Sublime Dreams of
Living Machines

Sublime Dreams of Living Machines

The Automaton in the
European Imagination

MINSOO KANG

HARVARD UNIVERSITY PRESS
Cambridge, Massachusetts, and London, England
2011

Copyright © 2011 by the President and Fellows of Harvard College
All rights reserved
Printed in the United States of America

Library of Congress Cataloging-in-Publication Data
Kang, Minsoo.
  Sublime dreams of living machines : the automaton in the European imagination / Minsoo Kang.
     p. cm.
  Includes bibliographical references and index.
  ISBN 978-0-674-04935-2 (alk. paper)
  1. Robotics—Popular works. 2. Robotics—Europe—History. 3. Robots in art—History. 4. Popular culture—Europe—History. I. Title.
  TJ211.15.K36 2011
  629.8'92094—dc22   2010019045

To my mother and my father,
with gratitude, respect, and love

Contents

	List of Illustrations	ix
	Introduction	1
1	The Power of the Automaton	14
2	Between Magic and Mechanics: The Automaton in the Middle Ages and the Renaissance	55
3	The Man-machine in the World-machine, 1637–1748	103
4	From the Man-machine to the Automaton-man, 1748–1793	146
5	The Uncanny Automaton, 1789–1833	185
6	The Living Machines of the Industrial Age, 1833–1914	223
7	The Revolt of the Robots, 1914–1935	264
	Conclusion	297
	Notes	311
	Acknowledgments	363
	Index	365

Illustrations

Hero of Alexandria, design for automaton figures   17

Hero of Alexandria, design for an automaton theater   18

Masahiro Mori, graph of the "uncanny valley"   49

Reactions to photographs morphing from robot to human (male)   50

Reactions to photographs morphing from robot to human (female)   51

Hero of Alexandria, design for an owl and other bird automata   82

Hero of Alexandria, design for an automaton archer that shoots a serpent   99

Jacques de Vaucanson's three automata   105

Salomon de Caus's Grotto automata   120

Giovanni Borelli, illustrations from *On the Movement of Animals*   126

Giovanni Borelli, illustrations from *On the Movement of Animals*   127

Wolfgang von Kempelen, the Turk chess-player   177

Wolfgang von Kempelen, the Turk chess-player exposed   181

Robert Seymour, *The March of Intellect*   236

George Grosz, *Republican Automatons*   274

George Grosz, *Daum Marries Her Pedantic Automaton "George"*   275

The female robot in *Metropolis*   290

Workers in *Metropolis*   291

Freder trying to help a worker in *Metropolis*   292

Honda robot Asimo pays tribute to Karel Čapek   309

Sublime Dreams of
Living Machines

Introduction

I offer you, myself, a venture into the ARTIFICIAL and its untasted delights!

—Villiers de l'Isle Adam, *Tomorrow's Eve*

O had I Father's gift I would breathe life
Into the lifeless earth, but who are we
To recreate mankind?

—Ovid, *Metamorphoses*

My story is similar to yours, but it's more interesting because it involves robots.

—The robot Bender, of the Fox television show *Futurama*

This is a historical study of the automaton, a machine that mimics a living being, as an idea in the European imagination. My purpose is to relate major instances of the object's manifestation in both actual and imaginary forms from premodern eras to the twentieth century, to explain its enduring presence and emblematic power. I discuss working automata in history, like those described by the ancient engineer Hero of Alexandria, the celebrated works of Jacques de Vaucanson presented in the 1730s, and the modern robot, but the central subject of this study is the automaton as a conceptual object. I examine how Europeans in different periods thought about the self-moving, life-imitating machine and how they used it for myriad intellectual and literary purposes.

A few months after I began research for this book, I saw something that brought home to me the true complexity of the ideas I was dealing with. A friend and I were walking down the Third Street Promenade in Santa Monica, California, packed with people on a Saturday evening, when we came across a street performer dressed in a silver suit, his face and hands painted the same color, pretending to be a robot. He entertained a crowd with slow mechanical movements of his limbs, hips, and head, until he suddenly stopped in midmotion. He remained still until a little girl approached him with curiosity. When she came close and made a face at him, the performer came alive again, sending the girl running with a delighted squeal, much to the amusement of the spectators.

"He's good," my friend said, "but the whole thing's kind of creepy, too, isn't it?"

I was immersed in thought of automata at the time, so I naturally found the scene of great interest. As I pondered the nature of the spectacle, three immediate thoughts came to me. First, from a conceptual point of view, the performer's act was of a complex order, as it featured a man pretending to be a machine that pretended to be a man. Second, even in this technologically advanced society in which all manner of machines, from the massively industrial to the conveniently portable, are ubiquitous and essential to the daily functioning of people's lives, there is something about the mimicry of machinery that can still enthrall people. And third, the performance was made possible by a ritualistic complicity on the part of the spectators.

The viewers knew full well that the performer was a human being pretending to be a machine that pretended to be a man, since only a child could seriously think that he was an actual robot. Yet they willingly suspended that knowledge for the peculiar pleasure of the play, to watch how well he executed the double act. When he stopped in midmotion, the performance was raised to another level as the viewers, aware that he was now pretending to be an inert object (a man pretending to be a machine pretending to be a man pretending to be a statue), waited for someone to approach him so he would move again with the pretense of frightening that person. The scene seemed like a reenactment of countless such moments from science fiction and horror movies in which people are terrorized by robots, statues, and dolls animated by malevolent forces. So the performance was one of harmless fun, but there was an undercurrent of uneasiness, in the pretense of danger by the young girl and in my friend finding the spectacle "creepy." Such expressions of anxiety at the thought of an inert object taking on characteristics of a living creature can be found not only in contemporary popular culture but also in literary works of the past.

In Alexander Pushkin's celebrated narrative poem "The Bronze Horseman" (1837), a disturbed man envisions the equestrian statue of Peter the Great coming alive to chase him through the streets of the czar's city.

> And all through that long night, no matter
> What road the frantic wretch might take,
> There still would pound with ponderous clatter
> The Bronze Horseman in his wake.
>
> And ever since, when in his erring
> He chanced upon that square again,
> They saw a sick confusion blurring
> His features. One hand swiftly then
> Flew to his breast, as if containing
> The anguished heart's affrighted straining;
> His worn-out cap he then would raise,
> Cast to the ground a troubled gaze
> And slink aside.[1]

In Prosper Mérimée's story "The Venus of Ille" (1837), the inexplicable death of a young man is believed by some to be the work of the unearthed statue of a pagan goddess; in E. T. A. Hoffmann's tale "The Sandman" (1816), a lover who finds out that his beloved is an automaton is driven mad by the revelation; and in the various versions of the Don Juan narrative, the libertine's damnation is presaged by the statue of his murdered victim coming alive to shake his hand.[2]

But such ambivalent feelings toward the performance at the Promenade (i.e., amusement as well as unease) were not just about the representation of an object that comes alive. The other component was the spectacle of a human being who behaves like a machine, an act that also arouses conflicting emotions in people. The complexity of such reactions can be discerned in the contradictory ways that certain types of people are described in mechanical terms in everyday language. On the one hand, referring to someone as a machine can be an expression of admiration, for the demonstration of great productivity at a task, unerring accuracy, and grace in its execution. To use some examples I came across in recent years in the media, we understand the characterization of the prolific novelist Stephen King as a "writing machine," the basketball star Michael Jordan as a "dunking machine," and even President George W. Bush as a "fundraising machine." On the other hand, when a person is called an "automaton" or a "robot" it is usually a derogatory comment denoting someone who is stiff and awkward in speech or movement; one who lacks imagination, emotion, spontaneity or a sense of humor; a fanatical follower of rules and regulations; a social or political conformist who is easily manipulated due to an inability to think critically and independently. In a recent issue of *Perspectives*, the news magazine of the American Historical Association, a history teacher objected to a Florida state education bill that asserted the view that American history is "factual, not constructed," by declaring that making students memorize information without teaching them analytic skills was tantamount to turning them into "little robots."[3]

One might jump to the conclusion that both the common as well as the contradictory nature of such mechanical descriptions of human beings are cultural expressions of our time, arising from anxieties over our dependence on all the machines of the digital era and the resulting desire to emulate the devices for their efficiency and productivity—but also from the visceral abhorrence we feel at the notion of the mechanization of humanity. But a survey of literary examples of people being described as machines reveals it to be a convention with a long history. In Charlotte Bronte's *Jane Eyre* (1847), Jane protests to Mr. Rochester that she is not "an automaton . . . a machine without feelings," while Honoré de Balzac, in *Père Goriot* (1834), describes the police informant Poiret as "a kind of automaton . . . one of the drudges of our great social mill . . . some cog in the machine of public business."[4] On the positive side, in Arthur Conan Doyle's story "A Scandal in Bohemia" (1891), Dr. Watson characterizes Sherlock Holmes as "the most perfect reasoning and observing machine the world has ever seen," and T. H. Huxley, in his essay "A Liberal Education; And Where to Find It" (1868), considers the well-educated man to be one who is firmly in control of his body, which "does with ease and pleasure all the work that, as a mechanism, it is capable" and one whose intellect is "a clear, cold, logic engine."[5]

To give one more interesting example of this ambivalence, in Henry James's novel *The American* (1877), Christopher Newman, a self-made man from the United States on a visit to Paris, has a conversation with the

French aristocrat Valentin de Bellegarde on the subject of Noémie Nioche, a woman who has transformed herself from an impoverished painter into an unscrupulous social climber. Valentin regards her as "a very curious and ingenious piece of machinery; I like to see it in operation." Newman, an industrialist who is familiar with actual machines, takes up Valentin's metaphor but points to the ruthless side of Noémie's personality. "Well, I have seen some very curious machines, too . . . once, in a needle factory, I saw a gentleman from the city, who had stepped too near one of them, picked up as neatly as if he had been prodded by a fork, swallowed down straight, and ground into small pieces."[6] Here we have a situation in which a person is admired as a machine but then is immediately cautioned against with the same metaphor.

For thousands of years, Westerners have been constructing, thinking about, and telling stories of automata, as many provocative examples of them, real and imaginary, can be found from ancient times to ours. As a concept, the automaton played a significant role in the world of ideas, the notion of the artificial entity appearing in some of the most important intellectual works in history. My larger purpose in this study is to demonstrate, by elucidating what the automaton represented in different contexts throughout the centuries, that it is a central idea in the Western imagination. The encounter with the street performer was a fortuitous one since it allowed me to articulate the two main themes pursued in this book: first, the nature of the automaton as an actual object, and the source of its ability to arouse powerful, conflicting emotions in people, and second, the consideration of human beings as machines, and the ambivalent reactions that elicits. The automaton is a particularly rich and complex subject of study for the historian, given its ability to take on an astounding array of meanings from one historical context to another. In addition to occupying an enduring place in Western thought, it has functioned as a kind of conceptual chameleon, embodying greatly varying sets of ideas and attitudes from one period to the next. At times its significance underwent drastic transformations in such short periods of time that it came to represent apparently contradictory notions. Two examples can serve as a preview of this protean nature.

During the Renaissance the automaton was regarded by Hermetic philosophers like Cornelius Agrippa and John Dee as an object of wonder, a marvel that could be achieved through the use of what they called "natural magic," or the manipulation of occult forces in the world. In the course of the seventeenth century, many natural philosophers of the scientific revolution, with their rationalistic worldview, rejected magical ideas, yet the automaton was taken up by such figures as René Descartes, Gottfried Leibniz, and Robert Boyle as the very image of the well-ordered universe. In the course of a single century, then, the automaton was transformed from being a wondrous object of natural magic in a pantheistic world to the central emblem of the mechanistic cosmos.

In addition, from the time Descartes described a human being as a combination of the material body, which is an automaton built by God, and the immaterial soul, the mechanistic view of humanity was a generally positive one, since there was a refrain of praising the Creator for having fashioned such an intricate, efficient, and beautiful machine as the body. Yet from the 1740s on, the automaton-man idea was used in a negative fashion by many late Enlightenment writers, denoting a person who lacks the principle of freedom, stunted in thought and spirit through either external oppression or witless conformism. Within a few decades, then, the image of the machine-man as an example of God's superb workmanship came to represent deeply flawed humanity, deserving only pity or contempt.

The easiest task for a historical study of the automaton is to establish its continuous presence in the Western imagination by enumerating its countless manifestations as both an object and an idea throughout the centuries. But four more essential and challenging subjects need to be elucidated for a full understanding of the object's significance: first, the origins and transformations of the two main themes of the automaton as a thing of enthrallment and as a representation of mechanistic humanity; second, the ambivalent emotions aroused by the object, from fascination and delight to contempt and horror; third, most important for the historian, how the idea functioned in specific periods and why its significance changed from one context to another; and fourth, the connection between two aspects of the object's power—the perennial (i.e., the essential reason it has been such an enduring object of interest through all periods of history) and the historical (i.e., how people reacted to and articulated the impact of that power in a specific era). I will show that what links all these themes together is the automaton's value as a conceptual tool with which Western culture has meditated on both the possibilities and the consequences of the breakdown of the distinction between the normally antithetical categories of the animate and the inanimate, the natural and the artificial, the living and the dead.

One of the complexities involved in the historical study of the automaton comes from the word itself, which underwent a number of changes and now has several meanings. The ancient Greek word *automaton* had a very general meaning of "self-mover," denoting an object or being that possessed an innate capacity to be mobile without the manipulation by an outside force.[7] There are instances in which it referred specifically to mechanical devices, but that was not of common usage in classical writings. Even Hero of Alexandria, who described numerous moving figures of humans and animals in his treatises, did not refer to them as automata, using the word *zoan* (image) instead. This originally rare sense of the automaton as a self-moving machine was revived during the Renaissance by Cornelius Agrippa, who used the word in his 1533 work *De Occulta Philosophia*.[8]

In the second half of the seventeenth century, in the most important moment in the history of the automaton idea, the word became prominent in European writings, as the self-moving machine came to play a crucial role in the philosophical, scientific, and medical discourses of the scientific revolution and the Enlightenment. In the latter part of the eighteenth century, a narrower definition emerged: the automaton as a self-moving machine built for the specific purpose of mimicking a living creature, as opposed to solely serving a utilitarian function. This was due to the great success of Jacques de Vaucanson's three works (two musician figures and a duck) and the subsequent automaton craze that lasted through the rest of the century. From then on the word became associated in the larger culture primarily with a mobile machine in the shape of a person or an animal.[9] To clarify the difference between the two definitions: the wristwatch I am wearing, despite its self-moving nature, is no longer thought of as an automaton since its purpose is not to imitate the actions of a living creature, but it would have been referred to as one in the seventeenth and eighteenth centuries. The word was used in both senses in the nineteenth century, but the newer meaning became the dominant one.

Most modern dictionaries provide three definitions of "automaton"—the older and more general one of any self-moving machine (including my wristwatch and any other device run by clockwork mechanism); the newer and narrower one of a self-moving machine built specifically to mimic a living creature; and a person who acts like a machine in some way. The last meaning is of complex significance since, as previously noted, a person who is called an automaton is often one who is easily manipulated due to the lack of individual will or thought. This, in a sense, contradicts the first two definitions of the word (i.e., a machine *capable* of independent motion versus a person *incapable* of independent action or thought)—as noted by Mark Seltzer, who pointed to the oddity of the fact that "automaton" can denote such opposing ideas.[10] These conflicting meanings emerged in the course of the eighteenth century, in the context of changing attitudes toward human nature, as I detail in Chapter 4. In the twentieth century, the word has been eclipsed to a great extent by the modern "robot" (I have often found myself explaining the subject of my research to people unfamiliar with it by informing them that "an automaton is the old-fashioned term for a robot").[11]

I will elaborate on the emergence of the varying definitions, since the rise and decline of meanings attached to the object are important to the story. Yet my primary—though not exclusive—interest is in the machine that mimics a living creature, the most provocative and beguiling form of a "self-mover." The central subject of this book can be specified, then, as the history of the self-moving, life-imitating machine and the role it played as an idea throughout European history.

The first chapter begins with an exploration of the perennial aspect of the automaton's symbolic power, to account for its enthrallment of the Western imagination and its capacity to arouse powerful and ambivalent emotions in people. After a discussion of some ideas that have already been advanced about the source of that power, I present a new theory, using works of history, anthropology, psychology, and art theory, to explain what exactly makes the automaton such a seductive and at the same time disturbing object. My ideas are based on how the human mind categorizes natural phenomena in building a picture of reality, and what occurs when that picture is violated through the appearance of an anomaly that defies attempts to fit it into an established conceptual schema. I revisit the theory throughout the book, illustrating its relevance with concrete examples and revealing the connection between the essential and the historical reasons for the automaton's uncanny attraction.

The second chapter is an overview of premodern automata, with a detailed look at two major manifestations of the object in the medieval and Renaissance contexts. For the earlier period, I consider the magical speaking head. A marvelous device built for oracular purposes, whether through diabolical sorcery or natural magic, it was associated in legends with the intellectual figures Gerbert of Aurillac (Pope Sylvester II), Albertus Magnus, Robert Grosseteste, and Roger Bacon. After a discussion of the ancient origins of the object, I elucidate its significance in terms of medieval attitudes toward unorthodox forms of knowledge like alchemy, astrology, and theurgy. For the Renaissance, the analysis centers on the significant lists of famous automata found in the works of Hermetic thinkers like Cornelius Agrippa, John Dee, and Tommaso Campanella, in which the objects were used to demonstrate the power of natural magic.

The third chapter begins with a general consideration of the life and works of Vaucanson, the most famous automaton-maker in history, and a survey of the automaton craze of the eighteenth century. I examine the success of his works as both a popular and an intellectual phenomenon, in the context of the mechanistic worldview that dominated European thought from the second half of the seventeenth century on. The automaton served as the central emblem of the era's views on the nature of the world, the state, and the body, as is apparent in the important role the idea of the self-moving machine played in advanced philosophical, scientific, political, and medical writings of the period. Given the rich and frequent deployment of the object in such works, I argue that this should be regarded as the golden age of the automaton.

The fourth chapter looks at the post-1740 period, when the image of the automaton-man underwent a drastic transformation into a highly negative representation of flawed humanity. In the works of radical thinkers, including Jean-Jacques Rousseau, Denis Diderot, Paul-Henri Thiry d'Holbach, and Louis-Sebastien Mercier, the notion of mechanized humanity was used to describe the sorry state of man lacking freedom through either oppression or conformism. In this context, the mechanistic metaphor shifted from the elucidation of bodily function to the characterization of certain types of personality. I analyze this change in the context of the rise of vitalistic ideas in the period that occurred in conjunction with the so-called revolution of sentimentality in the literate culture.

The fifth chapter takes the narrative to the Romantic period of the late eighteenth and early nineteenth centuries, when the automaton appears with conspicuous frequency in predominantly German works of fiction by Jean Paul (Friedrich Richter), E. T. A. Hoffmann, Ludwig Achim von Arnim, and others. The literary icon of the uncanny automaton appears in their dark stories as a living machine of uncertain nature whose mechanical functions seem to be tinged with the supernatural. While the mechanical man of the late Enlightenment was a figure of ridicule or pity, the uncanny version of the entity in fiction was an anxiety-inducing one that sometimes posed a threat to the sanity of the beholder. I examine the emergence of this image in the context of the worldview of Romantic natural philosophy, the rise of fantastic literature, and the resurgence of the magical aura attached to the mechanical object.

The sixth chapter looks at the new formulation of the automaton idea in the second half of the nineteenth and early twentieth centuries, especially in cultural responses to the Industrial Revolution. The uncanny nature of the living machine, inherited from Romantic literature, is very much present in the automaton works of this era, including those by Auguste Villiers de l'Isle Adam, Alfred Jarry, F. T. Marinetti, and Raymond Roussel. Yet they also feature new themes and ideas that emerged from complex attitudes toward the advancement of modernity. With new technology bringing about a rapid transformation of social and economic life in Europe, there was a sense in the larger culture that machines had taken on a life of their own, transcending their original purpose of serving humanity. This notion resulted in the common practice of describing the devices as living, superhuman beings driven by irrational will.

The seventh chapter deals with the interwar period, when the "robot" appears in such works as Karel Čapek's play *R.U.R.: Rossum's Universal Robots*, Romain Rolland's film script *The Revolt of the Machines*, and Fritz Lang's classic film *Metropolis*. These stories are distinct from those of the previous chapter, as they are pessimistic narratives of the mechanization of humanity, taking place in industrial settings and featuring major catastrophes that threaten to destroy all of civilization. These dire warnings of the fate of humanity in the machine age were informed by the experience of technological warfare in World War I, the Bolshevik revolution and its attempt to construct a high-tech utopia, and the controversial adoption of Taylorist regimes in European factories. This era is a fitting place to end the study proper, since it produced the most extreme fantasies of dehumanization, the machine society, and the living machine—the darkest visions of the West's centuries-long obsession with the automaton.

In the conclusion, I consider some current themes concerning the automaton, especially through an examination of how contemporary culture makes use of the artificial human idea. In our own time, we live with several competing ideas about our future in the digital world that express hopes and fears about the coming age of autonomous machines. Archetypal narratives of the robotic and cybernetic futures are told and retold in contemporary popular culture, in such films as *Blade Runner*, *The Terminator*, and *The Matrix* and in science fiction writings by Isaac Asimov, Philip K. Dick, William Gibson, and many others. Finally, I show that the age-old ideas of the living machine, including the magical automaton, the rational automaton, the witless automaton, the uncanny automaton, the superhuman automaton, and the rebellious automaton, are very much alive in the contemporary imagination, testifying once again to the emblematic power of the life-imitating device.

Previous works on automata have concentrated either on the technological aspect of the object (especially in the works of Alfred Chapuis and his various collaborators)[12] or on a particular period or episode, especially the automaton craze of the eighteenth century set off by the success of Vaucanson's creations.[13] I build on those works to present something new in automaton scholarship—a long-term study of the object as an idea, with special attention to the various shifts in meaning it underwent in transitional periods of history. While I take a panoramic look at the topic from ancient times to the twentieth century, it is not the aim of this book to be an encyclopedic history of the automaton, something impossible to do in a single volume. Instead, I alternate between providing a general overview of its conceptual usage in a given period and engaging in detailed analyses of the most significant instances of its appearances in intellectual and literary texts. The bulk of the work is concentrated on the early modern and modern periods since it was in the seventeenth century that the automaton first became a major concept in the West.

This book also addresses itself to two different audiences. For academic historians, especially cultural and intellectual historians of Europe as well as historians of science and technology, it is my purpose to provide a broad survey of the automaton idea. Given the millennia-long nature of this study, there are many concepts and themes I cannot elucidate to the fullest extent, not least because of the practical consideration of this book's length. It is my hope that other scholars will study them in greater depth in works with narrower topics, as I plan to do in the future. The central contribution I am seeking to make is to present historians with a general account of the major trajectories the concept of the self-moving machine has taken over the centuries as a basis for more detailed explorations of the subject. For the more general readership, especially for those interested in contemporary issues of robotics, artificial intelligence, and cybernetics, I unfurl a sweeping narrative of the West's inherent fascination with the automaton, with concise descriptions of the historical and intellectual contexts that informed the thoughts, fantasies, and uses of the object in each period. My purpose on this level is to demonstrate the enduring fascination with automata as it stretches back to the earliest period of Western civilization, and to show that the contemporary concern with artificial beings is only the latest development of a long-standing obsession.

This is essentially a work of intellectual history but with significant forays into cultural history in the analyses of public appearances of the automaton as an actual object and as a figure in popular literature and film. I also hope to explore the possibility of a new field I will call the history of the imagination, defined in the most general sense as the history of what people fantasized about. As I practice it in this book, the significance of a single entity or idea is traced through the examination of intellectual, cultural, and popular expressions in a given historical context to elucidate what the meanings invested in the subject reveal about the period's mentality in the nexus of thought, imagination, and emotional reaction. An interdisciplinary approach is necessary for such a purpose, as it seeks conceptual and thematic connections among works of the liberal arts and the sciences as well as popular narratives, polemics, and myths.

Ultimately, the real importance of the automaton idea lies in the crucial role it has played as a conceptual tool with which Western culture has pondered the very nature and boundaries of humanity. From ancient myths of inanimate objects coming alive to the creation of mechanical devices essential to modern life, philosophers, scientists, writers, and artists have struggled with myriad beguiling questions raised by the self-moving machine.

Is a human being a kind of machine? Does the mechanical represent the way to our empowerment and perfectibility or to debasement and the loss of what is vital and unique about being human? Is the image of the mechanized man that of a superhuman and a god or a slave and a monster? Can we recreate humanity, as humanity was created by a higher force? Why do we harbor the strange desire to do so? Is that desire a dangerous one of hubris that will take us to the dark realm of the unnatural and the inhuman, or is it a natural product of our innate yearning to know, to create, and to love? In these fantasies of continuing ourselves through recreated versions of ourselves, are we seeking the eternal and the infinite or are we plotting our own demise through the creation of what could surpass and replace us?

These are some of the questions that have haunted humanity before the image of the mechanical object that mimics life. And they haunt us still, in the sublime dreams of the living machine.

1
The Power of the Automaton

The First Automata: The Ancient World

The task of demonstrating that the Western obsession with the automaton is a long-standing one may seem simple, as many books on robotics have pointed to self-moving entities from the earliest eras of European civilization.[1] A recounting of such objects from the ancient period may also be the obvious place to begin this narrative, but an examination of the devices reveals a critical problem that is likely to create confusion over the exact nature of the subject rather than help elucidate its historical origin. Entities that are commonly referred to as examples of ancient automata, including the moving tripods in Homer's *Iliad*, the animate statues made by Daedalus, and the designs of Hero of Alexandria, are of such varying types of objects that a clear justification is needed for lumping them together under the same category. Many are thought of as automata only because they are reminiscent of contemporary robots, with the additional attraction of appearing at the dawn of European civilization, perhaps hinting at an innate desire in Western culture to create something that became feasible only in recent times. Such a view imposes a teleology that constructs a narrative of continuity from the ancient objects to modern creations, seeing in the former the dream of moving toward the mechanistic, technological, and automatic.[2] Even if such a dubious argument is to be made, a history of ancient automata must begin by establishing some sense of commonality among the great variety of objects. In view of such complexities it will be useful to make a brief foray into the ancient world to examine the exact nature of the entities that have been thought of as automata.

Objects that are frequently cited as examples of ancient automata can be put into four distinct categories. The first is that of mythic creatures that resemble modern robots only in appearance. In most cases, their makers are gods who created them through the use of supernatural power, in the same way they formed the first animals and humans (a universal trait of divinity being the ability to instill life in lifeless matter), with no reference to mechanical craft. We do not think of the biblical story of the creation of Adam and Eve as a technological one, so it would be a mistake to read the machine into the classical tales, no matter how much they seem to evoke the robot.[3] As examples, in the *Iliad*, the goddess Thetis travels to the forge of Hephaestus to commission armor and a shield for her son Achilles. There, she finds the creator god

> sweating as he turned here and there to his bellows busily, since he was working on twenty tripods which were to stand against the wall of his strong-founded dwelling. And he had set golden wheels underneath the base of each one so that at their own motion they could wheel into the immortal gathering, and return to his house: a wonder to look at.[4]

In *The Voyage of Argo* by Apollonius of Rhodes, the heroes returning from their quest for the Golden Fleece are attacked on Crete by a bronze giant, Talos, who hurls rocks at their ship. As is apparent from the following description of this fantastic creature, he is not a mechanical being but very much a living creature.

> A descendant of the brazen race that sprang from ash-trees, he survived the days of the demigods, and Zeus had given him to Europa to keep watch over Crete by running round the island on his bronze feet three

times a day. His body and his limbs were brazen and invulnerable, except at one point: under a sinew by his ankle there was a blood-red vein protected only by a thin skin which to him meant life and death.[5]

Some works even cite the legend of Pygmalion, the king of Cyprus who falls in love with a statue of his creation that is transformed into a real woman, but this example stretches the definition of automaton to an untenable degree. The object of his desire starts out as an ivory artifact, and the transformation is made through the divine power of the goddess of love, not by the mechanical ingenuity of man, and it is turned into a fully living woman and not some life-imitating machine.[6]

In the second category of classical objects are also mythic, self-moving devices, but of human rather than divine manufacture. Pindar, in an Olympic ode to a boxer from Rhodes, praises the people of the island for their skills in art and technology (their most famous achievement being the gigantic bronze Colossus, one of the seven wonders of the ancient world), which he says were a gift from the goddess Athena by means of which they filled their streets with images "in the likeness of beings that lived and moved."[7] And the Roman writer Aulus Gellius (second century CE), in his anecdotal collection *Attic Nights*, reports a story about the Pythagorean philosopher and mathematician Archytas of Tarentum (fourth century BCE) who once constructed "a wooden model of a dove with such mechanical ingenuity and art that it flew; so nicely balanced was it . . . with weights and moved by a current of air enclosed and hidden within it."[8] One can interpret these stories as being about technological achievements, but unlike the description of the dove, most of them do not spell out how exactly those works were achieved, whether through mechanical craftsmanship or magic.

The third category of automata are those that were either actually constructed or explicitly described in terms of their mechanical operation. The most famous name associated with such works in the ancient world is the engineer Hero (Heron) of Alexandria (ca. first century CE).[9] In his two surviving treatises, the *Pneumatics* and the *Automatic Theater*, he describes the workings of numerous automata, including singing birds, satyrs pouring water, a dancing figure of the god Pan, and a fully articulated puppet theater, driven by air, steam, and water power.[10]

The fourth category of automata are speculative objects, or ideas of

Hero of Alexandria, design for automaton figures. Reconstruction—illustration in *The Pneumatics of Hero of Alexandria*, trans. Bennet Woodcroft (London: Taylor, Walton, and Maberly, 1851).

possible self-moving devices. Aristotle in his *Politics* ponders the nature of property, including human beings (i.e., slaves). Beginning with a preliminary definition of a possession as "an instrument for maintaining life," he defines a slave as a "living possession" and property in general as a collection of possessions.[11] From this, he makes a rather startling jump to a fantasy about a device that, like slaves, "could accomplish its own work, obeying or anticipating the will of others." After he cites examples from myths, including the tripods of Hephaestus and the animate statues of Daedalus, he ponders the potential of such machines: "if . . . the shuttle would weave and the plectrum touch the lyre, chief workmen would not want servants, nor masters slaves."

Hero of Alexandria, design for an automaton theater. Reconstruction—illustration in *Herons von Alexandria Druckwerke und Automatentheater*, ed. Wilhelm Schmidt (Lipsiae: Teubner, 1899).

To summarize, the four categories of what are commonly regarded as ancient automata are:

1. mythic and of supernatural creation
2. mythic and of human creation
3. of actual human creation or design
4. speculative

Given the fundamental differences in their natures, is it useful or even possible to place these objects under the same rubric of automaton? The danger here is that in the absence of an exact definition, the word becomes an amorphous term that can encompass anything that is even vaguely reminiscent of the modern robot. It is crucial, then, for a sophisticated analysis, to determine the exact nature of an individual artifact (e.g., of divine manufacture or human; imagined or real; magical or technological) before specifying why the object should be considered an automaton. As noted in the introduction, the word was rarely used in the ancient world in the sense of a mobile machine, as it had the broad definition "self-mover." In that general sense, all the aforementioned examples would count as automata, but most of them would be excluded from the modern usage of the meaning that refers exclusively to the mechanical, a definition that arose in the seventeenth century in the context of the scientific revolution, when the word was established in direct opposition to the magical and the preternatural.

Given the disparate nature of the fantastic objects from the ancient world, they may represent a false start in the history of the automaton in the modern sense, but they point to two important aspects of that history. First, one abstract quality is, crucially, shared by all the great variety of (so-called) ancient automata: whether mythic or real, technological or magical, they all give the appearance of life (whether truly alive or only made to appear so) but are, to a substantial degree, made of matter we normally think of as inert or dead (metal, clay, stone, etc.). So as different as the mythic bronze giant Talos is from a water-driven bird doll of Hero or Aristotle's idea of a self-weaving shuttle, they resemble one another in their amalgamation of the animate and the inanimate. I will show how this aspect of the automaton as a hybrid entity that traverses the worlds of the living and the dead is at the heart of its attractive power.

Second, certain themes of remarkable endurance throughout the centuries can be explored without resorting to a simplistic teleology of essential tendencies. To give one extended example, in the celebrated legends of the master craftsman Daedalus and his mobile statues, we can discern ideas that resonate even in modern conceptions of the robotic, as it relates to notions of enslavement and rebellion.

In the classical world, Daedalus was as famous for his construction of self-moving statues (an example of the second, mythic-human category of ancient automata) as he was for his winged escape from the labyrinth of King Minos. The extent to which the story captivated ancient Greeks can be discerned in how often the tale was referred to in literary and philosophical works.[12] Aristotle in his treatise *On the Soul* speculates that Daedalus may have achieved the animation by pouring quicksilver into his creations, while the historian Diodorus Siculus provides a mundane explanation that the legend may have been inspired by the artist's great skill in making statues, as he was the first to represent human figures with eyes open and limbs in midmotion.[13] They seemed so lifelike to the viewers that stories were told of how he made living statues. Maurizio Bettini relates a legend of Herakles mistaking such a work for an actual person in the dark and striking it in fright, in one variation of which the statue actually comes alive and strikes him back![14]

The most interesting aspect of the Daedalus story is that the statues prove to be recalcitrant, as they constantly seek to run away, forcing their owners to physically restrain them. Plato mentions the tale in the *Meno*:

SOCRATES: It is because you have not observed the statues of Daedalus. Perhaps you don't have them in your country.

MENO: What makes you say that?

SOCRATES: They too, if no one ties them down, run away and escape. If tied, they stay where they are put.[15]

Socrates further explains, "If you have one of [Daedalus's] works untethered, it is not worth much; it gives you the slip like a runaway slave. But a tethered specimen is very valuable, for they are magnificent creations." So the self-moving object is useful as a slave, but it is also liable, like a slave, to try to escape bondage if it is not maintained under control. As previously mentioned, Aristotle explicitly draws a connection between the moving tripods of Hephaestus and the statues of Daedalus to human slaves. From this, one can infer that Daedalus's statues tried to run away because they wanted to be free.

Even in the modern period, a major way the idea of the robot has been articulated is in the notion of servitude, with such themes as the fantasy of owning robots as a metaphor for owning human slaves; the robot as the image for a slave or a dehumanized worker in the industrial system; and the robotic human as a representation of a person who lacks autonomy. Jean Baudrillard points out, however, that if the image of the robot is that of the slave, we must not forget that

> the theme of slavery is always bound up—even in the legend of the sorcerer's apprentice—with the theme of *revolt*. In one form or an-

other, robots in revolt are by no means rare in science fiction. And that revolt is implicit even when it is not manifest. The robot, like the slave, is both good and perfidious: good as a captive force; perfidious as a force that may break its chains.[16]

The connection between the automaton and the slave found in both ancient writings and contemporary themes of the robot is a significant one, but it would be a mistake to view the ideas of Plato and Aristotle as precursors or anticipations of modern concerns with automatic machinery. Such an ahistorical approach would ignore the fact that the Greek philosophers were writing in a cultural context in which the institution of slavery was largely unproblematic (i.e., neither thinkers were decrying the practice of slavery by linking it to automata) and that the contemporary robot symbolism comes out of concerns specific to the industrial era. But one can still discern a perennial theme in the ancient and modern works, from the beginning of Western thought a notion has persisted that if one grants life, even a semblance of it, to what began as a lifeless thing, there is always a possibility that it will go beyond one's control, to flee or to revolt. Adam and Eve disobeyed their creator, and many of their descendants went on to defy, offend, and challenge God; and according to Hesiod, the Olympian gods had to destroy the silver and bronze races of men (Talos being the last member of the bronze) because of their unruly and impious behavior, as the gods would surely destroy us when we inevitably degenerated as well.[17] So why should we expect our own lesser creations to remain obedient to us? Artificial beings, in other words, were seen from the beginning as inherently unstable, and that provides a clue to the automaton's power to fascinate throughout the ages.

This foray into the world of ancient automata, with its great variety of objects that combine the animate and the inanimate, and the delineation of enduring themes that connect the ancient entities to the modern robot, leads to the central purpose of this chapter which is to provide an explanatory framework for the perennial Western obsession with the automaton. The teleological view would see older instances of real and imagined automata as signs of an innate urge toward modern technology and the mechanistic worldview, and as rehearsals of cultural anxieties provoked by them. In contrast, I present a general theory of the fundamental source of the automaton's symbolic power, from which it draws its power to both fascinate and horrify. In the following chapters I will examine how these emotional reactions to the automaton were articulated in different periods, serving many conceptual functions from one cultural and intellectual context to the next. But before I proceed it will be useful to present an overview of some ideas on the significance of the automaton that have already been advanced. While all of them contributed in varying degrees to my own thoughts on this subject, I felt that a more comprehensive explanation of the object's captivating power was still needed.

The Uncanny and the Transcendent

In discussions on the significance of the automaton, the works cited most often are Sigmund Freud's essay "The 'Uncanny'" (1919) and John Cohen's *Human Robots in Myth and Science* (1966), both written from the perspective of psychology. Given their influence, it is necessary to consider their ideas at the outset. Freud's celebrated essay builds on an earlier psychological work by Ernst Jentsch, "On the Psychology of the Uncanny" (1906).[18] Jentsch defines the uncanny as a feeling that is aroused when one encounters an entity or finds oneself in a situation that is unfamiliar or unexpected, making it difficult to make sense of it through one's established worldview. The "psychic insecurity" caused by such an event translates into emotions ranging from anxiety to terror. An example of this is the reaction to uncertainty as to whether something one faces is an inanimate object or a living being, the insecurity being heightened when a thing not only looks like an animate creature but also behaves like one, like an automaton that plays a musical instrument or dances.[19] Jentsch points to E. T. A. Hoffmann as a writer who successfully makes use of this effect in his fantastic tales. Although Jentsch does not cite it, the Hoffmann work he obviously has in mind is the 1816 story "The Sandman," in which a man who finds out that the woman he is in love with is an automaton is driven insane by the revelation.

Freud uses the story to criticize Jentsch's ideas on two levels. He asserts, first, that Jentsch's definition of the uncanny is incomplete (i.e., not all things that are unfamiliar or unexpected necessarily arouse the feeling of the uncanny, and so it needs to be specified), and second, that given the ambiguity inherent in the etymology of the German word *unheimlich*, the cause of the uncanny is not the unfamiliar at all but something that is familiar but repressed.[20] In his analysis of "The Sandman," Freud dismisses the importance of the automaton in the story, seeing the essence of the uncanny in the recurring references to eyes and to the doppelgänger figure. He explains that the feeling arises from two closely related sources—castration anxiety (i.e., loss of eyes—loss of testicles) and a return to the mentality of early childhood. Young children have an animistic view of the world that is characterized by the idea that all things are alive and the magical belief in the "omnipotence of thought." As we grow up and progressively gain understanding of how our environment actually works, we shed such ideas in favor of more realistic ones. Even then, however, we sometimes find ourselves in situations in which we are taken back to the animistic, if only for a moment. For instance, if I wish that someone who annoys me greatly would just drop dead, and then I hear that the person died suddenly around the time I had the thought, I might experience a chill from the feeling, however fleeting, that it was my wishing that was responsible for the death. In addition, young children tend to treat their dolls and toys as at least potentially living beings.[21] The child psychologist Jean Piaget affirms the essentially animistic nature of children's worldview, and describes how between the ages of three and eight they gradually develop a sense of the distinction between the living and the inert through a series of stages.[22] So when we encounter an automaton, an object that acts as if it is alive, we momentarily return to the time in our lives before the boundary between the animate and inanimate was firmly established in our mind. Freud does not spell out why exactly this situation leads to the feeling of the uncanny, but presumably an adult encountering an intimation of the animistic is taken back to childhood feelings of being small, vulnerable, and frightened of all the unknown things in the world, which shakes all the confidence gained in the process of growing up. So the anxiety and terror that result are from the fear of losing the grip on reality and consequently being reduced to a powerless child.

Jentsch's and Freud's ideas on the uncanny shed significant light on the more unsettling aspects of the automaton figure but are inadequate as comprehensive explanations of its conceptual power. There is no doubt that the uncanny is a major component of people's reactions to automata, but while the object can be creepy and frightening under certain circumstances, it can also be fun, amusing, fascinating, and enlightening in others. What makes it such a rich subject of study is precisely the fact that it can arouse such a wide range of emotions. Freud's argument could be extended to the more positive side of our reactions to automata, so that their joyous and captivating aspect of the object also arises from childhood, a temporary return to the magical time when we derived enjoyment from playing with dolls and robots which we imagined were our friends. As Freud himself points out, "children have no fear of their dolls coming to life, they may even desire it."[23]

But what is lacking in the Jentsch and Freud essays is an explanation of how the automaton can switch from being a frightening, uncanny thing in one context to an amusing and fascinating object in another. This is crucial because both of their analyses, as they apply specifically to the automaton, raise the following questions. If the object is such an uncanny thing, why do we, as adults, keep making automata, writing about them, and producing movies that feature them, with so much pleasure? If the uncanny is the result of the return of the repressed, what is the source of the amusement and joy we also derive from automata? And what exactly are the conditions under which the pleasure and fascination we feel toward it turn into anxiety and horror, and vice versa? A comprehensive explanation of the symbolism must cover the full range of these emotions and explain their relationships to one another.

The psychologist John Cohen presents a more positive account of the automaton idea, seeing in it "man's never-ending struggle to achieve, first, a technical mastery of his surroundings . . . and, second, to become as one of the gods himself, by transcending both matter and himself."[24] In a later chapter on the motive of the robot-maker, Cohen presents a more detailed theory, derived from his view that human beings live with two contradictory worldviews simultaneously, what he calls the literal (also reductive, scientific, technological) and the symbolic (also mystical, magical, transcendent). The desire to create artificial life is derived from both sources, as the literal-scientific seeks technological control over nature and the symbolic-mystical aspires to overcome human limitations in the natural world. Despite the differences between these goals, the human mind, Cohen affirms, is "a unity which enfolds within itself in distinct but interrelated forms of consciousness."[25] One can deduce from this argument that what is behind the robot fantasy is the desire to heal the apparent rift between the two worldviews. This could be achieved by gaining a true understanding of the nature of life, which would allow us to obtain mastery over the environment and transcend to a higher realm of existence through the divine power of bringing life to the lifeless.

Just as Jentsch's and Freud's ideas do not provide an explanation of the pleasurable aspect of the automaton's symbolism, Cohen neglects to account for the uncanny, despite the fact that he looks at Hoffmann's "The Sandman" and other works that feature similar themes. In fact, he dismisses darker fantasies of the Frankenstein variety as distortions of "the potentialities of human effort," and even speculates that science would become more of a beneficent force if it was represented as such in the larger culture.[26] For Freud the sense of the uncanny experienced by an adult toward a lifelike automaton is about powerlessness, feeling oneself back in the animistic world of a child, while for Cohen the effort to create life from lifelessness is about the attainment of power, over the world and oneself. But what is the link between these two contradictory reactions in terms of power and powerlessness? What exactly are the circumstances under which a person feels like a frightened child or like a god before an automaton? Even if one combines Freud's and Cohen's theories, one still has no conceptual framework with which to determine how this movement occurs from the positive to the negative, the pleasurable to the uncanny, and the empowering to the emasculating.

One of the problems with many ideas on the significance of the automaton idea, even some of the more sophisticated and provocative ones, is that explanations that work for a specific historical and cultural context are often stretched to other eras, resulting in distortions. Even if one believes that there is a singular source of the perennial fascination with the automaton in the West, that source must be sought with the knowledge that this interest was articulated in vastly different ways throughout history. The problem arises when a scholar mistakes the idea of one period for the universal meaning of the symbol. For instance, Cohen asserts that of the two sources of the desire to create artificial life, the literal and the symbolic, the former is expressed in terms of the desire to master the environment through the use of technology. There is no doubt that the technological control of nature is a central theme in the modern period, and of the modern automaton idea as well, but it is inappropriate to see it as the central explanatory factor for automata in premodern contexts.

Likewise, another major theory of the automaton symbolism is one that ties it to issues of gender and sexuality, seeing in it an essentially male fantasy of attaining woman's power to give birth to new life. The psychologist Robert Plank in his essay "The Golem and the Robot" (1965) argues that the desire to create artificial life is "essentially a wish to circumvent the sexual act of creation. By being divorced from the sex act, the process would, as it were, become purified. The stain which by some is felt to be attached to anything sexual would be removed."[27] Similarly, Andreas Huyssen, in his analysis of the 1927 film *Metropolis*, asserts that the creation of the female robot reflects the male desire to procreate without a mother—

> but more than that, he produces not just any natural life, but woman herself, the epitome of nature. The nature/culture split seems healed. The most complete technologization of nature appears as re-naturalization, as a progress back to nature. Man is at long last alone and at one with himself.[28]

Huyssen's work, one of the most incisive analyses of the classic film, also sheds significant light on the social and political concerns of Weimar Germany, as I show in Chapter 7. There is no doubt that the linkage of technology to gender and sexuality is a major component of automata fantasies of the late nineteenth and twentieth centuries. The problem arises when Huyssen extends his analysis, especially the convincing connection he makes between the fear of female sexuality and the fear of industrial technology in *Metropolis*, to automaton symbolism in general all the way back to the eighteenth century. In a section entitled "The Machine-Woman: A Historical Digression," he starts by pointing to Julien de la Mettrie's 1748 essay "L'Homme machine," claiming that La Mettrie's mechanistic and materialistic ideas led to the notion of the world as a gigantic automaton, providing the worldview for the coming age of modern technology.[29] Huyssen further asserts that once the Industrial Revolution has begun, the great cultural anxiety over its impact can be discerned in fictional works in which dark, threatening automata make appearances. This point relates back to his notion of the link between woman and technology, as he notes that while actual eighteenth-century automata represented figures of both sexes, fictional automata of the nineteenth century tended to be female.

This narrative is problematic in several ways. As I show in Chapter 4, it was precisely at the time of La Mettrie's essay, which itself is a much more complicated work than its provocative title indicates, that the mechanistic worldview (which was established much earlier in the mid-seventeenth century) came into crisis, challenged by new vitalistic ideas in science, medicine, philosophy, and political theory. Even in the nineteenth century, mechanistic materialism was never unproblematic, as it was continuously challenged, first by Romanticism and later by revived vitalism in the last decades. It is, furthermore, inaccurate to say that there was a preference for female automata in fiction of the period. As examples of stories with female automata, Huyssen points to such works as Jean Paul's "Ehefrau als bloßem Holze" (1789), Ludwig Achim von Arnim's "Isabella von Ägypten" (1800), E. T. A. Hoffmann's "Der Sandmann" (1815), and Villiers de l'Isle Adam's *L'Eve future* (1886).[30] But among many examples of automaton stories from the same period with male automata there are Jean Paul's "Personalien vom Bedienten- und Maschinen Mann" (1798), E. T. A. Hoffmann's "Die Automate" (1814), Charles Nodier's "Voyage pittoresque et industriel dans le Paraguay-Roux et la palingénésie australe" (1836), Jules Verne's "Maître Zacharias" (1854), Herman Melville's "The Bell-Tower" (1855), William Douglas O'Connor's "The Brazen Android" (1891), and Ambrose Bierce's "Moxon's Master" (1893). So the preference for stories with female automata is Huyssen's, not that of nineteenth-century literature, since they presage the robot figure in *Metropolis* in a way that is consistent with his theory. Issues of gender and sexuality do play important roles in automaton stories of the modernist period, starting from the late nineteenth century, so Huyssen was right to discern them in the Lang film. In the earlier Romantic period, however, writers like Hoffmann and Arnim found male automata just as uncanny as female ones, and for reasons that have little to do with modern technology, since the Industrial Revolution did not commence in continental Europe until the 1830s.[31] Such distortions in Huyssen's "historical digression" are the result of trying to read a late nineteenth- and twentieth-century theme into a premodernist context. Recently, the robotics expert Rodney Brooks, in the Errol Morris documentary *Fast, Cheap and Out of Control* (1997), expressed skepticism toward the "womb-envy" theory by pointing out that there were just as many female researchers as male ones at his advanced robotics lab at MIT.[32]

Another theory, articulated by Linda M. Strauss in her rich essay "Reflections in a Mechanical Mirror," asserts the automaton's function as a double for humanity.[33] I largely agree with her assessment of the object's liminal nature as well as its value as a conceptual tool with which people in the West have explored the nature of humanity. But this notion of artificial humans as a mirror image of actual humans was something that did not arise until the late Renaissance, so it cannot provide a "universal account" of its power.

Despite such pitfalls, I do believe that it is possible to account for the automaton's essential power and to relate how that power has been explained, utilized, and reacted to in different ways in various periods. A theory to that effect has to explain how the automaton can arouse in the viewer such widely varying feelings as amusement, fascination, creepiness, and horror and the conditions under which the switch from one reaction to another occurs. What follows is an attempt at such a theory.

Categories, Anomalies, and the Liminal

The problem with overly determined theories of the automaton's power is that they present explanations that are too intellectual for what initially occurs at an instinctual level. A man strolls past an object he assumes to be an inert, immobile thing like a statue, until it suddenly moves, revealing it to be a robot or a person pretending to be a robot. This startles him greatly, despite the fact that the object poses no physical danger to him. The initial feeling of fright comes from a sense of the unexpected—of a thing he assumes to be one thing turning out to be another—but even after the moment passes and he understands what it is, a feeling of uneasiness persists in him. He may dispel it with laughter, assuring himself that what occurred was not a serious thing and he can be a good sport about it, or he may react with anger at the psychological disturbance it caused him. But the initial impact is felt at an immediate, visceral level, capturing his attention in a single instant, with the conscious articulation of that reaction coming afterward. For this reason the explanation of the automaton's power must start at the level of perception, cognition, and emotional reaction.

During the Enlightenment, philosophers from Leibniz to Kant reacted against the empirical notion, associated with John Locke, that the human mind begins as a tabula rasa, a blank slate on which impressions from the senses get imprinted, eventually forming ideas and understanding.[34] If the mind were such a passive receiver of phenomenal data from the outside world, it would become hopelessly lost in a chaotic swirl, overwhelmed at every moment by an avalanche of information flooding in from the five senses. As Kant elaborated in his *Critique of Pure Reason* (1787), the mind is equipped with innate tools that allow it to sort out the received data and shape them into a coherent picture of reality in which things are organized in terms of space, time, causation, unity, and other categories of understanding. From this automatic structuring of phenomena, one further develops a comprehensive sense of reality by imposing larger patterns of understanding on them.

Claude Lévi-Strauss in his anthropological works has examined how this individual process of constructing reality is related to how an entire society develops and maintains its worldview. The most common way both a person and a community put together a structure of reality is through the use of binary categories.[35] An individual makes sense of the world by organizing things in a series of dual oppositions such as day/night, human/animal, living/dead, man/woman, adult/child, safe/dangerous, and so on, which leads to the community's development of more abstract concepts like pure/impure, sacred/profane, natural/unnatural, normal/abnormal, sane/insane, moral/immoral. Lévi-Strauss thinks that the prevalence of such dualistic thinking has biological roots, perhaps in the structure of the human brain itself.[36] It is beyond the scope of this study to assess whether this tendency to view reality in binary terms is indeed rooted in biology or is linguistically determined or is socially constructed (in all probability a combination of all three), and the point is a controversial one in contemporary discussions on the topic.[37] Once an entire worldview based on such categories is set up, it is affirmed over time through ritual, custom, law, and education, eventually solidifying into tradition. Such a predetermined schema of reality provides the confidence people need to face the world, solve its problems, and explore its boundaries. Because that is essential to survival, a conservative bias sets in after a time, as people feel compelled to defend the schema against what threatens it.

The problem is that reality-in-itself is not composed of clear-cut categories, being essentially an amorphous, unstable, and ever-changing state that the human mind is constantly trying to impose a sense of coherence on.

Consequently, no matter how rigidly set and strongly supported by tradition and institutions a worldview is, it inevitably encounters entities, events, and situations that defy it and threaten to expose it as the arbitrary human construction that it is. Such a situation poses a danger to the community as a whole, as it threatens to undermine the foundations of its shared reality, potentially throwing it into a conceptual chaos in which its members become lost in an unfathomable world, resulting in the collapse of order and authority. Disturbances to the established schema appear in the form of entities that do not fit into either side of binary categories, sometimes in a terrifying fashion—for example a deformed baby that disrupts the categories of human/animal; a hermaphrodite—the categories of man/woman; or a solar eclipse—the categories of night/day. Certain forms of bigotry that are found even in advanced societies are also rooted in emotional reactions toward those who seem to cross categorical boundaries. Racists decry miscegenation that produces mixtures of ethnicities, religious fundamentalists attack homosexuals whose nature defies traditional notions of gender and sexuality, and anti-Semites look at Jews with suspicion because they are a people who are simultaneously a part of and apart from mainstream Western culture. As the anthropologist Mary Douglas explains,

> the yearning for rigidity is in us all. It is part of our human condition to long for hard lines and clear concepts. When we have them we have to either face the fact that some realities elude them, or else blind ourselves to the inadequacies of the concepts. The final paradox of the search for purity is that it is an attempt to force experience into logical categories of non-contradiction. But experience is not amenable and those who make the attempt find themselves led into contradiction.[38]

When a society comes face to face with what does not fit into the categories that make up its worldview, how does it deal with it, as it threatens to undermine its entire view of reality? Douglas describes methods that are negative ("we can ignore, just not perceive them, or perceiving we can condemn") and positive ("we can deliberately confront the anomaly and try to create a new pattern of reality in which it has a place") before outlining five specific ways different societies have traditionally handled them.[39] The first way is to introduce an interpretation that explains the anomaly in "normal" terms; second is to physically control the anomaly (as in destroying and removing the evidence of its existence); third is to establish rules of avoiding it (prohibitions); fourth is to label it as dangerous; and fifth is to assimilate it through its usage in special rituals. As an example of the third, a negative reaction, Douglas points to the dietary prohibitions in the Book of Leviticus in which all the creatures that are declared unclean are revealed to be ones that do not fit into the traditional classifications of birds (winged and two-legged), animals (four-legged, and hopping, jumping, or walking), and fish (scaly and swimming).[40] So insects, reptiles, shellfish, and other "borderline" creatures that do not conform to the categories are restricted by prohibitions. The pig is reviled not for hygienic reasons or for its scavenging behavior but because it is a cloven-footed animal that, unlike other domesticated beasts like the cow, sheep, and goat, is not cud-chewing. As an example of the fifth, a positive method, Douglas points to the Lele people of Central Africa, who practice a cult centered on the pangolin, or scaly anteater, an ultimate category-defying creature. By meditating on the oddity of its appearance and behavior, initiates are invited "to turn around and confront the categories on which their whole surrounding culture has been built and to recognize them for the fictive, man-made, arbitrary creations that they are."[41]

Whether a society decides to ignore, physically control, or condemn a categorical anomaly on the one hand or accept and assimilate it on the other depends on the specific nature of the culture (its openness to new ideas and the flexibility of its worldview, etc.) and the nature of the anomaly itself (the level of danger, conceptual or physical, it poses to that society). Zakiya Hanafi notes how different civilizations have reacted in a variety of ways toward the birth of a deformed baby—the ancient Egyptians, Indians, and Persians treating it with awe and respect, the Greeks, Romans, and other Europeans regarding it as an abomination, sometimes with dire consequences for the unfortunate mother.[42] Similarly, transsexuals, who are despised and persecuted in many cultures, are accepted in others as people of special status sometimes endowed with magical powers.[43]

What complicates this situation is that while categorical anomalies can disturb, frighten, and infuriate people, since they question their worldview and so threaten to cast them into confusion and chaos, some of them can also, under certain circumstances, bring pleasure and a positive sense of awe. There is no doubt that from the most basic level of perception to the most elaborate one of building an entire worldview, people absolutely need categories and schema. There is, however, a part of us that knows full well how arbitrary and artificial these necessary structures are and how restrictive they are to our unmediated experience of reality. They protect us from chaos, but they also impose rules that deny us the freedom of an infinitely protean world. As in Nietzsche's ideas on the Apollonian and Dionysian impulses, we need order but we also long to be released from it.

The complete overthrow of the communal order is, obviously, an unacceptable solution to this dilemma, so societies develop rituals in which people can indulge in a temporary release from the strictures of the normal, within controlled parameters. So attending a drama or a film in which actors play at experiencing great emotions in extreme, unusual, or fantastic situations provides the viewers with the catharsis described by Aristotle; the traditional European carnival, as analyzed by Mikhail Bakhtin, is an occasion when the rules governing the social order are put in abeyance and even made fun of; and the modern dance club is a space where people are allowed to move in certain rhythmic and sexually suggestive manners that are pleasurable but would make them look ridiculous, bizarre, or even offensive in everyday life.[44] Such rituals work only if they are temporary and remain within a preset boundary, since overstepping them can lead to the terror and confusion of total disruption (e.g., a cheering crowd at a sporting event turning into a riotous mob in the streets).

A particularly disturbing example of an object of transcategorical nature that every society must contend with is the recently dead body. It is one of the most vivid and terrifying examples of a thing of uncertain status—no longer alive but undergoing the activity of rapid transformation, changing form, color, and smell. It has left the world of the living but has not yet become a stable object of the inanimate world (i.e., a skeleton, which is a safe, inert object that can be displayed in a museum or a science classroom). The corpse is such an unsettling thing because it is in a transitional state from one part of a binary category to another and, as Douglas puts it, danger "lies in transitional states, simply because transition is neither one state nor the next, it is undefinable."[45] As a result, every society has developed elaborate rituals of disposal and mourning, to control and smooth over the disturbing event.[46]

On that point, Carlo Ginzburg in his essay "Representation: The Word, the Idea, the Thing" picks up on an issue raised by Ralph Giesey in his work on the royal funeral ceremony in Renaissance France.[47] The question concerns the display of the effigy, an artificial representation of the deceased, during funeral rituals, originally practiced by the nobility of ancient Rome.[48] It was discontinued with the onset of Christianity due to the object's association with idolatry but was revived on the occasions of the deaths of Edward II of England (1327), Charles VI (1442), and Francis I (1547) of France.[49] To explain the phenomenon, Ginzburg first elucidates the function of the effigy in the ancient funeral.

The death of any community member is a disturbing event, but the demise of a person of the highest status can be traumatic to a society, especially in the case of a ruler who functioned as the parent, order-giver, and, in some cases, spiritual leader of the people. The funeral ritual of such a person, therefore, must be elaborate enough not only to allow them to mourn and express their distress at the possible chaos that could ensue but also to assuage those emotions by demonstrating that the death is not an end but merely a transition. The spirit of the deceased is moving on to a different plane of existence, but his power and authority, the "eternal" part of the ruler (as elucidated by Ernst Kantorowicz in *The King's Two Bodies*), is transferred to a successor who is still among the living.[50] This is essential for both reassuring the people of the continuity of things and maintaining the legitimacy of the living power-holders. The difficulty lies in that the more elaborate such a ritual is, the more time it takes to complete it, and when the displaying of the body to the people is an essential part of the ritual, it can be problematic in societies that lack an effective embalming technique. In other words, the horrifying sight and smell of the decaying corpse could undermine the very purpose of the ceremony.

The purpose of the funeral effigy, then, was to represent that body as a stand-in during the rituals, providing a stable and reassuringly clean object of reference. In the case of Roman emperors, two funerals were performed, the first of the body itself, which was quickly cremated, the remains being taken to a tomb outside the city, and the second of a wax image of the deceased, which was displayed for a few days before it was transported to a temple (a sacred space that would have been defiled by the presence of an actual corpse) to be consecrated.[51] To further help the people get through the worrisome transitional period, elaborate playacting took place involving the effigy. Giesey tells the story from Herodian of the image of Emperor Septimius Severus being placed on a great bed where it lay for seven days, during which time physicians visited it every day, pretending to examine it, and then declaring that the emperor was getting worse; and the anecdote from Dio Cassius of the funeral of Emperor Pertinax, whose effigy was laid out on a bier, where a youth stood by to keep the flies away as if the figure were sleeping.[52] After the use of the funeral effigy was adopted by the French during the Renaissance, the image of the deceased Francis I was served meals and the uneaten food was given away to the poor afterward.[53] In a time of worrying uncertainty for a community, when the body of its leader is undergoing the harrowing transformation from a living being to a dead thing, the effigy played the crucial role of allowing people to go through the rituals without having to deal directly with the decaying body.

When Christianity became the dominant religion in fourth-century Rome, it brought with it the ancient Judaic prohibition against certain representative images, as in the Second Commandment: "Thou shalt not make unto thee any graven image, or any likeness of any thing that is in heaven above, or that is in the earth beneath, or that is in the water under the earth."[54] The practice of constructing an image of the deceased, displaying it during the funeral, and finally consecrating it in a temple smacked of idolatry, so it was discontinued. Ever since, Christianity has had a complex history of reaction to images of living creatures, as artistic flowering of sacred art has alternated with iconoclastic reaction.[55] But on the prohibition itself, what exactly was it about such images that disturbed the ancient Jews and early Christians? Was it solely about the concern over confusing the thing with what it represents and the resulting fall into idolatry, or was there a deeper, more primordial fear at work? This is a question worth pursuing since, as I will show at the end of this chapter, the attractive as well as disturbing power of the automaton is derived from its capacity to go beyond mere representation.

The purpose of an image such as an effigy, a statue, or a portrait is, of course, to represent things in the world, including a living being. The closer the resemblance to the subject, the more successful the object is at representing it. From a conceptual point of view, however, the more "like" an image is to what it represents, the more dangerous it becomes to the viewer. Not only is the beholder liable to confuse it with the real being, but the object itself seems to be on verge of coming alive, becoming—like a corpse—yet another category-defying, transitional thing. In other words, we all know that a work of art such as a statue or a painting is an object, but what is both interesting and unnerving about it is that it *seems* to be more than just a thing, *seems* to be alive, or about to come alive. For this reason the artist is often seen as a kind of magician, one who can seemingly create life out of lifeless material, like Daedalus the statue-maker.[56] It will be useful at this point to briefly consider the figure of the magician, whose significance will become clear in the course of this study, since the automaton-maker was often described in magical terms throughout the object's history.

Every culture in the world is familiar with the idea of a person who wields uncanny powers. The magician is often portrayed as a diabolical person in league with an evil force, but in premodern societies he occupies an important place in the community. As Marcel Mauss has described in his classic work on the subject, magicians in primitive tribes are typically people who either have dropped out of the community for a time or never quite belonged to it in the first place because of some unusual characteristic (e.g., deformity or mental illness).[57] Because of the quality that sets him apart, he lives on the margin of that society (both physically and conceptually), a space that is referred to in anthropological studies as the liminal. Victor Turner explains:

> Liminal entities are neither here nor there; they are betwixt and between the positions assigned and arrayed by law, custom, convention, and ceremonial. As such, their ambiguous and indeterminate attributes are expressed by a rich variety of symbols in the many societies that ritualize social and cultural transitions. Thus, liminality is frequently likened to death, to being in the womb, to invisibility, to darkness, to bisexuality, to the wilderness, and to an eclipse of the sun or the moon.[58]

As someone who dwells in the zone between the inside and the outside, the visible world and the hidden, the magician is thought to be capable of tapping into the chaotic power of the world beyond. A conceptual schema brings coherence to reality but also limits it, setting strict rules on what is possible and impossible, thinkable and unthinkable, controllable and uncontrollable. The world beyond the schema, usually described in supernatural terms, is regarded as a place of darkness and chaos but also of infinite power. So the magician is unsettling to people as a borderline creature and is never treated as a full member of their society, but he is also a source of fascination and revered as a healer, diviner, and rainmaker. Venturing beyond the boundaries of ordinary reality can bring one tremendous power, but to attain it one has to pay the price of becoming a perennial outsider, and risk as well the danger of that power going out of control in one's hands. And an ultimate form of such a power is the divine prerogative of bringing life to lifeless matter.

Given the conceptual framework I have built here, where does the automaton fit in? From what source does it draw its power to entrance and to frighten? To put it in the most general terms: the automaton is the ultimate categorical anomaly. Its very nature is a series of contradictions, and its purpose is to flaunt its own insoluble paradox. It is an artificial object that acts as if it is alive; it is made of inert material yet behaves like a thing of flesh and blood; it is a representation that refuses to remain a stable version of the represented; it comes from the inanimate world but has the characteristics of an animate creature; and, finally, it is a manmade thing that mimics living beings. What normal representative images only threaten to do, namely come alive, the automaton seems to actually realize. It appears to leave the hands of its mortal maker and take on a life of its own, animating itself to mock the idea that the power of creation belongs to God alone. The binary categories of living/dead, animate/inanimate, creature/object all break down in its wake, as it moves from one to the other, mesmerizing and terrorizing its beholders by turns. And its maker is not only an artist but a magician as well, perhaps an ambiguous creature himself who traverses the worlds to tap into uncanny power. One either admires his skill and his creations as divinely inspired or rejects them as things of the unnatural and the diabolic. Either way, the power exhibited by the self-moving object is far beyond that of inert images that can never literally reach out to stir up the deepest fears in the beholder. One can always turn and walk away from a still image that disturbs, but what if that image follows, speaks, and touches?

In 44 BCE, after the assassination of Julius Caesar, the murdered ruler's ally Marcus Antonius delivered his grand speech at his funeral, a scene made famous by Shakespeare. The Bard does not recount, however, a peculiar episode from the event related by Appian. When the speech was done,

> someone raised above the bier a wax effigy of Caesar—the body itself, lying on its back on the bier, not being visible. The effigy was turned in

every direction by a mechanical device, and twenty-three wounds could be seen, savagely inflicted on every part of the body and on the face. This sight seemed so pitiful to the people that they could bear it no longer. Howling and lamenting, they surrounded the senate-house, where Caesar had been killed, and burnt it down, and hurried about hunting for the murderers, who had slipped away some time previously.[59]

The purpose of the effigy, and the whole funeral ritual for that matter, is to allow people to calmly witness the passing of the deceased and to reassure them of the continuity of things despite the death. On this particular occasion, however, Marcus Antonius did not want to calm and reassure the people. He wanted to fill them with grief and anger, thereby arousing them to action against the conspirators. In order to achieve this effect, he hid the corpse and turned the effigy of his friend into an automaton. The thing that was supposed to be stable in its inert solidity became the most unstable thing there is, getting to its feet and showing off its fatal wounds.[60]

Despite the fact that this episode has been presented uncritically in some history books, it was in all probability imagined by Appian, as it is not mentioned in any other Roman accounts of the funeral, including in those of Cicero, Livy, and Suetonius (all of them earlier than Appian).[61] Nevertheless, it provides an interesting insight into the significance of the effigy and the automaton, for the psychological dimension of the story rings true. What it demonstrates is that the automaton is the diametric opposite of the effigy, or its dark twin. While the latter stabilizes the dangerous situation by standing in for the corpse that is going through the journey from the world of the living to the world of the dead, shielding the people from its frighteningly ambiguous nature, the automaton deliberately disturbs by pointing to its liminality, playing havoc with people's notion of what is alive and what is dead, what can move and what must remain still. One can imagine the terrifying, hair-raising moment when it appeared to the people that Caesar's dead body had risen of its own accord to cry out for vengeance, before that terror turned into unbearable sorrow and rage. What is finally significant about the effigy-automaton story is the audacious yet somehow convincing idea that the power of an automaton is such that if used at the right time and manner, it could arouse such powerful emotions in people and change the very course of history.

From this conceptual framework, we can finally examine in detail how the automaton causes such diverse emotional reactions as amusement, fascination, creepiness, and terror and under what circumstances it switches from one to another.

From the Uncanny Valley to the Eerie Peak

Let us start with the most innocuous reaction to the automaton—amusement. Given the fact that the automaton is an inherently unstable object of ambiguous status that can seem to threaten one's entire sense of reality, how can it also be a thing of playful fun? As Freud noted, this is unproblematic for children since they live in an animistic universe where the boundary line between the animate and inanimate is not yet set, a world before the schema of adult reality. And the sense of sheer fun and play even adults find in harmless automata afford temporary escapes from the strictures of the grown-up world. At several academic conferences where I gave presentations on the history of automata, I began by presenting a small windup toy to the audience, setting it in motion before them. Far from finding it creepy or frightening, people always reacted with smiles and laughter, some even commenting afterward that the object was "cute," as if it were an adorable animal.

Henri Bergson in his 1900 essay "Laughter" asserts that the essence of the comic lies in one's reaction to human action and thought that resemble those of a machine or "something mechanical encrusted on the living."[62] We laugh at people displaying clumsiness, inflexibility, absentmindedness, or literal-mindedness, because these make them look like automatic devices incapable of spontaneity, flexibility, and change. Laughter, furthermore, has a social function in that it seeks to correct such behavior, reminding people, sometimes through ridicule, what it is to be human. Bergson's work ultimately fails to provide a comprehensive account of what arouses laughter, perhaps an impossible task for such a complex phenomenon. Yet the essay is revealing as an incisive analysis of a specific brand of the comic and as a historical document that embodies some of the central concerns of European culture at the end of the nineteenth century. After decades of rapid industrialization, there was continued anxiety in Europe over the effect of the enormous changes it brought to both society and human identity, reflected in the agonizing debates over the question of adopting Taylorist regime in European factories.[63] As a central figure in a revived vitalist philosophy, Bergson emphasized that humans are living, supple, evolving beings of élan vital, even if they live in a world that is constantly trying to turn them into machines.[64] The social function of laughter, then, can be interpreted as a defensive measure by us to remind ourselves of our essential vitality. But how does this help to understand why an automaton can be an amusing thing?

According to Bergson, laughter is elicited when a person acts like a machine but clearly is not one. In other words, the spectacle is of an entity that acts like something it is not and, importantly, does a bad job of it. One can apply that logic to the other side of the equation as well—a machine acting as if it is a living being and doing it unconvincingly is humorous for the same reason. So a small device like the simple spring-driven one I presented at conferences is both laughable and cute because it seems to be playacting at being a little animal, when everyone can see what it really is, an insignificant thing made of metal and plastic. Our attention is arrested by the device because it tries to cross the boundaries of animate/inanimate, natural/artificial, living/dead, but does such a bad job of it that it ends up *reaffirming* our normal schema of reality. And so we laugh in relief at its failure and domesticate the object in terms of childhood playfulness, finding it amusing and cute. To put it another way, the very failure of the automaton to convince us that it is a living being works as the conceptual controlled parameter that turns a potentially disturbing thing into an amusing one, so we can enjoy the sense of return to childhood without being subject to its terrors.

So the amusement in response to the automaton has two related sources: first, the minor disturbance to our categorical worldview that is immediately corrected by the object's failure to convince us of its living nature, and second, the object's taking us back to the pleasure of childhood play, in a world before the adult schema, a world of infinite imaginative possibilities. From this, a first corollary of the human reaction to the automaton can be drawn, as follows:

> *The less powerful (often but not always because it is small) and more apparently mechanical an automaton is, the more amusing it is.*

A famous recent example of this is the robot duo R2-D2 and C-3PO of the *Star Wars* movies—despite exhibiting human-like characteristics and behavior, neither of them can possibly be mistaken for living creatures, and they obey their human masters absolutely (C-3PO is life-sized but made innocuous by its whimpering cowardice as well as its physical slowness, clumsiness, and stiffness). But even such figures of robotic fun hint at the dark side of such entities. The film scholar Daniel Dinello asserts that the representation of benign, servile, and silly robots represses people's essential fear of them. He makes a provocative point that people's "amusement at the groveling robot is even reminiscent of earlier racist laughter at black people forced to play submissive butlers and servants in pre–civil rights era movies."[65] In fact, in *Star Wars: Episode II—Attack of the Clones* (2002), the character Obi-Wan Kenobi (Ewan McGregor) says ominously that if droids (the *Star Wars* term for robots) "could think, there'd be none of us here." From this, I can proceed to the more negative emotions aroused by the automaton.

If the first corollary is right, then presumably the more powerful and lifelike an automaton is, the less amusing and more disturbing it is. This is true in a general sense, but this idea must be refined through clearer delineations of the emotional responses. When we have a negative reaction to an automaton, the disturbance can occur at two related but distinct levels. The object may arouse fear, which can turn into terror at an extreme point; or it may arouse the feeling of creepiness (the uncanny), which can turn into horror. But what is the difference between the natures of terror and horror, and what arouses one or the other? Terror arises in situations where our physical well-being is threatened, especially when our very lives are in danger. Horror, on the other hand, occurs in reaction to something that disturbs us psychologically, as in a categorical anomaly that could potentially undermine our grasp of reality. To clarify with concrete examples: if we encountered an armed criminal in a dark alley or a predatory animal in the wild, we would feel terror from the possibility of being hurt or killed, but there would be nothing innately uncanny about the experience. Likewise, we may find an entity or an event creepy to the highest degree of horror even if there's no *physical* danger: for example a harmless person with a deformity that we may find extremely disturbing, or the movement of something obscure in the dark that suggests a ghost.

What makes this distinction between terror and horror complicated is that the two can meet at an extreme point in a complementary fashion. The terror in the face of approaching death can give rise to psychological horror at the idea of the extinction of the conscious self. And the experience of horror in response to a highly disturbing event can lead to such a serious undermining of one's grasp of reality that it arouses the terror of finding oneself in a world that has become utterly alien and therefore infinitely dangerous to one's well-being. In fantastic literature and film, this connection between terror and horror is commonly exploited to create the maximum effect of disturbance, which is why dangerous villains, supernatural beings, and out-of-control robots are often presented as deformed (to be more precise, deformed versions of normal human beings and animals) as well as life-threatening, and sometimes even sexually depraved.

So what exactly are the circumstances under which the automaton stops being amusing and becomes terrifying or horrifying? What are the circumstances under which the automaton arouses terror as opposed to horror, and vice versa? And if the automaton can be such a disturbing object, why indeed do we keep entertaining ourselves with stories and movies in which they rebel and try to destroy us? It seems strange that we enjoy spectacles of our own endangerment, the deliberate arousal of our terror and horror in the face of the dangerous and the uncanny, as in the playacting of the young girl fleeing the robotic performer I encountered in Santa Monica. What is the nature of this attraction to the terrifying and the horrifying?

A significant insight into how the same object, event, or situation can be a source of both terror and pleasure under different circumstances is found in Edmund Burke's classic study *A Philosophical Enquiry into the Origins of Our Ideas on the Sublime and Beautiful* (1757). Burke notes the oddity of the fact that things that pose a danger to our self-preservation causes terror, yet when those same things are presented to us with the possibility of actual harm removed, we derive a peculiar pleasure out of it that is a species of the sublime.[66] For instance, we would be terrified if we encountered a wild beast or a murderous lunatic, but we enjoy looking at the same animal in a zoo or watching a drama about homicide. The crucial element here that turns intense fear into the pleasurable sublime is, to use a contemporary phrase, that of controlled parameter. If what frightens and disturbs to the highest degree is taken and secured in a safe environment, whether physically in a cage or as make-believe on stage or on film, it can give us a cathartic thrill. Aristotle, in fact, in his description of catharsis, points out that things that are "painful to see" in real life, including low animals and dead bodies, are pleasurable to gaze at if they are imitated in art.[67]

The gothic novelist Ann Radcliffe in her 1826 dialogue "On the Supernatural in Poetry" points out that Burke focuses his discussion of the sublime only on the feeling of terror, neglecting that of horror, which has to do with the uncertain and the obscure, when his argument would work just as well for the latter emotion.[68] In other words, people seem to experience the pleasure of the sublime when they are presented, within controlled parameters, with things or situations that threaten their psychological security as well as their physical well-being. I have noted in my discussion of Mary Douglas's ideas on categorical anomalies that the appearance of such entities can be traumatic because of the threat they pose to the reality schema of an individual or of an entire community; but they can also be a source of deep fascination. We are drawn to them because a part of us yearns to be free of the artificial and arbitrary strictures of our culturally imposed worldview. In societies that are willing to accommodate the anomalous to a certain extent, allowing people temporary escapes into the world beyond the conventional strictures of society without having them fall into anarchy and madness, the solution is the enactment of rituals in which dangerous entities are placed in strictly bounded zones. Within them they can be gazed at and interacted with, while they are prevented from unleashing their dangerous power into the community.

Applying this idea, one can theorize that what causes people terror and horror can be transformed into sources of the pleasurable sublime when placed within a controlled parameter, because that allows people to come face to face with the vastness, the danger, and the chaos of the world and existence itself without the possibility of actual physical or psychic harm. The pleasure comes from the feeling of freedom from the constricting patterns of normal, everyday life, the safety and predictability of which is made possible by the maintenance of limitations on possibilities. So the lethal power of the tiger behind the bars of a zoo provides us with a glimpse of the awesome world of the wild, while a horror movie about ghosts teases us with suggestions of the supernatural. Since most of us have an interest in living in the mundane world of civilized society, the pleasurable sublime works only under the condition that its subject remains within the controlled parameter, as an escaped tiger or the actual appearance of a ghost would cause terror and horror, respectively, perhaps to the point of utter panic and madness.

From this idea, a few more corollaries can be drawn about the human reaction to the automaton.

The more powerful (often but not always because it is bigger) a machine, the less amusing it is and the more sublime.

We feel awe rather than amusement in the face of a great, powerful machine, at its locomotive strength, efficiency, and relentless productivity, in the same way as at the sight of a grand view of nature.[69] One of the most famous descriptions of this feeling can be found in "The Dynamo and the Virgin," the famous chapter of Henry Adams's autobiography that describes his experience in the great hall of dynamos at the Great Exposition in Paris.

> As he grew accustomed to the great gallery of machines, he began to feel the forty-foot dynamos as a moral force, much as the early Christians felt the Cross. The planet itself seemed less impressive, in its old-fashioned, deliberate, annual or daily revolution, than this huge wheel, revolving within arm's-length at some vertiginous speed, and barely murmuring—scarcely humming an audible warning to stand a hair's-breadth further for respect for power—while it would not wake the baby lying close against its frame. Before the end, one began to pray to it; inherited instinct taught the natural expression of man before a silent and infinite force.[70]

The religious element at the end of this passage is of particular interest since the sublime is often expressed in terms of the otherworldly and the supernatural.

For the automaton, there is the added factor of being a liminal object, which heightens its aura of power. As already explained, the world beyond the adult schema of reality is one of infinite chaos and power, from which the magician draws his uncanny abilities, including the divine capacity to create life from lifeless matter. The automaton, as a thing that traverses the categories of animate/inanimate, living/dead, gives the appearance of being a product of such a power. This is why creators of artificial creatures, even in the modern period, are so often portrayed in the manner of magicians (solitary, half-mad beings dwelling in the margins of human community, as described by Mauss). To give a few examples, Mary Shelley's character Victor Frankenstein, before he secludes himself to create his creature, draws inspiration from the works of Cornelius Agrippa, Albertus Magnus, and Paracelsus (medieval and Renaissance authorities on magic who are also associated with artificial beings); the fictional Thomas Edison in Villiers de l'Isle Adam's novel *Tomorrow's Eve* also works in apparent isolation and dabbles in spiritualism; and the mad scientist Rotwang in the film *Metropolis* lives by himself in a gothic house with a prominently displayed pentagram (a traditional symbol of magic).[71]

The technological sublime turns into terror the moment the powerful machine appears to have gone out of human control.

As with a tiger that escapes from a zoo or a natural phenomenon like an earthquake or a volcanic eruption that can actually endanger people, the loss of the controlled parameter effectively cancels the pleasurable sublime. So the awe we might feel toward the machinery of an aircraft, a locomotive, or a factory turns into terror the moment we realize that due to some malfunction it is no longer under the control of its human masters and is operating under what seems like its own will with the possibility of causing us harm. It is significant that in films dealing with fantastic beings like robots (e.g., *Westworld* [1973]), cloned creatures (*Jurassic Park* [1993]), and giant animals (*King Kong* [1933, 1976, 2005]), the stories are precisely about the sublime awe turning into terror when such entities slip out of their constraints to wreck havoc on people.

The more lifelike an automaton, the greater the sense of the uncanny sublime.

Even if an automaton poses no physical danger to the viewer, our level of uneasiness toward it increases as it looks and acts more and more like a living creature, especially a human being. This is because increasing proficiency at the mimicry of life takes the object into the liminal zone, posing an ever greater danger to our reality schema based on the categories of the animate/inanimate, natural/artificial, and living/dead. What allows it to be a source of the pleasurable sublime or the sublime uncanny, without falling into horror, is the controlled parameter, which is defined in this case by our certain knowledge that it is indeed only a machine, no matter how good it is at pretending to be a living being. This understanding allows us to enjoy its spectacle, as it frees us temporarily from the conceptual boundary that separates us from the inanimate, inorganic world. So even if we feel the creepiness in response to the moving, talking thing, we are captivated by it through the sense of sublime awe at its mimetic effectiveness.

The sixteenth-century sculptor Benvenuto Cellini relates an interesting story in his autobiography of how he took advantage of this sense of the sublime aroused by an inanimate object that takes on an animate characteristic. On the occasion when he was to present a new work, a silver statue of Jupiter, to King Francis I of France, Cellini's enemy, Madame D' Étampes, conspired to undermine him by placing it in a gallery full of beautiful copies of antique statues. When the monarch came to view the piece, however, Cellini had his servant push his work toward the king (having placed four balls beneath its wooden plinth to make it mobile), which impressed him greatly because it "made the statue seem alive."[72]

At the Museum of Westward Expansion, located in a complex below the famous Gateway Arch monument in St. Louis, Missouri, several animatronic automata representing a Native American, a cowboy, a cavalryman, and other figures move and narrate the history of the American West. On my visit there I noticed that in every one of these displays an object, like a desk or a rock, was placed strategically in front of the automaton as a barrier between it and the spectators. These objects' ostensible purpose might be to prevent people from reaching out and touching the figures, but I think there must also be an element of protecting the people from the automata as well—or rather, easing their sense of the uncanny with the barriers between them symbolically marking the controlled parameter. The visitors might find it too creepy to have the figures moving and talking right in front of them with nothing to protect them if they were to go out of control and reach out to touch *them*. The artist Elizabeth King, in her research into the origins of a sixteenth-century monk-automaton, now at the Smithonian Institution, describes the stirring of her "animal flight urge" in response to the forward movement of the figure toward her.[73]

The uncanny sublime loses its pleasurable aspect and turns into horror when a human being turns out to be an automaton, or vice versa, in an unexpected way.

With no basis of safety from which to enjoy the transcategorical (i.e., the loss of the conceptual controlled parameter), the threat to one's sense of reality is at its maximum. This may last only an instant, as in the case of a man who is frightened by a statue that moves before he understands that it is really a robot or a person who was pretending to be a statue (the reestablishment of the controlled parameter of the knowledge of what the entity really is). Nevertheless, in that particular moment of the unexpected, it is the sense of complete uncertainty that brings about the shock. In fiction, characters fall into madness or are driven to the edge of it by such encounters. I have already mentioned Hoffmann's story "The Sandman," in which a young man loses his mind after witnessing his beloved being mishandled as an automaton. In a similar scene, the character of Lord Ewald in the Villiers novel *Tomorrow's Eve* experiences a moment of severe disorientation when his lover, Alicia Clary, confesses that she is, in reality, the android Hadaly; in Herman Melville's tale "The Bell-Tower," the people of an Italian village go on a frenzied rampage after they find the corpse of an engineer, with its head shattered, beneath a club-wielding automaton; and more recently, in Jeff VanderMeer's wonderful story "Dradin, in Love," a missionary whose mind is already on the verge of collapse from his sojourn in the jungles loses his sanity altogether after destroying a mobile mannequin he has mistaken for the woman of his dreams.[74] In other fictional works, including Ernst Jünger's *Glass Bees* (1957) and Philip K. Dick's *Martian Time-Slip* (1964) and *We Can Build You* (1972), characters experience mental breakdowns not from such sudden revelations but just from working in the robotics industry, as if prolonged exposure to artificial beings could be enough to eventually drive someone insane. One can speculate that if we are able to produce perfect robotic simulacra of living beings and sentient artificial intelligence in the future, there will be many who, from this sense of horror, will regard them as abominations and react violently toward them, while others will be forced to adjust their definition of life and consciousness in order to assimilate them into their worldviews.

The central attraction of the blockbuster film *The Terminator* is the spectacle of the technological sublime (as per the second corollary) exhibited by the awesome power and relentless drive of the time-travelling killing machine played by the former bodybuilder Arnold Schwarzenegger. The combination of the display of strength and the image of a machine that is completely successful at pretending to be human (as per the fourth corollary) was what made him an iconic figure in popular culture, even though the character was the villain of the story. The most uncanny scene in the film occurs toward the end when in its final battle against its human opponent, explosions tear off some of its skin, revealing shiny, metallic parts within. It is in that moment that it appears as a truly liminal, transcategorical monster. In fact, it becomes less frightening when all of its skin is burnt off to reveal a mechanical robot, since the complete loss of its human appearance takes away its liminal aspect, leaving only the terror of a life-threatening machine.

The various emotional reactions to machinery in general and the automaton in particular under different circumstances can be charted as shown in Table 1 on the following page.

In 1970, the Japanese robotics expert Masahiro Mori published his article "The Uncanny Valley," in which he theorized that people's feeling of "rapport" with machines increases as they become more lifelike—our affection toward toy robots rises as they better resemble and mimic life.[75] At a point in this rising level of rapport that is tied to resemblance to the living, Mori asserts, there is a sudden and precipitous drop in the comfort level, as we find uncanny the very robots that pretend to be living beings so well. In other words, the more lifelike a robot, the more at ease we feel with it, but when it reaches a certain level of being too lifelike, we suddenly find it creepy and horrifying. Then, as the robot reaches an even greater level of perfection in the mimicry of humanity, the rapport level goes back up, leaving behind what he calls the "uncanny valley," as expressed in the graph on page 49.[76]

On the basis of this theory, Mori advises robot-makers that when making their products lifelike, they should try to hit the first peak of rapport as closely as possible for maximum comfort level for the people who will interact with them, without falling into the uncanny valley.

This is, obviously, a highly provocative theory for the purpose of this study, but experts currently involved in robotics and cognition research have pointed to some of its problems. Karl MacDorman and Hiroshi Ishiguro have revealed that Mori's ideas were based on theoretical extrapolations from anecdotal experiences with robots, mannequins, and prosthetic limbs, not on empirical evidence from controlled testing, and David Hanson has questioned the simplistic notion of "realism" in human resemblance in the study.[77] In fact, when Mori published the essay, no robot existed whose resemblance to a human being was so perfect that one could verify whether it did pull itself out of the uncanny valley. The right end of the graph, then, is purely speculative, and Mori's reference to the traditional Japanese bunraku puppet is problematic since, as he admits, it is seen in a theater at a sufficient distance to make its size and deficiency in lifelikeness irrelevant.

Table 1. Shifts in Emotional Reactions to the Automaton and Other Machines

	Within Controlled Parameter		Beyond Controlled Parameter
Physical			
(Controlled parameter is defined by the knowledge that a machine can cause no harm because it is under control)	A machine that is inherently harmless and interesting because it is particularly useful or beautiful	A powerful machine that is potentially dangerous but is under control so that it can cause no harm	A powerful machine that is dangerous and can cause harm at any moment because it is no longer under human control
	FASCINATION →	**SUBLIME** →	**TERROR**
Conceptual			
(Controlled parameter is defined by the knowledge that the automaton is not really alive, no matter how good it is at pretending to be)	An automaton that imitates life but utterly fails to convince that it is really alive	An automaton that does an excellent job of imitating life though one can tell that it is not really alive	An automaton that does such a good job of imitating life that one cannot tell (even temporarily) if it is a living being or not
	AMUSEMENT →	**UNCANNY SUBLIME** →	**HORROR**

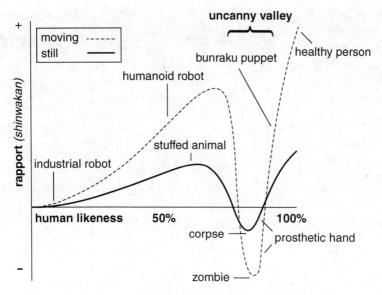

Masahiro Mori, graph of the uncanny valley. A simplified version of the original graph in Masahiro Mori, "Bukimi no tani" [The uncanny valley], *Energy* 7, 4 (1970): reproduced from Karl F. MacDorman and Hiroshi Ishiguro, "The Uncanny Advantage of Using Androids in Cognitive and Social Science Research," *Interaction Studies* 7, 3 (2006): p. 299, fig 2. Courtesy of Karl MacDorman.

Furthermore, the term "rapport" lacks a clear meaning. It apparently denotes comfort level, as Mori claims that what happens in the uncanny valley is that we notice flaws that in the machine's attempted human likeness that stand out so to give us a feeling of fright, as when one takes someone's hand and realizes through its hardness and coldness that it is a prosthetic. There is no explanation, however, of why at a certain level of resemblance to life, flaws that were unproblematic in less lifelike machines should suddenly become sources of discomfort. Perhaps the cause of the uncanny is not in the flaws but the conceptual uncertainty that they cause in the perceiving mind.

In a recent study to verify Mori's findings empirically, a group of Indonesian subjects (less likely to be familiar with robots than Westerners or Japanese) were shown a series of photographic images that morphed from a clearly mechanical robot to a robot resembling a human being to a true human being.[78] The participants were then asked to rate the photographs in terms of human likeness, familiarity, and eeriness. The following two graphs show the results for a male and a female image.

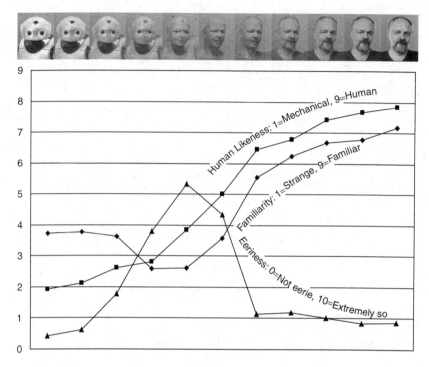

Graph of reactions to photographs morphing from robot to human (male). Reproduced from Karl F. MacDorman and Hiroshi Ishiguro, "The Uncanny Advantage of Using Androids in Cognitive and Social Science Research," *Interaction Studies* 7, 3 (2006): p. 305, fig 4. Courtesy of Karl MacDorman.

A few things stand out in this study. As is clear on the left side of the graphs, that which is unfamiliar is not necessarily uncanny (supporting Freud's criticism of Jentsch). Contrary to Mori's theory, the viewers did not find the least human-like robots eerie at all, even though they were moderately unfamiliar with them. In addition, instead of greater resemblance to human beings causing a steady rise in the sense of rapport, there is a noticeable dip in the middle in the transitional stage between the robot and the human. Most significant, what we have here is an "eerie peak," as opposed to the "uncanny valley," precisely in that middle range when there is a maximum level of uncertainty about the nature of the thing observed, when we are in the liminal zone between the artificial and the natural. So it seems apparent here that it is indeed uncertainty that is the operative concept in the arousal of eeriness.[79]

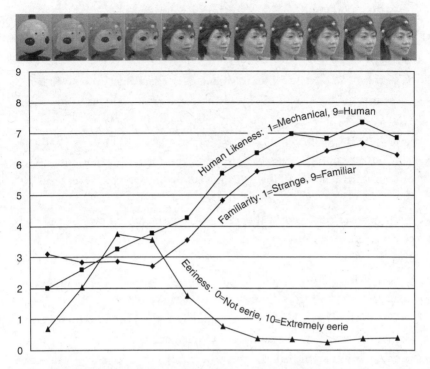

Graph of reactions to photographs morphing from robot to human (female). Reproduced from Karl F. MacDorman and Hiroshi Ishiguro, "The Uncanny Advantage of Using Androids in Cognitive and Social Science Research," *Interaction Studies* 7, 3 (2006): p. 305, fig 4. Courtesy of Karl MacDorman.

Mori also speculated in his study that the sense of uncanny might have something to do with death—that flaws we detect in the human-like robot reminds us of a cold, still corpse. MacDorman and Ishiguro explore this point further, questioning whether androids are frightening because their cadaver-like appearance reminds us of our own mortality.[80] In addition to these authors' empirical study of this issue, one can find many literary instances of this link between automata and death. In E. T. A. Hoffmann's story "The Automaton" (1814) the character Ludwig, shaken by viewing an uncanny automaton, decries such objects as those of "living death or inanimate life"; and in William Saroyan's novel *The Human Comedy* (1943) a little boy is entranced by an automaton named Mr. Mechano at a shop window, until night falls and he suddenly takes fright at the object: "He looked around, feeling a deep silent steady horror about all things—the horror of Mr. Mechano—Death!"[81] In the same vein, Gaby Wood asserts that the automaton is the very embodiment of death.

> Every time an inventor tries to simulate life mechanically, he is in fact accentuating his own mortality. He holds his creation in his hands, and finds, where he expected life, only the lifeless; the closer he comes to attaining his goal, the more impossible it reveals itself to be. Rather than being copies of people, androids are more like mementi mori, reminders that, unlike us, they are forever unliving, and yet never dead.[82]

Throughout this study I mention many instances where the object is spoken of in such mortal terms. Given the conceptual framework I have provided, however, I would like to assert a source of unease different from the corpse-like nature of the unmoving robot.[83] I believe the automaton's connection to death is not the reminder of mortality through the resemblance to a dead body but the evocation in our minds of dead things doing things they should not be doing: move, talk, and otherwise act like living creatures. A wax statue can be creepy because of its close resemblance to a person, but not to the extent of an automaton that can cause real uneasiness and horror through its animate behavior. The uncanny automaton is more akin to a zombie, a living mummy, or a vampire, as it crosses the categories of animate/inanimate, living/dead (as intimated in the passage from Wood above) and so threatens our reality schema in a particularly frightening way. So when a person finds an automaton particularly eerie, it is often expressed in terms of how one might feel if faced with an undead creature. The connection in this case is their common status as liminal entities, which as Victor Turner has pointed out are often likened to the dead and the monstrous.

If the source of both the fascination and discomfort in reaction to the automaton lies in its ambiguous nature that makes it a category-defying object threatening our normal schema of reality, one must question whether the same dynamic can be applied to all entities of uncertain status. I believe that any object, event, or situation that disrupts our normal worldview commands attention, but with differing levels of emotional reaction that are determined by how great a threat the phenomenon poses to our reality schema as a whole. The discovery of a new species of animal or plant that does not fit into the established scheme of zoological or botanical organization is likely to cause wonder but not horror, since scientists can easily assimilate such an entity by creating a new category or adjusting the method of categorization, whereas the appearance of an apparently extraterrestrial or supernatural being could be traumatic. Objects or events that seem to cause the greatest emotional reactions seem to be those that have directly to do with the nature of human identity itself. Because an automaton is a humanmade object, as opposed to one found in nature, and one that mimics life, it suggests all kinds of essential and disturbing questions about what exactly a human being is, raising doubts about our own place in the binaries of animate/inanimate, spiritual/material, soul/body. Are we also mere machines consisting of matter functioning according to a preset program, or is there also a nonmechanical and nonmaterial aspect of us?

What complicates the matter is that despite the prevalent nature of binary thinking in human culture, whether rooted in biology or not, we are certainly not condemned to it, as we are capable of understanding the arbitrary nature of its construction and of questioning our own prejudices based on it. So I can grasp the ideas that at a certain level of reality the distinction between matter and energy breaks down, that there is a variety of sexual orientations that are equally natural, and that varying but legitimate historical narratives can be written from differing perspectives. Likewise, an entire culture, given the right circumstances and influences, can become more open-minded and willing to critically examine its own values and prejudices without falling into confusion and anarchy. In fact, such a society is likely to be stronger and more enduring since its flexibility would allow it to handle change, contingency, and novelty much better than one holding rigidly onto its worldview.

At this point, I want to make absolutely clear my purpose in laying out the ideas in this chapter. I have pursued the reasons behind the perennial interest in the automaton throughout Western history and used terms such as *categories, schema, the liminal,* and *controlled parameter* in a deliberately abstract fashion. It is important to understand, however, that the conceptual framework presented here is not meant to be a suprahistorical explanation of how the automaton has functioned as an idea in every period. This framework points to an essential explanation of the automaton's captivating power, or why it is inherently an interesting entity and what has allowed people to make fruitful use of it for various cultural and intellectual purposes throughout the centuries. But the exact content of the meanings people have invested in the object must be examined historically in each specific context. In other words, the explication presented here of why the automaton is such a fascinating and disturbing thing in itself cannot account for what a medieval theologian, an eighteenth-century philosopher, or a twentieth-century filmmaker have seen in the object and what it represented in their works. For the historian it is essential to elucidate the meaning of the automaton motif in specific contexts since the ideas that people invested in the object and the feelings it aroused changed so radically from one era to another. Because the source of its ability to both enthrall and terrify originates from its ability to traverse the categories of reality, it is an inherently unstable object with a protean essence that can be fully explicated only within a given historical and cultural milieu. So beyond the abstract explanation of its essential power, the automaton has to be comprehended historically.

For a comprehensive understanding of the automaton idea, then, one has to examine how the levels of amusement, fascination, unease and horror at the object fluctuated in accordance to the beliefs, concerns and needs of each period. One also has to analyze the ways a particular culture articulated such reactions, revealing how the object threatened or contributed to that culture's ideas about the nature of humanity and the world. And so I turn to the life of the automaton as a historical being.

2

Between Magic and Mechanics: The Automaton in the Middle Ages and the Renaissance

In 968, Liudprand, bishop of Cremona and advisor to Holy Roman Emperor Otto I, was sent on a mission to Constantinople to negotiate a possible marriage between a Byzantine princess and Otto's heir. His account of the journey, *The Embassy to Constantinople*, is a litany of complaints about his maltreatment at the hands of the corrupt Byzantines, including the "fat-headed" and "piglike" Emperor Nicephorus II Phocas.[1] Liudprand portrays the Eastern state as a decadent and moribund civilization, putting under a favorable light the newly energetic West under the leadership of Otto.[2] Yet in his longer work the *Antapodosis*, a narrative of the political history of his time, he admiringly describes some wonders of Constantinople, including those in the great throne room of the emperor. In addition to the magnificent Throne of Solomon, which could be elevated to the ceiling, the chamber featured a bronze tree with automata in the forms of a lion and birds of different species, which roared and chirped as people approached.[3] Even as he denigrates the Byzantines, Liudprand marvels at the greatness they inherited from the ancients, including the artificial wonders they possessed—probably based on the designs of Hero of Alexandria, such sophisticated mechanical knowledge having been lost in the Latin world after the fall of the Western Empire.

Throughout the Middle Ages and the Renaissance many regarded such marvels and the arcane learning necessary for their construction as magical, with all the ambivalence that concept carried in the Christian world. So the automaton motif in the premodern context can only be explicated through an understanding of the complex link between the mechanical and the magical. The science fiction writer Arthur C. Clarke famously asserted that "any sufficiently advanced technology is indistinguishable from magic."[4] The problematic phrase here is, of course, "sufficiently advanced" since that has to be defined relative to the level of technological familiarity of a given society. A typical western European of the eighteenth century, for instance, would have been found nothing magical about a clock, but anxiety-inducing uncertainty about one would not have been unusual in the early Middle Ages, when any complex device of unknown purpose or design could provoke accusations of sorcery.

In fact, the distinction between what is magical, typically defined as the use of preternatural power to create extraordinary effects in the world, and what is technological, or the use of machines powered by natural forces, is a modern one that did not begin to be established until the seventeenth century. During the Renaissance, even after many ancient works on mechanics and mathematics were rediscovered and made use of, both these fields were often regarded as magical. As late as 1684, when John Wilkins, a key figure of the scientific revolution as one of the founders of the Royal Society, published his major work on mechanics, he entitled it *Mathematicall Magick*, "in allusion to vulgar opinion, which doth commonly attribute all such strange operations unto the power of Magick."[5] His denigration of the attitude reveals the prevalent linkage of the technological with the magical in the popular imagination that he was seeking to correct through the book. What makes this matter complex is the uncertain status of magic in general throughout the Middle Ages and the Renaissance, as can be seen in major intellectuals' attempts to distinguish different types of it according to their sources, efficacy, and legitimacy. It is crucial, then, that I first describe how magic, as it manifested itself in many different forms, was defined and organized.

When someone performs an uncanny act, for instance animating an inert object like a statue to make it move or speak, what are the possible explanations for how the feat was achieved? A modern person might consider that it was done either mechanically (i.e., a robot or a puppet) or through trickery, as in the statue turning out to be a human being in disguise. In a culture where there is a lack of familiarity with complex machinery and widespread belief in the ability of some to tap into otherworldly powers, magic can be a legitimate explanation for such a phenomenon. Then the question arises of what kind of magic is being utilized. From the medieval Christian point of view, a number of positions could be taken: that all forms of magic are products of pagan superstition and are fraudulent tricks; that some forms of magic are real and efficacious but should not be practiced because they derive their power from demons or Satan; that some forms of magic are real and divinely sanctioned because their power comes from angels or from God; that some forms of magic are real but neither diabolical nor divine in themselves, as they involve the manipulation of occult (i.e., hidden and not apparent to everyday perception) and impersonal forces in the natural world.

Questions on whether magic was real or not, and if real, whether it was always demonic or if there were also divine or neutral forms of it, were debated throughout the Middle Ages and were never satisfactorily settled in a consensus. The Bible itself offers no definitive answer since it condemns pagan magic as fraudulent but then relates stories like that of Moses competing with the magi of Pharaoh in turning sticks into snakes (showing that magic can be divinely sanctioned but also that pagan magic was real, just not as powerful as divine magic, as the Mosaic stick-snakes overcome those of the magi). Valerie Flint has demonstrated that in the early Middle Ages there were theologians who condemned all magic as false and others who decried the forms of magic associated with paganism but advocated the efficacy of what could be called Christian magic. The latter position was strategically advantageous for the purpose of establishing Christianity, in that it translated age-old magical practices into the language of the new religion instead of forcing people to abandon them altogether, thereby avoiding a destructive confrontation between faiths.[6] A Christian counterpart to virtually every type of pagan magical activity can be found, from saints and other holy men performing the same acts as sorcerers, through the grace of God or with the help of angels, to certain prayers and images of Christ having the same beneficial effects on sick people as magical formulas and talismans. And despite Christian condemnation of idolatry, we are familiar with stories reported even today of images of saints, Christ, or the Madonna moving, weeping, bleeding, and otherwise showing signs of life.[7]

In addition to this theological tension of either condemning all magic or Christianizing certain forms of it to make them legitimate, a new intellectual position was developed in the thirteenth century by such thinkers as William of Auvergne and Albertus Magnus on the Continent and Robert Grossesteste and Roger Bacon in England. William and Albertus not only affirmed that demons and angels could indeed aid one to magically cause harm or good in the world but also advocated what they called "natural magic," through which one could perform marvelous acts through the manipulation of forces inherent in the world. Such magic does not involve spirits and so is neither good nor evil in itself, though the end to which it is used could be to the benefit or detriment of humanity.[8]

This position was articulated following the expansion of intellectual life in the high Middle Ages that is sometimes referred to as the twelfth-century renaissance. At the end of the tumultuous period from the dissolution of the Western Roman Empire to the investiture controversy of the eleventh century, Europe entered a relatively stable period following the abatement of the conflicts over the latter issue and the commencement of the Crusades. The proliferation of institutions of learning, including universities in Paris, Bologna, and Oxford in the late eleventh and twelfth centuries, served the needs of a growing class of literate professionals in the expanding cities across Europe. In addition, renewed interest in ancient knowledge about the natural world, initiated by the reappraisal of Plato's *Timaeus* and Aristotelian works that were beginning to trickle back into Europe through Muslim Spain, set the stage for the revival of learning and art that is associated with such seminal figures as Peter Abelard, Suger of St. Denis, Bernard of Clairvaux, and John of Salisbury.[9] Crucial to the development of natural magic was the twelfth- and thirteenth-century transmission of Arabic works on alchemy and astrology that described such operations as transmuting metals, healing the sick, and making prophecies through arcane methods of observation, calculation, and manipulation of natural properties.[10] Intellectuals like Marsilio Ficino, Cornelius Agrippa,

John Dee, and Tommaso Campanella took up the advocacy of this neutral form of magic again in the fifteenth and sixteenth centuries, in the context of the revived interest in Neoplatonic and Hermetic ideas.

The difficulty was that even if one were to adopt the idea that some forms of magic were neutral and therefore legitimate means of investigating the natural world, there was no general agreement on which exact knowledge, practices, and methodologies counted as natural and which did not. This is a complex matter to unravel, especially for those who would see in natural magic certain ideas and practices that contributed to the scientific revolution of the seventeenth century. For example, Roger Bacon is often regarded as a protoscientist for his advocacy of the experimental method in studying nature and for his descriptions of the myriad technological marvels—such as a great ship that could be manned by one person, a horseless carriage that could travel at high speeds, a flying machine, and a vessel that could travel underwater—he thought could be created through what he called the study of art and nature.[11] Yet even as Bacon ridiculed magical practices as fraudulent activities of the superstitious and the gullible, he believed in the neutral efficacy of astrology for both medical and prophetic purposes and in the reality of demons that could cause mischief in the world.[12]

A particularly complicated case is that of image magic, including theurgy, which uses statues and other artificial figures to summon otherworldly beings. One of the most important Arabic works of magic that contains descriptions of such an operation is the *Picatrix*, a compendium of alchemical and astrological magic written in Muslim Spain probably in the eleventh century, a Spanish translation of which appeared in the thirteenth century.[13] William of Auvergne, the first major advocate of natural magic in the high Middle Ages, declared such practices both fraudulent and heretical since they dealt with demons, which in reality cannot be coerced by human beings to do anything, including being summoned into statues to animate them.[14] In the sixteenth century, however, Cornelius Agrippa, whose works on magic were informed by the *Picatrix*, cited William on the practice without mentioning his condemnation of it and even went so far as to assert that the operation was evil only if one dealt with demons but perfectly legitimate if one used it to summon angels.[15]

The continuing debates on magic throughout the Middle Ages and the Renaissance were bedeviled by such uncertainties arising from the shifting boundaries of religious and intellectual legitimacy. In times of relative social stability and intellectual freedom, such as the twelfth and thirteenth centuries, there was room for intellectuals to consider such unorthodox ideas on magic with impunity, but in a more repressive environment expressing such views led to serious consequences. In the period of the Counter-Reformation, when there was a real political need to affirm the Church's power and authority, advocates of natural magic were severely persecuted for their beliefs. The practice of the dark arts was one of the crimes for which Giordano Bruno was burnt at the stake, and Tommaso Campanella was repeatedly imprisoned and tortured, while John Dee was hounded by accusations of demonic sorcery throughout his life.

The appreciation of the uncertain status of magic throughout the centuries is crucial to understanding the automaton idea for several reasons. First, to return to Arthur C. Clarke's dictum, any machine whose operation or purpose was not apparent to the viewer was liable to be regarded as magical, especially for an object, like the automaton, that was evocative of theurgy.[16] Consequently, both fictional and actual automaton-makers were often regarded as sorcerers and necromancers. Second, the advocates of natural magic themselves regarded mechanics as a type of neutral magic, closely related to or subsumed under the rubric of mathematics. During the Renaissance, people like Agrippa, Dee, and others explicitly pointed to the automaton as a type of marvel that could be achieved through the use of natural or mathematical magic. Third, for reasons detailed in the last chapter, the automaton was such a captivating but unstable object in terms of its conceptual significance that it was deployed in the ongoing debates on the nature and legitimacy of magic. Through its innate power to both fascinate and horrify, the automaton occupied and exposed the liminal space between the legitimate and the heretical, the natural and the demonic, and the mechanical and the magical.

Lorraine Daston has shown that in the high Middle Ages all phenomena were categorized in connection with three discrete realms, as outlined by Thomas Aquinas—the natural, the preternatural, and the supernatural.[17] Entities and events of the first realm were those that occurred in accordance with the regular laws governing the world; those of the last two had extraordinary natures that appeared to defy the normal workings of nature. A supernatural phenomenon was easy to define: it was a miraculous act of

God that resulted from his unmediated intervention into the world. All other entities or events of unusual nature were placed in the amorphous middle category of the preternatural ("that twilight zone between the natural and supernatural"), including freakish natural phenomena, which sometimes had portentous value, and those caused by demons, astral intelligences, and other spirits, who brought them about through skillful manipulation of natural forces (the power to actually suspend worldly laws belonging to God alone).[18] Until the idea of the preternatural collapsed when mechanistic philosophers of the seventeenth century naturalized it, the category served as an explanatory concept for all events and entities of anomalous and ambiguous status. The idea is crucial to understanding the significance of the automaton in the Middle Ages and the Renaissance since the object was regarded as a powerful example of that kind of a strange, unusual, and marvelous thing.

As in the case of the ancient world, for a clear elucidation of the significance of automata, it is important to specify their different types. By adjusting for the medieval and Renaissance periods the four categories I presented in the last chapter, one can find examples of the following types of automata:

1. fictional and of otherwordly origin
2. fictional and of human manufacture but involving some form of magic
3. of actual human construction or design
4. speculative

In the medieval period there is no significant report of an actual automaton or mechanical description of one (the third category) in western Europe until the thirteenth century. In the notebook (ca. 1240) of the master of gothic architecture Villard de Honnecourt, filled with illustrations of wonders he witnessed, heard of, or imagined during his extensive travels, one finds many mechanical devices, including automata. The latter include a mechanical lectern in the shape of an eagle that can hold up a Bible and turns its head toward the book's reader by a pulley device and a sundial shaped like an angel that follows the movement of the sun.[19] Only in 1299, however, is there a record of actual automata being built: the "engines d'esbattement" at the castle of Hesdin in Artois, by the order of its lord, Count

Robert.[20] At the castle's park an enormous funhouse was built, at great expense, filled with trick devices like pipes that suddenly spouted water or flour at viewers and many animal automata, including an astounding troupe of playful monkeys. We have detailed knowledge about them through a 1432 account by Philip the Good, the Duke of Burgundy, who was obsessed with wonders and bought devices that had fallen into disrepair and renovated them.[21] In addition, during the 1377 coronation pageant for the eleven-year-old Richard II of England, the image of a golden angel set atop a tower bent down to offer a golden crown to the monarch-to-be.[22]

Such automata would have been difficult to construct in the early Middle Ages due to the loss of ancient mechanical knowledge after the fall of the Western Roman Empire. In the eleventh and twelfth centuries, in conjunction with the expansion of intellectual life in the period, the stabilization of the feudal system gave rise to what some scholars have called the medieval industrial revolution, which saw the expanded use of natural power and the invention of new techniques and devices like iron-casting, the tidal mill, the water-driven bellows, and the compound crank.[23] In general, the mechanical arts were traditionally regarded as low forms of knowledge unfit for learned gentlemen, medieval intellectuals having inherited from the ancients the distinction between the liberal arts, subjects associated with literate scholars and organized into the established educational systems of the trivium and the quadrium, and the servile arts, including mechanics, that were linked to the manual labor of the lower orders.[24] The situation began to change in the twelfth century, as intellectuals began to find a more respectable place for technological study in the schema of learning. The most notable case of this is Hugh of St. Victor, who included mechanics in his reclassification of the liberal arts.[25] The case of Hugh is a significant one since he was one of those who took a hard-line position on magic, vociferously denouncing it in all forms. For him, the knowledge of mechanics was beneficial to the people because it not only allowed them to construct useful devices but also made them less ignorant and superstitious and so less gullible toward the fraudulent tricks of false magicians.[26]

The loss of ancient mechanical knowledge in the early Middle Ages was not the only reason that no significant automaton construction took place until the thirteenth century. Early Christianity tended to take a negative view of representative images of animate beings, in accordance with the

Second Commandment, seeing in them the lurking danger of idolatry. Even a simple image of Christ crucified sometimes came with a warning like "This is God, as the image teaches, but the image itself is not God. Think upon this, but in spirit worship what you see in it," so that people would not mistake the representation for what it represented.[27] This was no idle concern since many people thought images of Christ, the Madonna, and saints were endowed with magical powers, much like pagan idols.[28]

In his study of the fourteenth-century revival of effigies in royal funerals, Carlo Ginzburg links the phenomenon with the thirteenth-century proclamation of the dogma of transubstantiation, which was a "decisively important event in the history of how images have been perceived," as it constituted "an extraordinary victory of abstraction."[29] In the high Middle Ages, then, perhaps in conjunction with the cultural revival of the period and the renewed interest in ancient philosophy, art, and architecture, the authorities' confidence in the people's ability to understand the notion of representation (crucial in the case of transubstantiation since in its ritual the worshipper is asked to see beyond its visual elements to their spiritual reality), making images of living beings less conceptually dangerous and anxiety-inducing for the society. Ginzburg notes that in the thirteenth century "the fear of idolatry begins to lessen. Ways are found of domesticating images, including those that have come down from pagan antiquity."[30]

In other words, the relative stability of western Europe in the high Middle Ages, along with the intellectual openness and technological innovations of the period, produced a culture secure enough to consider the categories of its worldview under a critical light. This was sometimes done by taking ideas, phenomena, and objects that had previously been condemned or labeled taboo because of their transcategorical nature and reassessing them in light of the preternatural. So just as maverick intellectuals sought to legitimize areas of learning that were seen by many as lowly (mechanics) or heretical (magic), there was a new appreciation and assimilation of representative images with their power to disrupt the boundary line between the animate and the inanimate—even automata that extended that power to an extreme degree.

The dangerously uncertain nature of the automaton (living or dead? magical or mechanical? diabolical or divine?) can be seen explicitly in its fictional manifestations (the first and second categories of automata).[31] The fantastic object appears in the romances and epic poetry of the twelfth and thirteenth centuries, some of them inspired by real-life travel accounts of wonders witnessed in exotic places, most significantly Liudprand's descriptions of the marvels of Constantinople and William of Rubruck's (ca. 1220–1293) narrative of his journey to the Mongol court in Karakorum. The latter was a Franciscan missionary who was sent by Louis IX of France to establish diplomatic relations with the Mongols for a possible alliance against the Muslims. At the imperial capital, William met a master goldsmith named William Buchier, originally from Paris, who had constructed for Mongke Khan a mechanical liquid-dispensing tree with automata of lions, serpents, and a trumpet-blowing angel.[32] In romances, automata were used to heighten the wonder of their marvelous environments where magic and technology were indistinguishable. But even as imaginary devices, they embodied all the anxiety-producing uncertainties manifested in the debates on the study of magic and the legitimacy of representative images. This can be seen in the two prime modes in which automata appear in medieval and Renaissance literature.

First, as provocative examples of the wonders of ancient civilizations with their glorious splendor and arcane knowledge, automata appear in exotic worlds that are portrayed in a manner designed to arouse awe and nostalgia, in the same way Liudprand described the marvels of Constantinople. These are, in other words, positive images of a bygone golden era or a faraway land where fantastic and magical things were or are possible. The twelfth-century *Pèleringe de Charlemagne*, for example, begins with a description of the revolving palace of Constantinople, which features youths of bronze who smile and blow their ivory horns when the wind from the sea comes up.[33] In the fascinating "Chambre de Beauté" episode in Benoît de Sainte-Maure's *Roman de Troi*, Prince Hector of Troy recovers from his wounds in a wondrous chamber that contains four moving statues, two fulfilling "social" functions of holding up a magic mirror and correcting misbehavior and two providing entertainment by playing musical instruments and performing acrobatic feats with other automata, including an eagle and a satyr.[34] Penny Sullivan describes the chamber as a "breathing-space" in which the reader is asked to ponder the horrific and senseless nature of war, which will lead to the destruction of Troy and the loss of a great civilization.[35]

Second, in contrast to those positive depictions, automata also appear as the products of diabolical magic practiced by evil witches and warlocks, especially in the form of "necromancy," a condemned form of magic associated with the raising of the dead and the summoning of demons. An example of this is found in the early thirteenth-century Arthurian romance *Perlesvaus*, in which the hero, Perceval, goes to a copper castle inhabited by an oracular image filled with an evil spirit that guards its realm with a pair of copper warriors armed with iron mallets who are "made through the necromantic art."[36] So the automaton appears in literature alternatively as the wondrous creation of ancient knowledge and as the diabolical work of heretical magic, the former meant to arouse sublime awe and the latter horror.

Even when the issue was not whether such an entity was the product of divine power, ancient knowledge, or demonic magic, the automaton was portrayed as an ambiguous, double-sided, and transcategorical thing. The most striking example of this in Renaissance literature can be found in the fifth book of Edmund Spencer's epic *The Faerie Queen* (1596), in which the goddess Astraea grants to Artegall, the knight of justice, an "yron man" by the name of Talus (first category of premodern automata) to help him in his tasks.[37] He is, of course, a revival of the ancient Talos, though in ancient literature he was the sole surviving member of the bronze, not iron, race of men. In the beginning the figure seems a positive character, an invincible, tireless, and obedient servant of the righteous knight, who slays villains with relentless power like a Renaissance Terminator. But as the story goes on, Talus proves to be single-minded and inflexible in his work to a disturbing degree, killing even those Artegall would have spared for mercy's sake. The hardness of his iron shape ultimately comes to represent the inhumanity of justice without mercy, as well as what Jessica Wolfe calls "political instrumentalism gone awry."[38] The eminently efficient and powerful creature proves to be an entity of ambivalent nature as it shifts from being a loyal servant to an implacable executioner.

E. R. Truitt has pointed out, significantly, that fantastic automata in medieval literature are found in "liminal spaces—thresholds, bridges, or tombs" and fulfill functions of surveillance and discipline, which points to not only the in-between nature of the automata themselves but also "the ways in which they enforce boundaries of epistemological legitimacy and morality."[39] This analysis is insightful, especially as it pertains to positive depictions of automata as products of ancient knowledge, but also incomplete in view of negative portrayals of the entities as well. While they do serve the conservative function of pointing to and guarding social, political, and ethical lines in many narratives, in others they—especially the ones depicted as products of diabolical power—deliberately confuse and transgress those very boundaries. In fact, what makes the automaton so enthralling and conceptually dangerous at the same time is the very fact that even as it supports the status quo it also disrupts it. As a preternatural object dwelling in liminal spaces, it can expose the arbitrary and artificial nature of human boundaries of legitimacy and morality, just as the acts of Talus point to the double-edged nature of justice. Jacques le Goff has also pointed to the uncertain nature of the object in his categorization of medieval wonders into six types—sites, humans and anthropomorphs, animals, *mischwesen*, objects, and historical personages.[40] Automata are placed in the category not of objects but of *mischwesen*:

> half-human, half-animal creatures, such as Melusian or the sirens or Yonee in Marie de France's *L'Oiseau bleu* or werewolves. . . . The griffin also falls under this class, as do automata. Ultimately we find creatures that are half-living, half-animate, as in the paintings of Hieronymus Bosch.[41]

Uncertainty toward the exact nature of the artificial marvel can be seen in a particularly vivid and entertaining form in a series of related medieval texts dealing with automata in the imperial court of the Mongol Empire.

Seventy years after William of Rubruck witnessed the marvelous works at Karakorum, the missionary Odoric of Pordenone (ca. 1286–1331) journeyed to the Far East, arriving in Taydo (i.e., Dadu, "Great Capital") at Cambalech, today's Beijing, where he stayed from 1324 to 1327. Among numerous wonders he witnessed at the palace of the great khan (Yesun Temur, r. 1323–1328), he noted "many peacocks of gold." When "any of the Tartars wish to amuse their lord, then they go after the other and clap their hands; upon which the peacocks flap their wings, and make they would dance."[42] The display impressed the European visitor, but since he had no idea how the artificial birds were made to move, he wondered if these works were achieved through "diabolical art, or by some engine underground."[43]

The same automata appear in John Mandeville's *Travels* (1357–1371), a popular fictional travel narrative based on various actual accounts, including that of Odoric, which was the main source of the section on the Mongol Empire.[44] Reiterating Odoric's experiences, Mandeville arrives at Gaydon (Taydo—Dadu), the capital of the "Great Caan," and witnesses countless marvels there, including

> peacock of gold and many other manners of fowls of gold, curiously and subtly wrought. And these fowls are so wonderfully made by craft of man that it seems as they leaped and danced and beat their wings, and played them on other divers wise; and it is right wonderful to the sight, and how that such things may be done.[45]

As with Odoric, Mandeville wonders if the work was done "by artifice or by necromancy,"[46] despite the fact that he has already described them as products of man's craft.

An even more fantastic device appears in the Squire's Tale in Geoffrey Chaucer's *Canterbury Tales* (late fourteenth century), set yet again in the Mongol court. The story itself is derived from an *Arabian Nights* tale involving a foreign knight's visit to the court of the great and benevolent monarch Cambyuskan (Genghis Khan).[47] The mysterious stranger arrives on a "steede of bras" that, along with a magical mirror, he presents as gifts from the kings of Arabia and India.[48] The brass horse can not only become invisible but also take its rider to any place in the world within a day by either running or flying, as the rider wishes, through the manipulation of a series of "pyns" located in the horse's ear. The crowd that gathers to gape at the wonder offers a number of nervous speculations on its nature and function, or how "it koude gon, and was of bras."[49] Some think that it is a "Fairye" creature, others think that it is the gold-winged Pegasus of old, and yet others fear that it is like the Trojan horse and quake at the thought that it may contain soldiers who are plotting to take the city at an opportune moment. The last theory is that it is a kind of magical trick of appearance, like that of jugglers at great feasts. The knight's own description of its workings does not clarify the matter, as he evades the king's questions about its workings, just giving him detailed instructions for its operation. Its mechanical nature is implied when he explains that the pins control its internal devices and "therin lith th'effect of al the gyn" (gyn—engyne—

engine or machine).[50] Yet the knight claims earlier in the narrative that the creator of the wonder ("He that it wrought koude ful many a gyn") waited until the right constellations were in place before he made it, which he knew how to do because he was familiar with many magical "seals" and "bonds."[51] In other words, the brass steed seems to be the product of both mechanics and natural magic involving astrology.

In all of these accounts of actual and fictional automata in the faraway land of the Eastern lords, there are worries about whether they were made through mechanics alone or through magic, diabolical or natural. While none of these accounts provides a definite answer, they all express awe and wonder at the marvelous objects, and the effect is heightened by the exact nature of their operations being left in doubt. This uncertainty is very much a part of their status as liminal entities, simultaneously entertaining and frightening their viewers as they blur, straddle, and play around with the categorical boundaries of the viewers' reality schema. Worries about the legitimacy of such a feat appear even in stories where explicitly magic and not mechanics is in operation in the animation of the inanimate. At the end of Shakespeare's *Winter's Tale*, when the statue of the wrongfully condemned Hermione comes alive (though the text leaves open the possibility that she did not die at all but was kept in hiding until she was brought forth pretending to be her own statue), her friend Paulina insists that her "actions shall be as holy as / You hear my spell is lawful." Even so, Hermione's repentant husband, Leontes, as he marvels at the miracle, hopes that if "this be magic, let it be an art / Lawful as eating."[52]

To explore these complex issues surrounding the automaton motif in the Middle Ages and the Renaissance in greater detail, I will examine two provocative instances of their appearance in legends and intellectual works. The first is a group of stories involving the construction of an artificial head for the purpose of divination, and the second is a list of wonders, first appearing in Cornelius Agrippa's 1533 magnum opus on occult philosophy, that was reiterated in many modified forms throughout the early modern era.

The Talking Oracular Head

The legend of the talking artificial head is a particularly rich and interesting one to examine for the purpose of further understanding the magical automaton in the medieval imagination because it has been told in a number of significantly varying versions from the twelfth to the sixteenth centuries. It was also attached to the names of some of the most important intellectual figures of the Middle Ages. In this section, I will look at four major variations dealing with Gerbert of Aurillac, Albertus Magnus, Robert Grosseteste, and Roger Bacon. I will also demonstrate how the fantastic object alternatively expressed the fear of dangerous knowledge with diabolical possibilities and the awe of the arcane learning that could be used to produce such wonders. As with attitudes toward the automaton in general, one must take into account both the positive and negative reactions to the marvel to gain a comprehensive understanding of its meaning in the premodern context.

Gerbert of Aurillac (945–1003), archbishop of Rheims and Ravenna and pope (Sylvester II), was a key political and intellectual figure of his time, as he was patronized by Emperor Otto I, was a tutor and advisor to both Otto II and Otto III, and played a crucial role in the rise of Hugh Capet as the king of France.[53] As one of the most learned men of his time, Gerbert effected a flowering of learning through his tireless reformist activities as schoolmaster, clergyman, and pontiff. He was lauded by scholars and political leaders alike for his immense erudition, ingenuity, and modesty, but over a century after his death, when the English historian William of Malmesbury gave a concise account of his life in his *Chronicle of the Kings of England* (ca. 1125), the picture William presented was of a ruthless and heretical sorcerer who ultimately got his comeuppance through his evil practices. The legend begins with Gerbert stealing a book of magic from a powerful Islamic philosopher in Spain, who gives him chase but the trickster is able to escape with the help of the devil, to whom he promises his soul.[54] In the fantastic tale that follows, he uses necromancy to find hidden treasures in Rome that are guarded by moving statues of gold,[55] and he constructs, "by a certain inspection of the stars when all the planets were about to begin their courses," an artificial head that can talk, answering prophetic questions with yes or no.[56] The device assures Gerbert that he will not die until he sings mass at Jerusalem, so he avoids going there, but he unwittingly attends in Rome a small church called Jerusalem and falls fatally ill.

The source of these slanderous stories was the eleventh-century writings of Cardinal Benno, a supporter of the archbishop Guibert of Ravenna (ca. 1029–1100), who was installed as the antipope (Clement III) by Emperor Henry IV during his struggle over the investiture controversy with the reformist Pope Gregory VII. As a partisan of the schism, Benno set out to portray the recent pontiffs of Rome, starting with Sylvester II, as evil magicians who had corrupted the papacy with their diabolical ways. Roland Allen speculates that Benno probably did not invent the stories wholesale but rather enlarged on preexisting rumors about Gerbert that arose not only from his reputation for great erudition but also from the fact that he had spent some time as a young scholar in Spain, after his brilliance was recognized by Count Borel of Barcelona, who patronized him.[57] There is no evidence that Gerbert visited Muslim-controlled places like Córdoba or Zaragoza, but he came into contact with Arabic learning, as he introduced elements of its mathematical knowledge into Christendom, including the nine Hindu-Arabic numerals and the use of the abacus. In the popular imagination of the period Spain was, Allen explains, "a land where every form of magic was thought to be rife, a land of astrology and of necromancy, a land too of pagans and of the black arts."[58]

Politically motivated slander and the magical reputation of Spain provide an explanation for the legends of Gerbert's unholy magic, but the significance of the artificial talking head has to be elucidated through its reappearance in stories attached to other major intellectuals of the Middle Ages.[59] Such a tale involving Albertus Magnus appears in a 1373 moral treatise, the *Rosario della vita* by Matteo Corsini, that features a very different attitude toward the marvel. In the section on the subject of *sapientia* (wisdom), Corsini presents a fantastic tale as a moral for how wisdom should be revered and honored.

> We find that Albertus Magnus, of the Black Friars, had such a great mind that he was able to make a metal statue modeled after the course of the planets, and endowed with such a capacity for reason that it spoke: and it was not from a diabolical art or necromancy—great intellects do not delight in such things because it is something that makes one lose his soul and body; such arts are forbidden by the faith of Christ. One day a monk went to find Albertus in his cell. As Albertus was not there, the statue replied. The monk, thinking that it was an idol of evil invention, broke it. When Albertus returned, he was

very angry, telling the monk that it had taken him thirty years to make this piece and "that I did not learn this science in the Order of the Black Friars." The monk replied, "I have done wrong; please forgive me. Perhaps I can make you another one?" Albertus responded that it would be thirty thousand more years before another could be made for him, as that planet had made its course and it would not return before that time.[60]

A few elements of this story stand out—first, in direct contrast to the stories of Gerbert, there is an unambiguous insistence that Albertus was not engaged in diabolical magic when he constructed the object; second, far from a cautionary tale of a man's downfall from the hubris of dabbling with the dark arts, the moral of the story lies in the lack of respect for wisdom on the part of the monk (identified as none other than Albertus's star pupil Thomas Aquinas in a fifteenth-century variation of the story)[61] that leads to the wrecking of Albertus's precious work—a tale, in other words, of the destructive nature of ignorance; and third, the fact that the statue (not just a head) was constructed with the use of astrology, just like Gerbert's head and the brass steed in Chaucer, shows that it is a product of natural magic.

Not long after the appearance of the Corsini story, the English poet John Gower (ca. 1330–1408) related a similar tale in his *Confessio Amantis*, but this time about Robert Grosseteste, the thirteenth-century philosopher and bishop of Lincoln. In a section on the dangers of sloth, Gower writes that Grosseteste, practicing the secret arts, spent seven years constructing a head of brass that could foretell the future but missed the crucial moment of its prophecy by falling asleep.[62] As with the case of Albertus, natural magic is implied, with no mention of diabolical magic, and the story is about the failure of a legitimate intellectual work, due this time to fatigue rather than ignorance. The late sixteenth-century prose romance *The Famous Historie of Fryer Bacon* combines elements of the Gerbert, Albertus, and Grosseteste stories, with Roger Bacon as the creator of the magical object, aided by a friar, Bungey. Bacon is already a great magician who can perform all kinds of wonders, but he wants to gain fame by creating a protective wall around England to make it safe from foreign invasions. To attain the magical knowledge necessary for the feat, he and Bungey construct a magical head of brass that could tell them the secret. They make it so that "in the inwards parts [of the head] there was all things like as in a naturall mans head," but they fail to give it motion, without which it obviously cannot speak.[63] In their frustration they summon the devil and entrap him into revealing the secret of animating the head, which consists of subjecting the device to "a continuel fume of six hottest simples" (medicinal herbs of alchemical medicine) for a month, after which it will speak at some unspecified time.[64] It is important to note here that while the knowledge comes from the devil, the operations needed for the animation itself are of natural magic. Because they do not know the exact time the head will speak, Bacon instructs his idiot servant, Miles, to watch over the thing while he and Bungey sleep. The head comes alive and utters the words "Time is," then "Time was," and finally "Time is past," but Miles sees them as too trivial to wake his master. With the opportunity to question the magical object gone, it destroys itself in an explosion, awaking Bacon, who laments the loss of the work, which cannot be replicated. The Elizabethan playwright Robert Greene retold this and other stories of Roger Bacon as a magician in his 1592 play *The Honourable History of Friar Bacon and Friar Bungay*, in which Bacon not only receives the help of the devil but uses explicitly diabolical magic as he admits to utilizing "necromantic charms" and "the enchanting forces of the devil" to achieve his purpose.[65]

Beside the fantastic image of the talking metal head, a number of common features tie these stories together. Gerbert is made out to be a diabolical figure, but the creation of the head itself seems to have been achieved through some form of natural operation, as in the cases of the Albertus and Grosseteste variations, though Bacon has to elicit the help of the devil. The object is made to work, in other words, not simply by casting a spell they learned, but through arcane knowledge that takes some effort over an extended period of time—seven years in the cases of Grosseteste and Bacon (only in the Greene play—the original romance makes no mention of how long it took Bacon and Bungey to construct the head, so it is possible that Greene took elements from the Grosseteste story in Gower).[66] For Gerbert and Grosseteste, the purpose of the work is to gain the knowledge of the future, but the whole project goes awry due to the bungling of an inferior, in the cases of Albertus and Bacon, and to fatigue, in Grosseteste and Bacon. Table 2 organizes these narrative elements.

Table 2. The Artificial Speaking Head from Gerbert of Aurillac to Roger Bacon

	Head or statue	Reference to alchemy/astrology	Involvement of the devil	Construction time	Reason for failure
Gerbert (William of Monmouth)	Head	Yes	Yes	Unknown	Misinterpretation of words / hubris
Albertus Magnus (Corsini)	Statue	Yes	No	Thirty years	Ignorance of an inferior
Robert Grosseteste (Gower)	Head	Yes	No	Seven years	Fatigue
Roger Bacon (Anonymous and Greene)	Head	Yes	Yes	Unknown (anon.), seven years (Greene)	Fatigue + ignorance of an inferior

Obviously, this group of legends took elements from one another, but they also tell stories with varying messages, some of the hubris of using the dark arts but others of great and useful knowledge that is lost due to sloth or the incompetence of an ignorant.

So what exactly does the talking head represent? And why are the figures of Gerbert, Albertus, Grosseteste, and Bacon associated with it? The first step toward answering these questions is to understand what the real-life figures of the four characters had in common. Besides the fact that they were all important intellectuals of their times who articulated some of the most advanced ideas on nature and philosophy, they dealt in different degrees with two areas of knowledge—magic and mechanics. Gerbert, in addition to introducing Arabic mathematical knowledge into Christian Europe, invented numerous devices, including several improved organs and dials and globes for the study of astronomy and geometry, which might have been based on objects he had encountered in Spain.[67] In a period when the distinction between astronomy and astrology was virtually nonexistent, Gerbert also believed in divination through the observation of stars.[68] In the early medieval context there was little general familiarity with mathematics, geometry, and machinery, so Gerbert's association with them, in addition to his connection to Spain and Arabic learning, made him a figure ripe to be turned into a sorcerer of questionable legitimacy at best.

Similarly, Albertus was one of the first major intellectuals to advocate the study of natural magic, especially in such a neutral, nonspiritual area as alchemy. As exemplified in works like his *Book of Minerals* and *Book of Animals*, he was deeply interested in natural subjects, and there is evidence that he was directly involved in experiments to verify his ideas.[69] On the matter of alchemical transmutation of base metals into gold, for example, Albertus claimed to have tested the methodology but found that the gold that was produced was a noticeably inferior quality to natural gold and tended to decompose after several firings.[70] This means that, like Gerbert, he must have been in possession of or had access to experimental equipment that would have intimidated or even horrified people who were unfamiliar with them.[71]

The English philosophers Robert Grosseteste and Roger Bacon (the latter probably studied with the former during his student days at Oxford) are regarded as the two pioneering advocates of the experimental method in studying nature.[72] Grosseteste's greatest area of interest was optics, and his ideas and findings on the subject based on experimentation and observation point to the fact that he must have used equipment necessary for the purpose.[73] Likewise, Bacon once claimed to have spent the rather fantastic amount of 2,000 pounds on "books, instruments, experiments and other things necessary for the pursuit of wisdom."[74] Bacon railed against magic in his famous *Letter Concerning the Marvelous Power of Art and of Nature and Concerning the Nullity of Magic*, considering most forms of magic fraudulent or arising from ignorant superstition—making his later magical reputation unfortunate. Yet both he and Grosseteste considered legitimate the subjects William of Auvergne and Albertus Magnus categorized under natural magic, including alchemy and astrology.[75] Like Gerbert and Albertus, these two English figures were renowned for their great learning, were interested in mathematics, mechanics, and some forms of natural magic, and were probably in possession of arcane instruments used for experimental purposes. These aspects of their careers set them up to be transformed into figures of legend wielding uncanny preternatural powers.

As for the oracular talking head itself, Arthur Dickson has pointed to a number of ancient sources for it, including Greek myths about divine automata, Egyptian animated idols, and Arabic stories featuring the use of a corpse's head for prophetic purposes, also involving astrology.[76] All of these undoubtedly contributed to the development of the medieval stories, but another major source was the Hermetic tradition. The Latin book *Asclepius*, written probably in the second century CE, is a central work of the mystical pantheistic philosophy of Greco-Egyptian origin.[77] A controversial passage in this text deals with the making of statues. In the dialogue between the divine Hermes Trismegistus and his beloved disciple Asclepius, the former describes how "humanity persists in imitating divinity, representing its gods in semblance of its own features, just as the father and master made his gods eternal to resemble him."[78] When Asclepius asks if he is talking about statues, Hermes answers, "I mean statues ensouled and conscious, filled with spirit and doing great deeds; statues that foreknow the future and predict it by lots, by prophecy, by dreams and by many other means; statues that make people ill and cure them, bringing them pain and pleasure as each deserves." Later on, he further describes how such living statues were constructed by his ancestors after they "discovered the art of making gods."

> To their discovery they added a conformable power arising from the nature of matter. Because they could not make souls, they mixed this power in and called up the souls of demons or angels and implanted them in likenesses through holy and divine mysteries, whence the idols could have the power to do good and evil.[79]

He elaborates on the method of bringing those beings into statues:

> It comes from a mixture of plants, stones and spices, Asclepius, that have in them a natural power of divinity. And this is why those gods are entertained with constant sacrifices, with hymns, praises and sweet sounds in tune with heaven's harmony: so that the heavenly ingredient enticed into the idol by constant communication with heaven may gladly endure its long stay among humankind.
>
> Do not suppose that these earthly gods act aimlessly, Asclepius. Heavenly gods inhabit heaven's heights, each one heading up the order assigned to him and watching over it. But here below our gods render aid to humans as if through loving kinship, looking after some things individually, foretelling some things through lots and divination, and planning ahead to give help by other means, each in his own way.[80]

Some three centuries after the writing of the *Asclepius*, Augustine of Hippo was incensed enough about the passages on god-making to dedicate two entire chapters of his *City of God* to denouncing them: while images are the products of men's labors, "Hermes asserts that the visible and tangible idols are in some way the bodies of gods; certain spirits have been induced to take up their abode in them, and they have the power either to harm, or to satisfy the wants of those who offer them divine honours and obedient worship."[81] But Augustine does not necessarily think that those idols were mere dead objects. They may have been inhabited by crafty demons, mistaken by Egyptians as intermediaries between gods and men, who pretended to bring benefits to men in order to do the greater harm of drawing them into fellowship with them. But Christianity will eradicate those dangerous and "deceitful images with an irresistible finality corresponding to its truth and holiness, so that the grace of the true Saviour may set men free from these man-made gods." In the most provocative sentence of the chapter, Augustine eloquently summarizes the situation, concluding with a phrase that could be the motto of all future antitechnological attitudes:

> And yet in some way because of that 'darkening of the heart' [Hermes] sank low enough to wish men to remain forever subject to gods who, on his own showing, are the creations of men, and to bewail the prospect of their extirpation at some future time, as if there were any unhappier situation than that of a man under the domination of his own inventions.[82]

The celestial nature of the spirits drawn into the statues, their prophetic function, and the fact that the operation requires a great deal of arcane knowledge and technical work (both in the construction of the object and its maintenance) marks the god-making passage as the essential source of the medieval talking head. And Augustine's denunciation of it was a major referent for condemning idolatry and theurgy in medieval theology, as even for the thirteenth-century advocates of natural magic this form of sorcery involving the animation of statues was clearly in the territory of the illegitimate, the heretical, and the demonic. William of Auvergne, who was careful to distinguish natural from spiritual magic, describes in his work *De Legibus* the theurgical operation in *Asclepius*, including the "observation of the hours and constellations when the image is cast or engraved or fabricated"— the same type of consultation of the movement of the stars found in the legends attached to Gerbert and Albertus.[83] William goes on to deprecate the whole thing as impossible and superstitious, as he does not believe that spirits can be lured into inanimate material.

In an incisive article on the subject of the talking head in English literature, Kevin LaGrandeur has analyzed the stories essentially as cautionary tales about the dangers of dancing "along the limits between the innovative and frightening exercise of human ingenuity," which could "imperil the social order with the disruptive danger of the independent thinker's ambitions, and with the unchecked power of the novel ideas such instruments [as the brazen head]" represents.[84] This notion of the talking head as a symbol of dangerous knowledge is certainly appropriate, especially when applied to the stories in which the devil gets involved and the protagonists get their comeuppance in the end. This interpretation, however, is incomplete in the sense that it emphasizes the conservative element of the stories' moral, in the same way that Truitt concentrated on the boundary-enforcing role of automata in medieval literature at the expense of their disruptive, critical, and subversive potential. In Corsini's narrative of Albertus's talking statue, there was nothing diabolical or dangerous about the object, which is the product of the philosopher's great knowledge and wisdom (in Gower's Grosseteste story as well). It is an inferior monk, probably the model for Miles in the Bacon stories, who destroys something he does not understand. In other words, far from a tale of a man's intellectual hubris going awry to pose a threat to the world, it is ignorance that is shown to be the danger, one that can undo decades of wondrous work by a learned man.

Taken as a whole, the talking head represents both the danger and the glory of arcane knowledge, with the potential to arouse both horror and sublime awe. It is able to do so because the types of knowledge that the real-life as well as fictional Gerbert, Albertus, Grosseste, and Bacon dabbled in were forms of learning associated with the preternatural realm, located between the stable order of traditional scholarship and the chaos of uncontrolled but infinitely powerful knowledge beyond. The magical automaton of the oracular head served to express both popular anxiety and fascination with magic and with those who practiced it. Even as many people were intimidated and frightened by such figures, they were also deeply attracted to them, in the way magicians in primitive societies were both feared and revered. In the medieval and the Renaissance imaginations, then, the talking head was a general symbol of liminal knowledge, with the potential to both empower and destroy, enlighten and confuse, enthrall and horrify, just as did the fantastic object of the automaton itself.

The story of the talking statue or head occupied a powerful and enduring place in the early modern imagination, as demonstrated by its numerous appearances in Renaissance discussions on mechanics and wonders. Further variations of the story were also introduced in such debates—Gabriel Naudé, in his 1625 debunking work on magic, claimed that Thomas Aquinas destroyed Albertus's creation not because he thought it was diabolical but because "he could not endure its excesse of prating," while John Wilkins wrote that Thomas did so in order to boast that "in one Minute he had ruined the Labour of so many years" (this story probably resulted from Wilkins's misreading an earlier variant of the story, as he has transformed

Albertus's lament in the original Corsini text to Thomas's boast).[85] The talking head that also appears in a list of traditional marvels in a work of natural magic by Cornelius Agrippa, is the subject of the next section.

The List of Mathematical Wonders: Automata in the Renaissance

I will begin considering the role of the automaton motif in the Renaissance with a brief survey of actual automata that were designed or constructed from the fifteenth century on. I will then analyze a sixteenth-century list of artificial wonders that is of supreme importance to the entire history of the automaton idea in the early modern period. This list appears in a popular book by the German alchemist and philosopher Cornelius Agrippa and reappears numerous times in modified forms in the following centuries, reviving the ancient Greek work "automaton" and providing following scholars with a referent for marvels that could be achieved through the understanding and manipulation of occult forces.

The crisis period of the fourteenth and early fifteenth centuries saw the ravages of the Black Death, the bloody cycles of the Hundred Years War, the schism of the Church as it was ruled by two and then three popes, and the militaristic expansion of the Ottoman Empire, which captured Constantinople in 1453, ending the moribund state of Byzantium. As western Europe began to recover in the second half of the fifteenth century, the merchant city-states of Italy led the economic, political, and cultural revival that was the Renaissance. The rediscovery of ancient works on philosophy, nature, and magic created an intellectual environment that sought a more dynamic alternative to the Aristotelian synthesis achieved by Thomas Aquinas in the late thirteenth century. Renewed interest in classical learning in turn accelerated the search for more lost knowledge, creating a cycle of recovery, study, and assimilation. The most important site of such activities was the household of the Medicis in Florence, where the philosopher Marsilio Ficino, working under the patronage of Cosimo de Medici and his grandson Lorenzo, rendered the works of Plato, and those of the Neoplatonic philosophers Porphyry, Iamblichus, Proclus, and Plotinus into Latin. Ficino's translations, commentaries, and philosophical works initiated the emergence of a new worldview of the Renaissance version of Neoplatonic pantheism, with a heavy dose of Hermetic magic, which thinkers from

Ficino to later figures like Giambattista della Porta and Tommaso Campanella sought to make compatible with Christianity.

Also recovered during the period were ancient writings on the related fields of mathematics, geometry, and mechanics, as Latin translations of works by Euclid, Archimedes, Hero, Pappus, and Vitruvius appeared. Of particular importance to technological development was the widely read treatise *Mechanical Problems*, which was attributed to Aristotle (contemporary scholars consider it the work of his student Strato).[86] Jessica Wolfe has pointed to the courts of Frederico and Guidobaldo da Montefeltro in Urbino as the center of learning about ancient mechanics in late fifteenth- and early sixteenth-century Italy.[87] And the historian of technology Pamela Long has shown that a number of new works on mechanics appeared in Germany and Italy in the fifteenth and sixteenth centuries—by such figures as Conrad Kyeser and Giovanni Fontana—that built on ancient knowledge with designs for original devices of mainly military purpose.[88]

In this vastly improved state of technological knowledge, quite a few actual automata were constructed, some of them extant and in working order today. The fourteenth century was a crucial period in the history of horology, as large, elaborate, and mostly weight-driven clocks proliferated, beginning in Italy and later in France and Germany.[89] Among the greatest achievements of the era's technology was the enormous astronomical clock at the Strasbourg cathedral that was completed in 1354.[90] Along with a moving calendar and an astrolabe pointing to the movements of the sun, the moon, and the planets, this device featured automata of the Three Magi that bowed before the Virgin and a cock that opened its beak, thrust out its tongue, flapped its wings, and crowed three times at noon. Many such clocks with automata were constructed throughout the following centuries.[91]

I have already mentioned Philip the Good, the Duke of Burgundy, who renovated the late thirteenth-century automata of Artois as a part of his great collection of marvels.[92] At a feast Philip gave in 1454 to drum up support for his crusade, numerous fantastic automata were displayed on the banquet table, representing various animals. In the following century, such wonders were collected not only by royalty and nobility but also by learned men in the context of the upsurge of intellectual interest in the natural world.[93] Many such collections, especially the *Wunderkammer* (chamber of wonders) of the seventeenth century, featured artificial wonders like automata along with natural ones.[94] One of the most avid collectors was Emperor Rudolph II, who was also deeply interested in natural magic.[95] Among the many important intellectuals who came to his court in Prague, including Tycho Brahe, John Dee, and Rabbi Judah Loew, the astronomer Johannes Kepler was particularly impressed by a drum-playing automaton in the collection that suggested to him the model of the universe as a vast machine.[96]

Two of the oldest automata that are both independent (i.e., not attached to a larger device like a clock) and in working order are a lute-playing lady that can now be seen at the Kunsthistorisches Museum in Vienna and the rather astounding walking and praying monk figure that is at the Smithsonian Museum in Washington, D.C., probably the oldest automaton currently in the United States.[97] Both are dated to the sixteenth century and attributed to Juanelo Turriano (ca. 1511–1585), an Italian engineer who was served as clockmaker to Emperor Charles V.[98] Leonardo da Vinci, both an artist and an engineer, designed and constructed several automata. According to Giorgio Vasari, when the king of France came to Milan,[99] he "begged Leonardo to make something unusual, and so Leonardo made a lion which walked a few steps before its chest opened, revealing it to be filled with lilies."[100] He also left plans for an armored knight designed to sit up, wave its arms, and move its head and jaw.[101]

Also popular in the sixteenth and seventeenth centuries were the so-called hydraulic automata that could be found in the great gardens of Western Europe. Following the rediscovery of Hero's *Pneumatica*, engineers used its technological knowledge to decorate artificial caves and grottos with moving and water-spouting statues. During the extensive enlargements made to the royal château of Saint-Germain-en-Lay from 1589 to 1609 by the order of Henry IV of France, a Florentine architect and engineer, Tommaso Francini, and his brother Alessandro were brought in to embellish a series of terraces with grottoes. Among their creations were elaborate fountains and other waterworks and a group of mythological figures in the grottoes, including Neptune, Diana, Orpheus, and a sword-wielding Perseus that swooped down from the ceiling and killed a dragon rising out of the water, all of it powered by the flow of the Seine.[102] The Francinis went on to create such works at the parks of Fontainebleau and Versailles.

Salomon de Caus (1576–1626), a French Huguenot engineer in the service of the Elector Palatine Frederick V, built similar works during the renovation of Heidelberg castle and its grounds, including grottoes that featured automata of classical gods, monsters, nymphs, and satyrs. In 1615 he published an engineering treatise on such devices entitled *The relations of Motive Forces, with various Machines as useful as they are pleasing.*[103] Montaigne visited such a garden at the Villa d'Este of Tivoli in 1581 and marveled at its wonders, including hydraulic musical instruments and an owl automaton, which originated from a design of Hero.

> In another place you hear the song of birds, which are little bronze flutes that you see at regals; they give a sound like those little earthenware pots full of water that little children blow into by the spout, this by an artifice like that of the organ; and then by other springs they set in motion an owl, which appearing at the top of the rock, makes this harmony cease instantly, for the birds are frightened by his presence; and then he leaves the place to them again.[104]

The flowering of art that accompanied the expansion of knowledge in the Renaissance marks the climax of a process that began in the thirteenth

Hero of Alexandria, design for an owl and other bird automata. Reconstruction—illustration in *Herons von Alexandria Druckwerke und Automatentheater*, ed. Wilhelm Schmidt (Lipsiae: Teubner, 1899) p. 93, figure 17.

century—the appreciation of the power of representative images through the language of wonder and legitimate animism. The art historian Frederika Jacobs has shown that the creation of "living" images was an explicit goal for artists and mechanics alike in the period, who constantly spoke of their craft as one of breathing life into their works.[105]

Yet even in this period when mechanical wonders proliferated, often collected and displayed by the powerful as a demonstration of their glory, all the ambivalence and confusion about the nature of such objects was very much present.[106] For example, Anthony Grafton analyzes the works of the fifteenth-century engineer Giovanni Fontana, who had an interest in automata, illustrating in his treatises a flying bird, a fire-farting rabbit, and devils with mobile facial features, arms, and wings.[107] In his exploration of Fontana's motivation in designing such works, especially the frightening devils, Grafton encounters the same complex attitude toward magic and mechanics found in the Middle Ages. Fontana had contempt for the fraudulent tricks of magicians and may have exposed the inner mechanical workings of the diabolical automata in order to deflate stories of sorcery. But he also affirmed the reality of certain forms of natural magic, including theurgical operations.

In the realm of ideas, such lingering questions about the connection between magic and mechanics were explicitly dealt with in scholarly works on Hermetic magic, in which the automaton idea was employed in significant ways. Frances Yates in her classic work *Giordano Bruno and the Hermetic Tradition* recounts that around 1460 Cosimo de Medici received a collection of texts from Macedonia that included the *Corpus Hermeticum*, writings containing the teachings of Hermes Trismegistus (thrice-great), a priest, philosopher, and prophet, identified with the Egyptian god Thoth and the Greek Hermes, who was thought to have been a contemporary of Moses and Zoroaster.[108] Excited by this acquisition, Cosimo ordered Ficino to put a hold on translating Plato and study the new manuscripts.

Hermetic thought, a diverse and fragmentary group of animistic, alchemical, and Neoplatonic ideas, had been present throughout the Middle Ages when the Latin *Asclepius* and other writings like the enigmatic *Emerald Table* were available.[109] But the introduction of the *Corpus* in the intellectually adventurous atmosphere of the Renaissance vitalized the tradition, which became a major philosophical strain until well into the seventeenth century, even after the texts' origin in far antiquity was debunked by Isaac

Casaubon in 1614.[110] The Renaissance Hermetic worldview was derived from many disparate sources, including Neoplatonic philosophy, Arabic magic, and Jewish Kabbalah, featuring the notion of the world as a living entity, with a system of sympathetic emanations that link human beings to celestial and supercelestial creatures, up to God.[111] While Hermetic philosophers believed in the existence of angels and demons, they differed on the question of the legitimacy of using them for magical purposes, as they generally adopted the strategy of thirteenth-century advocates of natural magic in favoring the study of nonspiritual occult forces in the world. As in the Middle Ages, however, the boundary between legitimate and heretical magic was never clear, as some philosophers wrote of communicating with supernatural beings and even attempted to do so themselves.

In 1531, the German philosopher and alchemist Henricus Cornelius Agrippa von Nettesheim published his *De Occulta Philosophia* in Paris, and in 1533 a greatly expanded three-volume version in Cologne. A monumental survey of magical knowledge of his time, it was organized under a scheme that was thoroughly informed by Hermetic and Neoplatonic ideas.[112] The comprehensive nature of the work made it the central text of reference on all things magical during the Renaissance; becoming the most popular book on the subject. The section dealing with a list of artificial wonders in Book 2 is of enormous importance to the history of the automaton idea, for two reasons.

First, in the text Agrippa uses the obscure Greek word "automata," quoting Aristotle in his *Politics*, who was in turn quoting Homer in his reference to the self-moving tripods of Hephaestus.[113] After the word was revived, it came into ever greater usage in European writings in the course of the following century. Not long after the appearance of the book, François Rabelais introduced the word "automate" into the French language (1534) by having his character Gargantua amuse himself on a rainy day by constructing "several little automatic machines [plusieurs petitz engins automates], that is to say machines that moved by themselves."[114] Given Rabelais's famed erudition, it is possible that he may have picked up the ancient word independently from his reading of Aristotle or some other Greek source, but it is also plausible that he took it from Agrippa since he was familiar with his works[115]—paying tribute to him in *Pantagruel* as the character Herr Trippa, who could predict "all future events by the arts of astrology, geomancy, cheiromancy, metopomancy, and other sciences of that kidney."[116] An important aspect of the word's usage in Agrippa and Rabelais is its signifiying an artificial and mobile device (as opposed to, in the original Greek meaning, any self-moving entity). Agrippa had no intention of endowing it with this narrower definition, since he was only quoting Aristotle, but due to the great intellectual interest in his list, "automaton" became popularized in that sense. Furthermore, the list of artificial wonders itself, an example of Agrippa's great cataloguing work, became a popular reference for automata that played a significant intellectual role in the early modern period. The list was repeated, referred to, and modified numerous times throughout the next century in important texts dealing with cutting-edge ideas on magic, mathematics, mechanics, and natural philosophy. Through a series of direct and indirect transmissions, echoes of it can be discerned even in the Enlightenment.

In *De Occulta Philosophia*, Agrippa describes three different types of magic: "natural," dealing with the four basic elements and the variety of matter in the world; "celestial," concerned with mathematics, mechanics, music, and astronomy; and "religious," having to do with spirits, sacred rituals, and miracles (a hierarchy of worldly matter, universal laws, and spiritual revelation) and devotes a volume to each category. The book on celestial magic begins with a discussion of mathematics or, as the description of the first chapter declares, "of the necessity of mathematical learning and of the many wonderful works which are done by mathematical arts only."[117] He elaborates on the idea:

> The doctrines of mathematics are so necessary to, and have such an affinity with magic, that they that do profess it without them, are quite out of the way, and labour in vain, and shall in no wise obtain their desired effect. For whatsoever things are, and are done in these inferior natural virtues, are all done, and governed by number, weight, measure, harmony, motion, and light. And all things which we see in these inferiors, have root, and foundation in them.

It is clear from this passage that by "mathematical" learning, Agrippa is speaking not only of the study of numbers but of mechanics as well. Though the latter involves manipulating "inferior natural virtues" (inferior to spiritual virtues) and the products of such an activity do not partake in "divinity," its understanding is crucial to the study of the natural world. Within this realm, he argues for a naturalistic explanation of artificial wonders. And what better way to illustrate such marvels than through a list of famous automata as examples of great works that could be achieved through the use of mathematical magic:

> such as were those which amongst the ancients were called *Daedalus* his images, and *automata*, of which *Aristotle* makes mention, viz. the three-footed images of *Vulcan*, and *Dedalus*, moving themselves, which *Homer* saith came out of their own accord to exercise, and which we read, moved themselves at the feast of *Hiarba* the philosophical exerciser: also that golden statues performed the offices of cup-bearers, and carvers to the guests. Also we read of the statues of *Mercury*, which did speak, and the wooden dove of Archita, which did fly, and the miracles of *Boethius*, which Cassiodorus made mention of, viz. *Diomedes* in brass, sounding a trumpet, and a brazen snake hissing, and pictures of birds singing most sweetly.

This is followed by a list of geometric and optical illusions that were achieved through the use of mathematical magic. Toward the end of the chapter he mentions one more famous, and perhaps the most dangerous, example: "And so images that speak, foretell things to come, are said to be made, as *William of Paris* relates of a brazen head made under the rising of Saturn, which they say spake with a man's voice."[118] William of Paris is William of Auvergne, who was the bishop of Paris from 1229 to his death in 1249. It is significant that Agrippa makes reference to the talking head in William's *De Legibus* without mentioning that William condemned the practice as both fraudulent and heretical, creating an interesting complication for others who referred to the object.

I have already identified the sources of most objects in Agrippa's list:

1. "*Dedalus* his images": Plato, Aristotle, Diodorus Siculus, etc.
2. "*automata* of which *Aristotle* makes mention . . . which *Homer* saith came out of their own accord": Aristotle in his *Politics*, quoting Homer in the *Iliad*
3. "images . . . that moved by themselves at the feast of *Hiarba*"

4. "the Statues of *Mercury*, which did speak": Mercury = Hermes = Hermes Trismegistus, the *Asclepius*

5. "the wooden Dove of Architas": Archytas in Aulus Gellius, *Attic Nights*

6. "the miracles of *Boethius*, which *Cassiodorus* made mention of"

7. "images that speak . . . as *William of Paris* relates": William of Auvergne in *De Legibus*

The two items I have not yet mentioned are those attached to the names Hiarba (no. 3) and Boethius (no. 6).

The self-moving images at the feast of Hiarba are described in an episode in *The Life of Apollonius of Tyana*, written by the sophist Philostratus (late second to early third century CE) under the patronage of Julia Domna, the wife of the Roman emperor Septimius Severus. Neo-Pythagoreanism, which elevated the mathematical-minded Greek philosopher to the status of a master magician, was the most significant intellectual movement of philosophical magic in the Roman Empire of the first centuries CE.[119] The central figure for the movement was Apollonius of Tyana, a first-century ascetic who, according to Philostratus, traveled all the way to India to converse with the Brahmin philosophers. The reference to "Hiarba the philosophical exerciser" is misleading since the latter title is a creative rendering in the 1598 English translation by J. F. (probably John French) of the word *gymnosophistae*, (naked philosophers), the ancient Greek word for Indian philosophers of the dualistic religion of Jainism.[120] Hiarba is surely Iarchas, a native of India whom Apollonius befriends and at whose feast

> four tripods stepped forth like those which advanced in Homer's poem, and upon them were cupbearers in black brass resembling figures of Ganymede and of Pelops among the Greeks. . . . And dried fruits and bread and vegetables and the dessert of all seasons came in, served in order, and set before them more agreeably than if cooks and waiters provided it.[121]

The late Roman philosopher Boethius (480–524) had a magical reputation in the Middle Ages. Stories of his supposed powers originate from the misreading of a letter written to him by his kinsman Magnus Aurelius Cassiodorus (490–585), a Roman statesman who served the German king

Theodoric as rhetorical draftsman and legal advisor. The collection of his letters, most of which were written in the name of Theodoric and his successor, is an important source of knowledge on Italy under Ostrogoth domination. Around 506 Cassiodorus wrote to Boethius commissioning from him the construction of two clocks, one operated by water and the other by the light of the sun, to be sent as gifts to King Gundobad of the Burgundians. As Cassiodorus explains, Boethius is eminently qualified for this task by the great erudition in mathematical knowledge he has gained from studying in Athens and his Latin renderings of Greek works.

> For it is in your translations that Pythagoras the musician and Ptolemy the astronomer are read as Italians; that Nichomachus on arithmatic and Euclid on metaphysics and Aristotle on logic in the Roman tongue; you have rendered Archimedes the engineer to his native Sicilian in Latin dress.[122]

What follows is a description of great works that have been achieved through the use of such knowledge. Medieval readers misread Cassiodorus as saying that the marvels described, including Hero's automata, which are mentioned at the end, were the creations of Boethius rather than general examples of ancient technology. Such mathematical and mechanical knowledge

> labours to display events that men may wonder at: altering the course of nature in a wonderful way, it takes away belief in the facts, despite displaying images to the eyes. It causes water to rise from the deep and fall headlong, a fire to move by weights; it makes organs swell with alien notes, and supplies their piped air from outside, so that they resound with great subtlety. By its means, we see the defences of endangered cities suddenly arise with such solidity that machinery gives advantage to a man who despaired at their lack of strength. Waterlogged buildings are drained while still in the sea; hard objects are disintegrated by an ingenious device. Objects of metal give out sounds: a bronze statue of Diomedes blows a deep note on the trumpet; a bronze snake hisses; model birds chatter, and those that had no natural voices are found to sing sweetly.[123]

So Agrippa's reference to the miracle of Boethius is really to Cassiodorus describing Hero's automata.

What is apparent in lists of wonders similar to Agrippa's is the linkage of mechanics and mathematics with magic. And the automaton, a provocative example of an artificial marvel, served as a vivid example to illustrate the power of natural magic. Its use was particularly effective in the context of the Renaissance flowering of visual arts, in which people of culture, through the contemplation of representative images, pleasurably pondered the blurring of the boundary between the natural and artificial, the animate and inanimate, and the living and dead. One can also make sense, in this context, of the idea of the homunculus in the writings of Agrippa's contemporary Paracelsus, who gave a detailed formula for the alchemical creation of a smaller but very much living version of a human being. William Newman has shown that this idea of a chemically produced artificial being has a long tradition with roots in both Greek and Arabic lore, but it is significant that Paracelsus's new formulation captured the imagination of his time.[124]

Agrippa's list of wonders reappeared in a modified form in the writings of John Dee, the most prominent English Hermetic thinker of the period, who wrote the introduction to the first English translation of Euclid's *Geometry* (1570), by Henry Billingsley. Dee's short essay, no less than the translation itself, played an essential role in vitalizing mathematical studies in England. In it, he divides mathematics into two parts, arithmetic and geometry, and then further distinguishes "Art" and "Art Mathemticall Derivative," which are analogous to the modern categories of pure and applied. The bulk of his introduction is dedicated to describing different categories of the latter. Of the derivative mathematical arts of geometry, he counts no less than nineteen types, revealing the very broad nature of the mathematician's interests—Perspective, Astronomie, Musike, Cosmographie, Astrologie, Statike, Anthropographie, Trochlike, Heliocosophie, Pneumatithmie, Menadrie, Hypogeiodie, Hydragogie, Horometrie, Zographie, Architecture, Navigation, Thaumaturgike, and Archemastrie.[125] Thaumaturgike, or natural marvels, Dee describes as "that Art Mathematicall, which giveth certaine order to make straunge workes, of the sense to be perceived and of men greatly to be wondered at. By sundry meanes, this *Wonderworke* is wrought."[126] What follows the passage is a list of automata and devices of

"Perspective" that is close to the one in Agrippa's *De Occulta Philosophia*, probably a direct transmission since Dee was familiar with the work.[127] Among the automata, the recurring objects are "the Imagies of Mecurie" (item 4 on Agrippa's list), "the brasen head ... which dyd seme to speake" (item 7), "Boethius was excellent in these feates" (6), "The Dove of wood" (5), "of *Daedalus* straunge images" (1), and *"Vulcans Selfmovers"* (2). To these Dee adds several new objects. The examples he gives of the "wonderworks" include some that were achieved through the use of Pneumatithemie—"as the workes of *Ctesibius*, and *Hero*." Dee also reports a recent example:

> Mervaylous was the workemanship, of late dayes, performed by good skill of *Trochlike &c.* for in Noremberge, A flye of Iern, beyng let out of the Artificers hand, did (as it were) fly about the gestes, at the table, and at length, as though it were weary, retourne to his masters hand agayne. Moreover, an Artificiall Egle, was ordred, to fly out of the same Towne, a mighty way, and that a loft in the Ayre, toward the Emperour comming thether: and followed hym, beying come to the gate of the towne.

Throughout the sixteenth and well into the seventeenth centuries, the free imperial cities of Nuremberg, Strasburg, Ulm, and Augsburg were centers of mechanical crafts.[128] As they supplied many of Europe's clocks and other intricate mechanical devices, the ingenuity of their craftsmen became legendary. As for the stories of the iron fly and the mechanical eagle, Dee could have gotten it from the *Scholarum Mathematicarum* (1569), by the pioneering mathematician and logician Petrus Ramus, who credits the inventions to Regiomontanus. The latter—the astronomer and mathematician Johann Müller of Königsberg ("Regiomontanus" in Latin)—was a resident of Nuremberg from 1471 and was well known for various devices he constructed there, including a clock, a sundial, and several astrolabes, which may account for the legend of the improbable automata.[129] The "Emperour" Dee mentions would be Friedrich III, who visited the city on two occasions (1471 and 1474) when Regiomontanus was there, but as the astronomer's biographer Ernst Zinner points out, the tale may have originated from the Nuremberg custom of hanging a large double image of an eagle in a main street whenever the emperor came.[130] Dee also mentions a wonderwork he witnessed himself—"Of the straunge Selfmovying, which at Saint Denys, by Paris, I saw, ones of twise." The royal abbey of Saint-Denis possessed the greatest collection of treasures and wonders in all of Europe, as described by its twelfth-century abbot, Suger of Paris.[131]

Immediately following his list of wonders, Dee makes an enlightening "Digression Apologeticall" in which he ardently defends interest in such works against the accusation of diabolical magic, claiming it all for the glory of God, to reveal his great works in the world.

> And for these, and such like Marveilous Actes and Feates, Naturally, Mathematically, and Mechanically, wrought and contrived: ought any honest Student, and Modest Christian Philosopher, be counted, & called a Conjurer? Shall the folly of Idiotes, and the Mallice of the Scornfull, so much prevaile, that He, who seeketh no worldly gaine of glory at their handes: But onely, of God, the threasor of heavenly wisdome, & knowledge of pure veritie: Shall he (I say) in the meanse space, be robbed and spoiled of his honest name and fame? He that seketh (by S. Paules advertisement) in the Creatures properties, and wonderfull vertues, to finde juste cause, to glorifie the Æternal, and Almightie Creator by: Shall that man, be (in hugger mugger) condemned, as a Companion of the Helhoundes, and a Caller, and a Conjurer of wicked and damned Spirites?[132]

This was a particular sore point with him, as he was suspected of sorcery throughout his career, first when he constructed a machine for a flying scarab in a theatrical production of Aristophanes's *Peace* as a fellow at Cambridge and later, more seriously, when he was arrested in 1555 after an accusation that he had cast spells against Queen Mary on behalf of Elizabeth.[133] He was cleared of the charge, but the reputation stayed with him, prompting him toward the end of his life to make an appeal to the newly crowned James I to clear his name, which was rejected by the monarch who was hostile to magical studies.[134]

Dee was very much in the tradition of thinkers who distinguished demonic from natural magic and advocated the study of the latter for gaining a better understanding of nature and the manipulation of its forces for the benefit of humanity. The issue that complicates the matter yet again is the uncertainty of the boundary between legitimate and illegitimate magic. Despite Dee's indignation at the accusation of conjuring spirits, in 1582 he met with Edward Kelley, a supposed "skryer," or spiritual medium, and became intensely interested in spiritual magic, attempting to communicate with angels through Kelley.[135] All the major advocates of natural magic, from both the Middle Ages and the Renaissance, believed in the existence of spiritual entities, but they differed on such questions as whether such beings could affect things in the natural world, whether they could be coerced to do so, whether all practice of magic based on communication with them was heretical, and whether there were some forms of it that were legitimate. Interestingly, the Renaissance debates on those issues were often centered on the question of theurgy, in which the image of the talking head reappears.

Ficino, in his own philosophical works and commentaries on ancient works, attempted to interpret Hermetic and Neoplatonic ideas in ways that could be reconciled with Christianity. As D. P. Walker points out, however, the very nature of the worldview he was dealing with led to a kind of "astrological polytheism to which even the most liberal Catholic could not admit."[136] And in his efforts to somehow elaborate on the aspects of Hermetic philosophy that were particularly problematic from the orthodox point of view, Ficino had to resort to rather convoluted arguments, as when dealing with the notorious god-making passage in the *Asclepius*. In order to argue that the ancients were not in fact engaged in idolatry when they spoke of bringing spirits down into statues, he claimed that the images represented not gods but actual people who were born under the signs of powerful planets like Jupiter and Saturn and that the objects were used for astrological operations that had nothing to do with demon worship.[137] It was only later that wicked and ignorant priests used the practice to introduce idolatry and superstition.

When the Counter-Reformation commenced in the second half of the sixteenth century and it became particularly dangerous to express unorthodox ideas, the item of the oracular head in lists of wonders was either denounced or explained away in naturalistic terms. In Italy, the list shows up in the 1584 work *Trattato dell'arte della pittura*, by the mannerist artist Giovanni Paolo Lomazzo, who turned to writing art theory after he went blind.[138] In a section on the depiction of actions, gestures, situation, decorum, motion, spirit, and grace, Lomazzo discusses "mathematical motions" and then proceeds to give the list, whose first part is obviously taken from Cornelius Agrippa, as even the order in which they are presented is identical to his: the works of Daedalus; Vulcan's tripods from Aristotle; tripods at the feast of Iarbas; and the statues of Mercury.[139] Lomazzo includes at this point Leonardo da Vinci's flying machine and the walking lion mentioned by Vasari, before continuing to recount the items on Agrippa's list: the dove of Archytas and Cassiodorus's description of Hero's automata. Lomazzo concludes with the final item of the talking head, but an interesting variation occurs at this point, as he makes no mention of William of Paris but refers to the legend of Albertus Magnus building the object only to have Thomas Aquinas destroy it. Why the substitution, when all the other items on the list, except those of Leonardo, match those in Agrippa's? Although any speculation on this point runs the risk of reading too much into this minor difference, it should be pointed out that the reference to William in Agrippa speaks of "the rising of Saturn," whereas Lomazzo's description of the Albertus legend comes with a denial of the involvement of magic. Thomas Aquinas destroyed the object, Lomazzo claims, "because he thought it the Devil, whereas indeed it was a mere mathematical invention (as is most manifest)."[140]

The possibility that Lomazzo may have felt compelled to avoid mentioning any specific form of magic is supported by how other Italian writers of the period dealt with the item in their works. Giambattista della Porta, the most rigorous skeptic among the proponents of natural magic, discusses the speaking head in a section on pneumatic experiments in his popular book *Natural Magick* (expanded edition 1589) where he considers the question "whether material Statues may speak by any Artificial way."[141] He mentions the example of Albertus Magnus but says "to speak the truth, I give little credit to that man, because all I made trial of from him, I count to be false, but what he took from other men." He also rails against the superstition of people who believe that the talking head was created through the use of astrology, thus explicitly criticizing the position Agrippa took when he described its celestial operation: "I wonder how learned men could be so guld; for they know the Stars have no such forces." Yet he considers how the trickery of such an object could be achieved mechanically, through the use of a speaking tube:

> But I suppose it may be done by wind. We see that the voice of a sound, will be conveighed entirely through the Air, and that not in an

instant, but by degrees in time . . . if any man shall make leaden Pipes exceedingly long, two or three hundred paces long (as I have tried) and shall speak in them some or many words, they will be carried true through those Pipes, and be heard at the other end . . . and when the mouth is opened, the voice will come forth, as out of his mouth that spake it . . . I am now upon trial of it.[142]

It is possible that Porta may have actually heard of such a device, since modern archaeologists have found statues with speaking tubes in both Egypt and Greece.[143]

Tommaso Campanella, who was familiar with the works of Porta, makes a similar move in his discussion of automata in his *Magia e grazia* (early seventeenth century). He provides his own shorter list of wonders, which consists of Archytas's dove, the Nuremberg automata mentioned by Dee, and the statue of Daedalus. The last item is the talking head, on which he declares "I do not hold that to be true which William of Paris writes, namely that it is possible to make a head which speaks with a human voice, as Albertus Magnus is said to have done."[144] What is apparent here is that Campanella took the story from Agrippa, not directly from William of Auvergne, since he was unaware that the latter also denied its possibility. This passage is followed by a mechanistic speculation, possibly taken from Porta:

It seems to me possible to make certain imitation of the voice by means of reeds conducting the air, as in the case of the bronze bull made by Phalaris, which could roar. This art however cannot produce marvellous effects save by means of local motions and weights and pulleys or by using a vacuum, as in pneumatic and hydraulic apparatuses, or by applying forces to the materials. But such forces and materials can never be such as to capture a human soul.

The bronze bull Phalaris was a horrifying instrument of execution built by the tyrant Phalaris (ca. 570–549 BCE) of Acragas in Sicily. As the story goes, he had a certain Perilaus build the metal beast equipped with a firebox and a door through which people were thrown in to be roasted alive. The figure's throat and mouth were designed to make the victim's screams sound like the bull roaring.[145] So even as the automaton was continually being used to demonstrate the power and wonder of natural magic, the most dangerous item on this list was denied magical nature and domesticated as a trick device.[146]

In the period of the Counter-Reformation, writers from Lomazzo on had good reason to avoid writing of explicitly spiritual or demonic forms of magic, opting to interpret the workings of marvels like the talking head in mechanical terms.[147] But beyond the political consideration, Lorraine Daston and Katharine Park have noted, in the course of the sixteenth and seventeenth centuries there was a general shift from magical explanations of wonders to wholly naturalistic ones.[148] It is fitting, then, that this development is narrated in an episode involving the talking head in the great satire of medieval romance, Miguel de Cervantes's *Don Quixote*. In the second part of this epic (1615), the would-be knight is patronized by the nobleman Don Antonio Moreno, who is delighted by his madness and keeps him around to play a series of pranks on him. One of them involves showing Quixote into a dark room where Moreno reveals a bronze head he says was made by a great enchanter who "studied the stars . . . watched favorable moments, and at length brought the head to perfection."[149] The wonder astonishes Quixote, his squire, Sancho Panza, and other guests of Don Antonio by answering their questions. The knight is quite convinced of its power, but it is revealed that the whole thing is a hoax—a hollowed-out object fitted with a tin tube that the nobleman has made "for his own amusement and to astonish ignorant people." He has his nephew speak through the tube after giving him information about the guests who will address it. When word gets around that he has an enchanted object in his house, "Don Antonio, fearing it might come to the ears of the watchful sentinels of our faith, explained that matter to the inquisitors, who commanded him to dismantle it and have done with it, lest the ignorant should be misled."[150] Written in the same period as Porta and Campanella's naturalistic explanations of the talking head, this episode that features the medieval-minded Quixote's gullible awe, Don Antonio's mechanical trickery, and the fear of persecution by the humorless religious authority marks the definitive end of the legends of the talking head as an object of theurgical magic.

The shift from the magical to the naturalistic explanations of wonders also occurred as a prelude to the seventeenth-century campaign by many mechanistic philosophers of the scientific revolution to debunk, denigrate, and denounce all forms of learning that smacked of sorcery or superstition, including the "enthusiastic" approach to knowledge, which they associated with the religious fanaticism that caused Europe to fall into the carnage of the Thirty Years War. In the following chapter I will demonstrate the important role the automaton idea played in the intellectual transition from the Renaissance to the scientific revolution, to elucidate the intriguing situation of the persistent interest in the object throughout the seventeenth century. As I have shown, the automaton was regarded in both fiction and philosophy as a magical object, though there were disagreements on what kind of magic was involved in its construction. As a magical entity, then, it might have been denounced by the mechanistic thinkers for its connection to Renaissance wonders and animistic-pantheistic ideas of Neoplatonic and Hermetic variety. What happened, on the contrary, was that the natural philosophers took up the automaton, purified it of its magical aura through completely mechanical explications of its function, and elevated it as the central emblem of their newly emerging worldview, which was based heavily on the metaphor of clockwork machinery. The transition from the automaton as an object of natural magic to that of the mechanistic worldview is demonstrated in the reappearance of the list of wonders in the scientific works of the seventeenth century. I will elaborate in the next chapter how the conceptual use of the automaton by the Hermetic philosophers of the Renaissance and the mechanists of the scientific revolution alike points to significant areas of common interest among them.

The philosopher René Descartes began writing in 1628 his *Traité du Monde* (Treatise on the world), a detailed explication of his natural philosophy, including a mechanistic description of the human body in the second part entitled *Treatise on Man*. But five years later, with the condemnation of Galileo for heresy, Descartes decided not to publish the work since it was filled with Copernican ideas that could land him in trouble with the authorities. As a result, the *Treatise on Man* was not published until 1662, twelve years after his death, and in a Latin translation made in the Netherlands. The text in the original French appeared two years later, and the complete *Traité du Monde* in 1677. The Latin version was edited and translated by Florentius Schuyl, a professor of philosophy at the University of Leyden who was deeply interested in botanical and medical subjects.[151] He wrote a long introduction for the work, dealing mainly with mechanistic physiology, which was translated and included in the first French edition by Claude Clerselier, the most ardent advocate of Cartesian ideas in

France.¹⁵² As a result, the Schuyl introduction became an important document in the ensuing debates on medical matters.

One of the many ways Schuyl sought to defend the notion of bodies as organic automata was to show how it was possible to create machines in the shape of animals. He does this by presenting the reader with, as you may have guessed, a list of automata from ancient times to the present.¹⁵³ The items he mentions are "that famous dove made by Architas of Tarente" (item 5 on Agrippa's list), "the wooden eagle of Regiomontanus" (from Dee's list), "the admirable head of Albertus Magnus" (7—Schuyl cites Giambattista della Porta as his source), and "the Venus of Daedalus" (1). He also refers to moving and speaking machines that are described in the books of "Coelius Rodginius in his book of antiquities, Kircher and several others." The antiquarian Ludovico Caelius Rhodiginus mentions the Egyptian idols in his *Lectorum Antiquarum*, as does the Jesuit polymath Athanasius Kircher in his 1652–1654 work on ancient Egyptian civilization *Oedipus Aegypticus*, in which he also debunks the speaking statue, the trickery of which he himself has demonstrated by attaching a tube to a statue and talking through it.¹⁵⁴ Schuyl also mentions an interesting legend of an iron statue that came alive and went to see the king of Morocco to ask for the freedom of its maker.¹⁵⁵

I have already mentioned John Wilkins, one of the founders of the Royal Society. In addition to popularizing the astronomical ideas of Copernicus and Galileo, he wrote a great deal on the experimental and mechanical sciences, urging natural philosophers to learn the practical mathematical arts.¹⁵⁶ The major example of such writing is his 1684 treatise on Archimedean mechanics, *Mathematicall Magick: Or The Wonders that may be performed by mechanical geometry*, in which he expounds on the need for a study of this kind. As noted, however, the word "magick" in the title is an ironic reference to the association of mechanics with magic in the popular imagination, which Wilkins seeks to correct in this work.

The second part of the book, which deals with what today is called applied science, is entitled "Daedalus," and chapter 4 is on automata, which Wilkins divides into two types: gradiant ("such as require some Basis or Bottom to uphold them in their Motions") and volant (flying).¹⁵⁷ Of the former he gives the following descriptions (the first two are the initial items on Agrippa's list):

Such were those strange Inventions (commonly attributed to *Daedalus*) of self-moving Statues, which (unless they were violently detained) would of themselves run away. *Aristotle* affirms that *Daedalus* did this by putting Quicksilver into them. But this would have been too gross a way for so excellent an Artificer; it is more likely that he did it with Wheels and Weights. Of this kind likeness were *Vulcan's Tripods* celebrated by *Homer*, that were made to move up and down the house and fight over one another. He might as well have contrived them into Journey-man statues, each of which with a Hammer in his Hand should have worked at the Forge.

To these he adds a legendary automaton of a statue holding a golden apple that shoots an arrow if the fruit is touched, taken from a work by the Renaissance philosopher Girolamo Cardano but based on a design of Hero, and the magnetically operated toys described by Athanasius Kircher in his 1631 work on magnetism, *Ars Magnesia*.[158] Later in the chapter, he adds two items I have already discussed:

> There have been some Inventions which have been able for the Utterance of articulate Sounds, as the Speaking of certain Words. Such are some of the *Egyptian* Idols related to these. Such are the Brazen Head made by Friar Bacon, and that Statue, in the framing of which *Albertus Magnus* bestowed 30 Years, broken by *Aquinas*, who came to see it, purposely that he might boast, how in one Minute he had ruined the Labor of so many years.[159]

In chapter 6, Wilkins describes volant automata.

> The *volant*, or flying *Automata*, are such Mechanical Contrivances as have a self-motion, whereby they are carried aloft in the open Air like the flight of Birds. Such was that Wooden Dove made by *Archytas*, a Citizen of *Tarentum*, and one of *Plato's* Acquaintances: And that wooden Eagle framed by *Regiomontanus* at *Noremburg*, which by way of triumph, did fly out of the city to meet *Charles* the Fifth. The latter Author is also reported to have made an Iron Fly,—which, when he invited any of his Friends, would fly to each of them round the Table, and at length (as being weary) return unto its Master.[160]

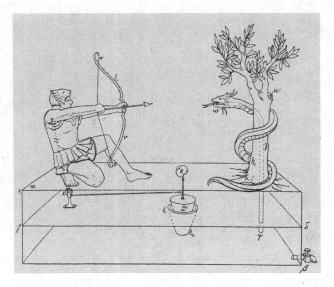

Hero of Alexandria, design for an automaton archer. Reconstruction—illustration in *Herons von Alexandria Druckwerke und Automatentheater*, ed. Wilhelm Schmidt (Lipsiae: Teubner, 1899).

The irony of his book's title notwithstanding, Wilkins's comprehensive reiteration of items from previous lists points to links between the natural magic of the Renaissance and the mechanistic ideas of the scientific revolution.

One can find faint echoes of the items from the lists of wonders even in Denis Diderot and Jean le Rond D'Alembert's eighteenth-century *Encyclopédie*. In the articles "Automate" and "Androide," the two traditional automata mentioned are the wooden dove of Archytas and the work of Albertus Magnus.[161] In the case of the former, it is probably the most innocuous item since in the original passage by Aulus Gellius the bird's operation is described in mechanical terms as using "weights and moved by a current of air enclosed and hidden within it,"[162] yet even here the reader is invited to consider whether it was just a fable or not. The origin of the word "android," defined as an automaton specifically in the shape of a human being (as opposed to other living creatures), is obscure, but it is a medieval coinage from Greek roots ("andros," man, and "eides," species) and is commonly linked to Albertus Magnus. The Renaissance historian Paolo Giovio (Latin name Paulus Jovius) in his work *Vitae Illustrium Virorum* (1549–1557) writes:

> Having become master of the magical sciences, Albertus began the construction of a curious automaton, which he invested with the powers of speech and thought. The Android, as it was called, was composed of metals and unknown substances chosen according to the stars and endowed with spiritual qualities by magical formulæ and invocations, and the labor upon it consumed thirty years.[163]

I am not arguing here for a case of direct transmission in the *Encyclopédie*, although the reference to Archytas was probably taken from Porta's *Natural Magick*, since that work is referred to in the article. This is, rather, evidence of the importance of Agrippa's original list, how the items he brought together in it and those in later modified lists became central referents of mechanical wonders for Renaissance magicians and natural philosophers of the scientific revolution alike. The intellectual impact of Agrippa's original list is charted in Table 3.

The enduring presence of the automaton motif throughout the early modern period not only gives evidence of the object's captivating power and conceptual flexibility, in its ability to take on such diverse ideas, but also points to an important aspect of the intellectual scene of the sixteenth century. As I will elaborate in the following chapter, scholars in recent decades have questioned the simplistic view of the scientific revolution as the struggle and eventual triumph of scientists over theologians and magicians. In more nuanced approaches to the intellectual history of the period, historians have pointed to significant overlaps in areas of interest and methodology in the works of Hermetic thinkers and mechanistic philosophers. The persistent use of the automaton for conceptual purposes in the transitional era points to two major continuities: the elevation of mechanics as a respectable field of scholarship and the further elaboration of the experimental methodology in the study of nature. For that reason, the automaton idea did not fade with the end of the Renaissance but became a significant concept in the mid-seventeenth century.

For reasons detailed in this chapter, the automaton was an object of highly ambiguous status in the medieval and Renaissance imagination because of not only its inherently transcategorical nature but also its association with preternatural knowledge that dealt with occult forces. Even as a product of purely mechanical craft, it could both fascinate and terrorize

Table 3. The List of Artificial Wonders from Cornelius Agrippa to Florentius Schuyl

Agrippa (1533)	Dee (1570)	Lomazzo (1584)	Campanella (early 17th century)	Wilkins (1648)	Schuyl (1662)
Statues of Daedalus	*	*	*	*	*
Tripods of Hephaestus	*	*		*	
Tripods of Iarchas		*			
Egyptian idols	*	*		*	*
Dove of Archytas	*	*	*	*	*
Miracles of Boethius		*			
Talking heads (William of Paris)	*	(Albertus Magnus)	*	*	*
Works of Ctesibius and Hero					
Wonders at St. Denys					
Nuremberg automata					
		Leonardo da Vinci's Works			
				Fighting automata in Cardan	
				Magnetic toys in Kircher	
					Iron statues

viewers in an age when the technological was often linked to the magical. The resulting confusion over whether it was a product of necromancy or natural magic, diabolical or legitimate art, was solved in the course of the seventeenth century when the philosophers of the new world order transformed the automaton into an object of mundane machinery that operated on the basis of regular, eternal, and immutable laws of nature. Once denuded of its magical aura, the automaton was then elevated as the single most important emblem of the emerging worldview that envisioned the world, the state, and the body in mechanical terms. So as rich and varied as the roles the automaton played in the intellectual and imaginative writings of the Middle Ages and the Renaissance, it entered the true golden age of its celebration, popularity, and significant use as both an actual and a conceptual object in the following period.

3
The Man-machine in the World-machine, 1637–1748

Jacques de Vaucanson and the Automaton Craze

In April 1738, *Mercure de France* reported on the latest craze in Paris:

> At the Hôtel de Longueville, rue St. Thomas du Louvre, for about two months all of Paris has been going to see with admiration a phenomenon of mechanics, the most singular and at the same time the most pleasing that has perhaps ever been seen.[1]

The object of fascination was the "fluteur automate"—a life-sized mechanical statue that could play fourteen airs, "all different in character, in a variety of notes and tempos," on a flute by emitting wind from its mouth and moving its fingers. According to one contemporary, some of the spectators refused to believe that it actually played the instrument, speculating on a hidden musical device, but even the most incredulous were convinced, as the "machine was submitted to the most minute examination and to the strictest tests. The spectators were permitted to see even the innermost springs and to follow their movements."[2]

Its creator was Jacques de Vaucanson (1709–1782), a mechanic from Grenoble who had come to Paris five years earlier to continue his studies and to raise money for his automaton project. Despite the fact that the price of viewing was 3 livres, about a week's wage for the average Parisian worker, thousands came flocking to marvel at the wonder.[3] When attendance began to wane in the following months, Vancanson presented two more works— another musical figure of a fife-and-drum player and, the most astonishing and famous automaton of the period, a mechanical duck that could flap its wings, drink water, swallow grain, and even excrete little pellets from its rear. As Joseph Spence, an Oxford literary scholar, described it in 1741:

> If it were only an artificial duck that could walk and swim, that would not be so extraordinary: but this duck eats, drinks, digests and sh-ts. Its motions are extremely natural; you see it eager when they are going to give him his meat, he devours it with a good deal of appetite, drinks moderately after it, rejoices when he has done, then sets his plumes in order, is quiet for a little time, and then does what makes him quite easy.[4]

The three automata were displayed together in the spring of 1739 with further public and critical approbation. After their inventor made a small fortune, he sold them to a consortium of businessmen from Lyons, who took them to London in 1742, where they were displayed at Haymarket Theatre to great success. The works were subsequently sent on a tour all over Europe, reaching as far as St. Petersburg and ending up in Germany, where in 1805 Goethe found the flute-player and the duck in a dilapidated condition at the house of a doctor named Gottfried Christoph Beireis in Helmstadt.[5] In 1839 the remains of the duck came into the possession of Johann-Bartholomé Rechsteiner, a Swiss mechanic, who made several working models of it and presented them for the viewing of Louis I of Bavaria at Munich in 1847.[6] The remains of the work disappeared some time after that.

The success of the Vaucanson trio set off a veritable automaton craze that lasted through the rest of the century, as a series of increasingly creative and sophisticated works made their appearance. The pinnacle of achievement in

Jacques de Vaucanson's three automata—the fife-and-drum player, the duck, and the flute-player. From Jacques de Vaucanson, *Le Mécanism du Fleutreu Automate, An Account of the Mechanism of an Automaton or Image Playing on the German-Flute*, trans. J. T. Desaguliers (Paris: J. Guérin, 1738).

the construction of clockwork automata was reached by the Swiss father-and-son team of Pierre and Henri-Louis Jaquet-Droz and their collaborator Jean Leschot with their 1773–1774 group of three automata: a writer, a draughtsman, and a musician.[7] The latter was an especially impressive work of a female figure that not only played an actual miniature harpsichord with fully articulated fingers but moved in time with the music while simulating breathing with the heaving of its chest. These automata were first displayed in the mechanics' native town, Chaux-de-Fronds, a center of clock manufacturing, and then, like the Vaucanson automata, sent on a tour all over Europe. In France they were demonstrated for the pleasure of Louis XVI and Marie Antoinette at Versailles. They can be seen today at the Museum of Neuchâtel in full operation, while another draughtsman-writer figure by Henri Maillardet, a sometime collaborator of the Jaquet-Drozs team, is at the Franklin Institute in Philadelphia.[8] There was also a series of speaking automata—a group of four by the Viennese mechanic Friedrich von Knauss that was exhibited in 1770; a ceramic head of 1778 by the Frenchman Abbé Mical and its follow-up, a pair of heads that exchanged sentences praising the king, demonstrated in 1783; and two others by Wolfgang von Kempelen and Christian Gottlieb Kratzenstein that were displayed in 1780.[9] In London, a jeweler named James Cox opened his celebrated museum of curios in 1772 with many automata, including a screeching, tail-spreading peacock, a silver swan with a moving neck swimming on "artificial water," and a pineapple that opened up to reveal a nest of chirping birds.[10] Most of the works there were the creations of a Belgian in Cox's employ with the suggestive name John Joseph Merlin, who eventually set up his own "Merlin's Mechanical Museum" in the 1780s.[11]

Vaucanson's career flourished after he sold off his automata. He was offered a position at the court of Frederick II of Prussia, who had an interest in them.[12] But Vaucanson only used the offer—by making it known to Cardinal de Fleury, the powerful chief minister of Louis XV, to gain the lucrative and influential position of royal inspector of silk manufacturers in 1741. Set to the task of overseeing the silk industry centered in Lyon and making it competitive with its rival in Piedmont, Vaucanson recommended sweeping reforms that would modernize production. In 1744, with the approval of the government for his proposals, he traveled to Lyon, only to be met with fierce opposition by the workers, resulting in the single most serious strike of eighteenth-century France, which was violently suppressed.[13]

After barely escaping the city with his life, Vaucanson returned to Paris to concentrate on inventing better silk-making machines, including an automatic loom powered by a single source, a draw-loom for brocade and figured silk, and a throwing mill for converting fibers reeled from the cocoon into raw silk.[14] In 1746 he was admitted into the Académie Royale des Sciences on the strength of his automatic loom.

In the following decades, he and a team of assistants at his workshop in Hôtel de Mortage produced many useful devices, including a water-driven saw, an improved dock, and a machine that made an endless chain. In 1757, he beat Diderot in the competition for the post of associate mechanic in the Académie. There were a few major failures among his many grandiose plans, including a model silk-producing factory he designed that was built in Aubenas in the 1750s. It was the first industrial plant in the modern sense, predating Richard Arkwright's cotton-spinning mills in Derbyshire by two decades, but it failed and went bankrupt in 1775.[15] There were also his intermittent attempts to construct an "anatomie mouvante," an automaton that replicated all the vital organs of the human body, which he began in 1741 at the suggestion of Louis XV and worked on over the decades, in competition with the famed surgeon Claude Nicolas Le Cat, which was never completed.[16] Nevertheless, he died a wealthy and celebrated man in 1782 as his eulogy was written by the Marquis de Condorcet and his obituary notice by Friedrich von Grimm.

The case of Vaucanson's works is essential to understanding the cultural significance of the automaton in this period for three related reasons. First, he was the single most famous automaton-maker in the history of the object, especially for his so-called defecating duck, and his name remained a household word into the early twentieth century. Second, the success of his automata was not just a popular phenomenon but an intellectual one as well, as some of the most advanced thinkers of his time commented on and made conceptual use of them in their works. Third, as an intellectual phenomenon, the automata embodied some of the central philosophical, scientific, and medical ideas that were being hotly debated at the time. Each of these points needs to be elaborated on at some length to demonstrate the crucial importance of Vaucanson's works in the history of the automaton.

On the first point of Vaucanson's fame as an automaton-maker, as he went through his careers as a government official, industrialist, and technological inventor, automata made by others quickly outstripped his in sophistication. Yet none of their creators reached his level of fame (with the Jaquet-Drozs a distant second), as it was his name that was repeatedly evoked for well over a century after 1738 in references to automata. On his achievements as an industrialist, Jean-Antoine Chaptal, who would go on to become Napoleon's most innovative minister of the interior, praised Vaucanson for his ability to combine mechanics with economics.[17] Likewise, Karl Marx, in *Capital*, used Vaucanson's technological works, along with those of Richard Arkwright and James Watt, as examples of crucial inventions that arose in the context of available labor in the early phase of the Industrial Revolution.[18] It is significant, however, that the vast majority of posthumous references to Vaucanson are to his automata, sometimes employed for high intellectual purposes. Immanuel Kant in his *Critique of Practical Reason* (1788) criticizes Leibniz's notion of the preestablished harmony of the world by asserting that freedom within such a system would be illusory and every human being would be "a marionette or an automaton like Vaucanson's, fabricated and wound up by the Supreme Artist."[19] Thomas Carlyle in his 1829 essay "Signs of the Times" mentions Vaucanson's duck, which "seemed to eat and digest," in the context of materialist-mechanistic philosophy, and the physicist Hermann von Helmoltz points to the automata of Vaucanson and Jaquet-Droz in his 1854 essay "On the Interaction of Natural Forces," calling the artificial duck the "marvel of the last century."[20] T. H. Huxley, in his 1874 work "On the Hypothesis That Animals Are Automata, and Its History," presents the case of a French sergeant who was brain-damaged in a battle and sometimes involuntarily makes the movements of a soldier in a skirmish. Huxley questions whether the man is dreaming that he is fighting or is "in the condition of one of Vaucanson's automata—a senseless mechanism worked by molecular changes in his nervous system."[21] The latest instance I have found of Vaucanson being referenced with no explanation of his identity (i.e., assuming the reader's familiarity with his fame) is in *The Sexual Life of Our Time and Its Relations to Modern Civilization* (1906), by the pioneering sexual psychologist Iwan Bloch, who points to mechanical constructs of human bodies made of rubber and plastic that are used for sexual purposes, calling their creators "true Vaucansons in this province of pornographic technology."[22] In literature, his duck and music-players (along with those of the Jaquet-Droz team) appear in the mechanical zoo and concert hall of the ridiculous

Machine Man in Jean Paul's 1789 story "Machinenmann nebst seinen Eigenschaften" (The machine-man and his characteristics), while the fictional Thomas Edison in Auguste Villiers de l'Isle-Adam's 1886 novel *Tomorrow's Eve* boasts that his electric android will be so superior to the automata of the past that he will make Albertus Magnus, Vaucanson, and others look like "barely competent makers of scarecrows."[23] The duck made a recent appearance as a fictional character in Thomas Pynchon's picaresque novel of the eighteenth century, *Mason & Dixon* (1997), in which it comes to colonial America and boasts of its fame in France:

> "Here,"—producing from some interior Recess a sheaf of Notices in print, clipp'd from various newspapers and Street-bills,—"here,— *voilà*, with the Flauteur, and the Tambourine-Player? in the Center, 'tis *moi, moi* . . . Listen to what Voltaire wrote about me, to the Count and Countess d'Argental,—'. . . *sans la voix de la Le More et le Carnard de Vaucanson, vous n'auriez rien que fit ressouvenir de la gloire de la France,*' all right?[24]

As for the second aspect of the importance of Vaucanson's works, his creations were regarded as significant even at the highest intellectual level. In fact, what is particularly interesting about the success of his automata was that it was both a popular and an intellectual phenomenon, as many important thinkers of the eighteenth century noted their ingenuity and utilized the object as an idea in their works. When Condorcet wrote in his eulogy to Vaucanson, he claimed that his name will be celebrated for a long time by the vulgar for his "ingenious productions," no doubt his automata, and by enlightened men for his "useful works."[25] What is surprising about this assertion is that it was demonstrably untrue even during Vaucanson's own time, as it was his automata rather than his subsequent inventions of practical value that captured the imagination of intellectuals. When the first of his works was displayed, the literary critic Pierre Desfontaines celebrated it in a long article on the exhibition in his journal *Observation sur les écrit modernes*, calling it "a masterpiece of mechanics, a prodigy of genius, and a miracle of art."[26] Smiliarly, Antoine François Prévost, the author of the novel *Manon Lescaut*, praised the automaton in his literary periodical *Le Pour et le contre* as "the most marvelous piece of mechanics that has appeared to this day."[27] In the same month, members of the Académie Royale des Sciences came to the Hôtel de Longueville to view the automaton, after which Vaucanson presented his *Mémoire descriptif* of his inventions to the august body.[28] Bernard le Bovier de Fontenelle, the permanent secretary of the Académie, who was also an ardent advocate of public education in science, wrote a flattering certificate to be included in the *Mémoire*, in which he attested to "the intelligence of its author, and his great knowledge of different aspects of mechanics."[29] Voltaire, who advised Frederick II to offer Vaucanson a position at his court, praised him in his 1740 poem "Discours sur l'homme": "the bold Vaucanson, rival of Prometheus / Seemed in imitating the forces of nature / To take fire from heaven to animate his bodies."[30] Julien de La Mettrie also evoked the name of the creator Titan in his controversial materialist treatise *Man a Machine* (1748) when mentioning Vaucanson in a discussion of machine-body analogy, speculating on a talking automaton that would be much more difficult than the flute-player and the duck to construct, but not impossible, in the hands of "a new Prometheus."[31] Diderot named Vaucanson the archetype of the mechanical genius in his dialogue *D'Alembert's Dream*,[32] and added a note to D'Alembert's article "Android" in the *Encyclopédie*, pointing out the flute-player "finesse in all its details" and the "delicacy in all the parts of this mechanism."[33] The article itself consists mostly of a detailed description of the flute-player, much of it taken from Vaucanson's *Mémoire descriptif*. Likewise, the anonymous article "Automate" is taken up almost entirely with a flattering account of his fife-and-drum player and duck.[34] And Louis-Sebastien Mercier in his panoramic work *Le Tableau de Paris* (1781–1788) complains of the bureaucracy of clerks who give themselves airs when their only talent lies in knowing how to fill out forms. Mercier thinks that if "Vaucanson instead of a mechanical musician had made an automatic clerk, this latter would have been more use in the world."[35] So even as those Condorcet called the vulgar came in droves to marvel at the ingenious productions, thinkers like Voltaire, La Mettrie, Diderot, and others praised them in their enlightened works. But this raises questions about how exactly the objects were utilized in their writings and what ideas they represented—leading to the third aspect of Vaucanson's importance in the history of automata.

As seen above, much of the intellectuals' praise of Vaucanson centered on his mechanical skill, representing him as a technological virtuoso, quite appropriate in an age when there was a widespread interest in mechanics, as is reflected in the detailed descriptions of all kinds of machinery in the *Encyclopédie*. Yet to many others, including Voltaire and La Mettrie, his automata suggested specific philosophical and medical ideas. Joan Landes has linked the Vaucanson works not only to anatomical models of the period but also to physiological theories prevalent in the medical discourse of the time.[36] Vaucanson himself in his *Mémoire descriptif* sought to highlight the scientific contributions of his works beyond their mere entertainment value by claiming that in the motions of his automata he replicated as closely as possible the function of natural bodies. In his description of the duck's wings, for instance, after a discussion of its bone structure, he asserts: "Inspection of the Machine will better shew that Nature has been justly imitated."[37] Earlier in the passage, he also claims that in "the Mechanism of the Intestines which are employed in the Operation of Eating, Drinking, and Digestion: Wherein the Working of all the Parts necessary for those Actions is exactly imitated."[38] He goes onto describe a small "Chymical Elaboratory" in the duck's gut that allows it to digest food and expel excrement through a pipe to the anus. What is interesting about the description of the duck's internal function is that Vaucanson was perpetrating a fraud when he wrote it. The automaton featured no such chemical laboratory or digestive system of any kind, but only a simple trap device released excrement-like pellets some time after the duck was fed.[39] His claim was debunked in the 1780s by Christian Friedrich Nicolai, an important figure of the German Enlightenment, who saw the duck during its tour.

Despite the deception, such descriptions in the *Mémoire descriptif* reflected Vaucanson's effort to bring intellectual respectability to his creations through the demonstration of not only his mechanical talent but also his knowledge of physiology. The didactic value of the automata lay in their representation of ideas from mechanistic medicine that used machines as conceptual devices for the elucidation of the natural body's operation. When Joseph Spence wrote to his mother from Paris in 1741 after having seen the duck, he began with a discussion of Descartes and his notion that "all animals were nothing but so many ingenious pieces of clockwork" and that "a good artist might make an animal in clockwork, that should do everything the same real animal can do."[40]

I begin this chapter with Vaucanson because the popular as well as intellectual success of his works represents the ultimate culmination and apotheosis of the intellectual golden age of automata that began in 1637, with the publication of Descartes's *Discourse on the Method* and ended in 1748 with La Mettrie's *Man a Machine*. The link between the automata and mechanistic physiology is further elaborated in the course of this chapter, but there is a larger contention to be made about their significance in the intellectual and cultural context of western Europe in the second half of the seventeenth and early eighteenth centuries, namely that the automaton emerged as the central emblem of the entire mechanistic worldview that was dominant in the period.

It is well known that the natural philosophers of the classical Enlightenment envisioned the universe as a great machine created by an engineer God who set it in regular motion according to his rational laws. Political thinkers of the period also described the state as a machine, with the enlightened sovereign as its mechanic, and medical thinkers considered the natural body in terms of machinery as well. In the context of what E. J. Dijksterhuis famously called the mechanization of the world picture that occurred in the course of the seventeenth century, in which machine-people lived in a machine-state in a machine-cosmos, the automaton emerged as the most powerful and conspicuous intellectual emblem of the era.[41] Starting from the mid-century it appears with notable frequency in many of the advanced philosophical, political, scientific, and medical writings of the time. In these works, "automaton" denotes specifically a self-moving machine, completely mechanical in function with no place for magical or any other kind of preternatural force. The word was then used to describe the detailed workings of the natural body, the government, and the entire world that was constructed by the mechanically oriented God. The automaton, in other words, became the central metaphor of the age.

To fully explore these issues, let us return to the point where the last chapter ended: the end of the Renaissance and the culmination of the scientific revolution.

From Living Nature to the Nature-Machine

Scholars of the late Renaissance period have noted an essential tension in the culture of the late sixteenth and seventeenth centuries that was exacerbated by the deterioration of the political situation in central and western

Europe.[42] The intellectually adventurous atmosphere of the period gave rise to a tremendous number of innovations and discoveries that laid the foundation for the scientific revolution. Yet the overturning of so many traditional ideas about nature, man, and society also brought about a crisis of uncertainty for intellectuals, especially during the religious wars that culminated in the Thirty Years War and the chaos of civil war in England. As a result, there was a marked concern among them with the establishment of a worldview that would provide order and stability while assimilating newly discovered knowledge. For many thinkers the answer lay in a mechanistic philosophy that saw the world as a rational construct of regular movements, immutable laws, and predictable actions, with no place for notions of occult powers, demonic beings, and miraculous intercessions that could give rise to the "enthusiasms" of ignorant superstition, religious fanaticism, and political strife.[43] In this context the language of machinery proliferated in the philosophical and scientific works of the period.

In explaining the decline of wonders as a subject of scholarly study in the seventeenth century, Lorraine Daston and Katharine Park have pointed to a shift in the idea of nature itself at the end of the Renaissance.[44] In the earlier view of such thinker as Girolamo Cardano, Nature was seen as a living, sentient being, a handmaiden of God who obeyed his commands (allowing God to affect his will on the world without getting his hands dirty in it) but with inventive touches of her own that manifest themselves in variations and anomalies. So wonders were important subjects of scholarly study since they revealed the creative workings of Nature, at times carrying messages as portents. In the new mechanistic model, this middle-woman of Nature between God and the world was removed, making the natural order directly subservient to the deity's laws. "In the name of simplicity, uniformity, and universality, not only nature but also God lost the spontaneity that made for surprises in the established order of things."[45] In other words, Nature, as a creative being, was replaced by a machine that operated strictly according to its original programming. Robert Boyle in *A Free Enquiry into the Vulgarly Received Notion of Nature* (1686) praises God for having created "so great and admirable an automaton as the world, and the subordinate engines comprised in it."[46] Consequently, anomalies and other forms of natural marvels became subjects beneath scholarly interest since they were nothing more than occasional irregularities that occurred in a generally regular system, like rare defective products from a factory that produces uniform goods.

Wonders remained things of popular interest, as magicians, tricksters, and displayers of the strange and the unusual continued to entertain the common people, but their status as subjects of intellectual interest underwent significant changes. Already in the late sixteenth and throughout the seventeenth century a process was under way of naturalizing the preternatural, of providing mundane explanations for marvels without any references to otherworldly beings or forces. Daston and Park have also demonstrated that some intellectuals of the period used the sense of wonder itself to combat the enthusiastic approach to knowledge, distinguishing between good wonder that leads to calm admiration for and curiosity about the mechanistic order of the world and bad wonder that leads to superstition and fanaticism.[47] By the early eighteenth century, however, wonders and marvels were rejected altogether, as Enlightenment savants "did not so much debunk marvels as ignore them," leading to the statement in the *Encyclopédie* that "whatever one says, the marvelous is not made for us."[48] This last point on the supposed lack of interest in wonders during the Enlightenment has been controversial.[49] Mary Terrall in her biography of the Enlightenment scientist Pierre-Louis Maupertuis discusses his interest in unusual characteristics of creatures like the salamander and the scorpion, claiming that strange natural phenomena such as regeneration and parthenogenesis pointed to new kinds of wonder that the natural philosophers of the period co-opted by means of their view of regular, predictable nature.[50] In other words, while Daston and Park are right in noting a trend in the eighteenth century involving thinkers who took a hard line in rejecting all wonders and wondrous attitudes in the study of nature, there were others who continued the project of interpreting wonders in the language of the new natural philosophy.

The automaton as a conceptual object provides a particularly vivid case of this. What made this object such a captivating entity in both the medieval and the Renaissance imaginations was the uncertainty about its exact nature. People in premodern periods articulated their alternating fascination and horror in the face of the inherently unstable, transcategorical, and liminal device in their anxious discussions on how the thing was achieved, either through mundane craft or some kind of magical power, "by artifice or by necromancy," as John Mandeville put it. As the mechanistic model of the world was being established and the preternatural was being naturalized, the automaton was likewise systematically deprived of its magical nature and completely rehabilitated as a product of human mechanics that promised the technology necessary for the control of nature.

Major thinkers of the period, like John Wilkins, René Descartes, Robert Boyle, and Thomas Hobbes, who made fruitful use of the automaton idea did not try to deprive this mechanical marvel of its wondrous aura. They took advantage, on the contrary, of that very attractive power to draw attention to the beauty, intricacy, precision, and power of the machine. There was no question now of any kind of otherworldly agency in its function, but they took up the automaton precisely for its ability to arouse a powerful sense of awe. And once deprived of its magic, it became a source of the good kind of wonder that inspired a decorous desire for the discovery and understanding of the worldly order. This also allowed thinkers to use the device as a conceptual and didactic object with which to illustrate the wonders of the world-machine, the state-machine, and the body-machine.

The adoption and rehabilitation of the automaton as a fully mechanical object by the natural philosophers of the seventeenth century also points to another important aspect of the transition from the Renaissance pantheistic worldview to the mechanistic one. In *Discourse on the Method* (1637) Descartes lays out his scientific ideas, contrasting his views from those of both the Aristotelian Scholastics and the Hermetic philosophers, railing against "the promises of an alchemist, the predictions of an astrologer, the tricks of a magician, or the frauds and boasts of those who profess to know more than they really do."[51]

Elucidating the position of mechanistic philosophy by distinguishing it from both the Peripatetic and the Hermetic positions was a strategy that was employed by many like-minded thinkers of the period. But the rhetoric led to the simplistic narrative of the scientific revolution as the triumph of scientists over theologians and magicians that has been questioned by historians of science in recent decades. Since the publication of Frances Yates's seminal work on Hermetic philosophy in 1964, various scholars have demonstrated the presence of alchemical and other Hermetic notions in the works of such central figures as Francis Bacon, Robert Boyle, and Isaac Newton,[52] while others have traced the impact and persistence of such ideas on seventeenth- and eighteenth-century natural philosophy, for instance in the relationships between alchemy and chemistry, astrology and astronomy.[53] There are fundamental differences between the animistic-pantheistic worldview of the Renaissance Hermetic magus and the mechanistic one of the seventeenth-century natural philosopher, but the "transitional" narrative of the intellectual shift, in contrast to the "oppositional," emphasizes the continuities and the transformations of key ideas and methodologies. For instance, Keith Hutchinson has shown that rather than completely rejecting the notion of "occult" or hidden qualities of natural elements, a central concept in Hermetic philosophy, early modern scientists altered it to fit it into the new worldview, as is evident in the debates between Newtonians and Cartesians over the nature of gravity.[54] Two of the most important areas of such continuity are the experimental methodology of studying nature, which the mechanistic philosophers directly inherited from the alchemists but then transformed into a public enterprise that abandons the magi's cult of secrecy, as well as the deep interest in mechanics and mechanical crafts.[55] On the latter, as shown in the last chapter, thinkers like Cornelius Agrippa and John Dee considered mechanics an important and respectable form of natural magic and used automata as examples of the wonders that could be achieved through that knowledge. The automaton's elevation as the central emblem of the mechanistic worldview provides a clear case of this continuity in technological interest.

To fully examine how the automaton as a conceptual object was utilized in various intellectual fields, I will look at its appearance in medical and political discourses before examining some of its manifestations in literature.

The World-machine and the Body-machine: Descartes

In the fifth part of *Discourse on the Method* Descartes outlines his ideas on mechanistic physiology, namely that the living body should be regarded basically as a complex machine. He claims that this idea

> will not seem at all strange to those who know how many kinds of automatons, or moving machines, the skill of man can construct with the use of very few parts, in comparison with the great multitude of bones, muscles, nerves, arteries, veins and all the other parts that are in the body of any animal.[56]

In this passage, he introduces two of his most famous and controversial ideas: first, that a human being is an amalgam of two distinct elements—the body, which is a God-constructed automaton made of matter, and the soul, an immaterial entity that provides consciousness, reasoned thought, and the ability to communicate, from its seat in the pineal gland in the brain (i.e., Cartesian dualism), and second, that animals, unlike human beings, have no souls, being pure mechanical constructs (i.e., animal-automatism).[57] On the latter idea, which was furiously debated for centuries to come, Descartes admits to a rudimentary form of intelligence in animals that allows them to carry out tasks like seeking food and shelter, procreating and nurturing, and even obeying simple commands. That intelligence, however, arises out of the mechanics of the body-machine and is of a significantly inferior type to the rational intelligence of man, which comes from his soul. To further elaborate on the essential difference between a human being and an animal, he points to two specific ways in which automata and animals are more limited than man.[58] One could conceivably construct an automaton that could utter words, just as one can train magpies and parrots to do the same, but none could be taught to converse with a person in a meaningful manner. In addition, automata and animals can do certain tasks very well, some even better than a human being, but they are limited in the range of things they can do, lacking the intelligence to learn new skills not granted them by their organic makeup.

That passage in the *Discourse* is a summary of the detailed description of mechanistic physiology in his *Treatise on Man*, the second part of his most comprehensive work of natural philosophy, *Treatise on the World*. Before looking at the central role of the automaton in the work, it must be pointed out that while Descartes used the machine-body analogy to a greater extent than anyone before him (with the possible exception of Gómez Pereira, a little-known Spanish doctor who had already taken the radical route of denying souls to animal-machines in his sixteenth-century *Antoniana-Margarita*),[59] he was not the first to do so. A brief overview of such ideas before Descartes's time will be useful at this point.

Aristotle, in his treatise *Movement of Animals*, compares animals to

> automatic puppets, which are set going on the occasion of a tiny movement (the strings are released, and the pegs strike against one an-

other); or with the toy wagon. . . . Animals have parts of a similar kind, their organs, the sinewy tendons to wit and the bones; the bones are like the pegs and the iron; for when these are slackened or released movement begins.[60]

Thomas Aquinas also draws an analogy between the movement of animals and that of a manmade construct like a clock, asserting that "artificial works are to human art as all natural things are to divine art. And so, like the things made by human ingenuity, the things moved by nature display order."[61] In contrast to Cartesian animal-automatism, however, the Scholastic position was that animals possessed "sensitive" souls (derived from the Aristotelian notion of "sensitive" intelligence in animals), which were incapable of reason but were still immaterial and transcendent.[62] And both Aristotle and Galen, the two central authorities of ancient medicine, affirmed the clear distinction between living entities and inanimate matter, seeing them as operating under varying principles and forces. In the passage quoted above from *Movements of Animals*, right after Aristotle makes the comparison between the mechanics of animal bodies and that of automatic toys, he spells out their essential difference on the basis of growth and change:

> in puppets and the toy wagons there is no change of quality, since if the inner wheels became smaller and greater by turns there would be the same circular movement set up. In an animal the same part has the power of becoming larger and now smaller, and changing its form, as the parts increase by warmth and again contract by cold and change their quality. This change of quality is caused by imagination and sensations and by ideas.[63]

Sylvia Berryman has pointed out two aspects of the role of automata in ancient medical writings that complicate the matter. Even as Aristotle draws a clear distinction between the animate and the inanimate, the automaton as a conceptual object makes the distinction problematic since one of the defining features of the animate is its ability move itself without being moved by a force beyond it.[64] So living beings are defined in contrast to the nonliving precisely as self-movers or automata (in the original Greek sense of the word), even as Aristotle clearly denied that they were like automatic devices in essence. In addition, Galen's criticism of rival schools of medical thinkers who had a penchant for using the machine-body analogy points to the existence of a mechanistic school of medical thought in the ancient world, possibly of Hellenistic doctors in Alexandria who followed the ideas of Erasistratus (310–250 BCE) and others.[65] Heinrich von Staden has asserted that the position of Alexandria as the center of technological arts in the ancient world must have given local doctors the idea of using various pneumatic and mechanical devices in operation there, including the inventions of Hero, as conceptual objects in describing the workings of the natural body.[66]

It was Descartes, however, who utilized the machine analogy to the fullest extent, in the context of the mechanistic turn in natural philosophy. The *Treatise on Man* opens with an invitation to the reader to imagine a hypothetical creature, a being that is an amalgam of two distinct parts, an immaterial soul and a material body that operates like a machine.

> We see clocks, artificial fountains, mills, and other similar machines which, even though they are only made by men, have the power to move of their own accord in various ways. And, as I am supposing that this machine is made by God, I think you will agree that it is capable of a greater variety of movements than I could possibly imagine in it, and that it exhibits a greater ingenuity than I could possibly ascribe to it.[67]

Since the treatise is a detailed description of how this machine works, the obvious inference is that the actual human body is that machine endowed with a soul. In the second part on the motion of the body-machine, he refers to devices that are moved by water in the "grottoes and fountains in the royal gardens."[68] The details he gives of what can be seen in such a grotto, including automata figures of Diana, Neptune, and a sea monster, make it apparent that he is referring to the works featured in the artificial cave at the royal château of Saint-Germain-en-Lay, constructed by the brothers Tommaso and Alessandro Francini between 1589 and 1609.[69] Descartes likens the "fountaineer" of those marvels with the rational soul that resides in the brain and controls the body machinery. This particular case of machine-body analogy is significant not only for Descartes's use of real-life hydraulic automata for illustrating his physiological ideas but also as an example of the naturalization and mechanization of wonders that was in full progress at the time.

Grotto automata from Salomon de Caus, *The relations of Motive Forces, with various Machines as useful as they are pleasing* (1615). Salomon de Caus, *Les raisons des forces mouvantes avec divers machines tant utiles que plaisantes* (Francfort: I. Norton, 1615).

When hydraulic automata like the ones at Saint-Germain-en-Lay became popular in the Renaissance, Salomon de Caus wrote his treatise *The Relations of Motive Forces, with various Machines as useful as they are pleasing* (1615), detailing their construction.[70] Such a technical explication of the mechanics served a useful purpose for engineers, but the ultimate purpose of automata was to be wonders that alternatively awed, frightened, delighted, and astonished the viewer. This could be achieved only by hiding their machinery, keeping the audience in suspense as to whether it was done by, to quote Mandeville again, "artifice or necromancy." The ancients were also aware that the spectacular effect could be achieved only through the secrecy of its function, Aristotle pointing out that "automatic marionettes" seem wonderful to people as long as they do not know how they work, and Hero advising engineers to keep the workings of the devices hidden for the same reason.[71] What Descartes did was to appropriate the Saint-Germain-en-Lay automata for a fully rationalistic-mechanistic purpose, conceptually wrenching them out of their original context of magical enchantment and focusing on their inner workings.[72] Opening up the devices to reveal their mundane machinery is indicative of the general process of disenchantment of the world that was taking place in the period. His philosophical exposure of the body-machine leads directly to Vaucanson's physical exposure of the workings of his flute-player to prove that nothing more than pure mechanics was going on in it. And while the Renaissance engineers built their automata to look like devils, pagan gods, and monsters to enhance their magical effect, Enlightenment mechanics like Vaucanson and the Jaquet-Drozs made them in the shapes of ordinary animals and people engaged in the everyday tasks of writing, drawing, and playing musical instruments. While the objects retained their wondrous aura, they represented the marvels of the machine in the everyday world, devoid of preternatural magic and supernatural entities.

Descartes apparently found the automaton such a useful idea that he used it for other conceptual purposes in many of his philosophical works. In *Meditations on First Philosophy* (1641) he illustrates the intellectual nature of "judgments," even ones seemingly based on empirical evidence alone, by posing the following situation.

> [I]f I look out of the window and see men crossing the square, as I just happen to have done, I normally say that I see the men themselves.... Yet do I see any more than hats and coats which could conceal automatons? I *judge* that they are men. And so something which I thought I was seeing with my eyes is in fact grasped solely by the faulty judgment which is in my mind.[73]

In *Principles of Philosophy* (1644) he demonstrates that the worthiness of a man is tied to his freedom in action by pointing out:

> We do not praise automatons for accurately producing all the movements they were designed to perform, because the production of these movements occurs necessarily. It is the designer who is praised for constructing such carefully-made devices; for in constructing them he acted not out of necessity but freely.[74]

And in *The Passions of the Soul* (1641) he characterizes the difference between the body of a living man and that of a dead one as the difference between, on the one hand, a watch or other automaton (that is a self-moving machine) when it is wound up and contains in itself the corporeal principle of the movements for which it is designed, together with everything else required for its operation; and, on the other hand, the same watch or machine when it is broken and the principle of its movement ceases to be active.[75]

A fascinating indicator of the impact of Descartes's mechanistic physiology, his frequent use of the automaton in explicating his ideas, and the era's general interest in the object can be seen in stories that began to circulate after his death that he had constructed actual automata. In Descartes's earliest notes—the "Cogitationes privatæ," extant only in fragments—is an interesting passage in which the young thinker considers the possibility of making an articulate statue with metal parts that could be moved by magnets.[76] Twenty years after his death, Nicolas Poisson, an early editor of his selected works, wrote in his *Commentaire ou Remarques de la Mèthode de René Descartes* (1670) that he had seen manuscripts in which the philosopher claimed that in order to illustrate his ideas on the soul of animals (i.e., its nonexistence), he had created various automata, including the magnet-operated man dancing on a cord, as well as a flying pigeon and a partridge chased by a spaniel.[77] Poisson admits that he has never seen the automata, but he thinks them plausible, considering the works of Torrez (i.e., Gianello Turriano, clockmaker to Emperor Charles V). Dennis Des Chene has speculated that Poisson deliberately misread Descartes's discussion of a speculative automaton as a claim to have actually built one in order to boost the philosopher's reputation as an experimenter.[78]

The most famous story of this type is that of his ill-fated sea voyage with an automaton he named Francine. As the tale goes, Descartes constructed a life-sized doll in the shape of a young woman, which he put in a box and took with him aboard a ship. During the journey, the curious captain discovered the automaton, the movement of which frightened him and compelled him to throw it into the sea, fearing that it was a devil. This story has been retold a number of times, including in Anatole France's fantastic novel *La Rôtisserie de la Reine Pédauque* (1893) in which the automaton is transformed into the mythic salamander.[79] Recently, there has been a proliferation of the story in both popular and scholarly works due to its connection with contemporary interest in robotics and artificial intelligence. Stephen Gaukroger opens his intellectual biography of Descartes with a consideration of the story, while Gaby Wood gives a rather dramatic rendering of the tale at the beginning of her book.[80] The source of this tale however, reveals a specific agenda to its telling that exposes it as a legend with no basis in fact.

The story originates from a book of literary and philosophical anecdotes, *Mélanges d'Histoire et de Litterature* (1700), by the Carthusian *moraliste* Bonaventure d'Argonne, who wrote under the pseudonym Vigneul-Marville. In it, he claims to have been informed by a "very zealous" Cartesian that the story in Adrien Baillet's first biography of Descartes (1691) that he fathered an illegitimate daughter named Francine was a lie concocted by his enemies and that the name belonged to an automaton he created and subsequently lost at sea.[81] The truth is that Descartes, after a dalliance with a maid at the house of a friend in the Netherlands, did have a daughter named Francine, who died in 1640 at the age of five, so the fantastic story was obviously told in a futile attempt to save his reputation.[82] Interestingly, Julian Jaynes speculates that he may have named the girl after the brothers Tommaso and Alessandro Francini in his admiration of their construction of the Saint-German-en-Lay automata—suggesting a series of provocative connections from these Renaissance works to Cartesian mechanistic physiology to Francine Descartes and finally to the legendary automaton Francine.[83]

We can also gauge the impact of Descartes's dualist vision by considering its appearance in the works of other great philosophers of the era. Hobbes opens his *Leviathan* (1651) by comparing the natural creations of God and the artificial products of man in terms of life as motion.

> Nature (the Art whereby God hath made and governes the World) is by the *Art* of Man, as in many other things, so in this also imitated, that it can make an Artificial Animal. For seeing life is but a motion of Limbs, the beginning whereof is in some principall part within; why may we not say, that all *Automata* (Engines that move themselves by the springs and wheels as doth a watch) have an artificiall life? For what is the *Heart*, but a *Spring*; and what the *Nerves*, but so many *Strings*; and the *Joynts*, but so many *Wheeles*, giving motion to the while Body, such as intended by the Artificer.[84]

Robert Boyle, in *A Free Enquiry*, refers to human bodies as "living automatons" and claims that "though the body of a man be indeed an engine, yet there is united to it an intelligent being (the rational soul or mind)."[85] He even elaborates on the specific type of machinery that the body can most aptly be compared with: "I look not on a human body as on a watch or a hand mill—i.e., as a machine made up only of solid or at least consistent parts—but as an hydraulical, or rather hydraulo-pneumatical, engine, that consists not only of solid and stable parts, but fluids and those in organic motion."[86]

In addition, Leibniz asserts in his *Monadologie* (1714) that "the organic body of each living being is a kind of divine machine or natural automaton."[87] A major aspect of Cartesian dualism that he departed from was that of the interaction of the soul and the body through the former's assertion of its will from the pineal gland. Leibniz could not accept the idea that the immaterial, transcendent entity of the soul could affect dead matter in any way, so he resolved the difficulty through his notion of preestablished harmony. God sets in advance the two parts of a human being in motion and in such a way that the soul's will and the body's actions are perfectly synchronized though they do not interact at all. In other words, when I will myself to get up from my chair, it seems to me as though I am controlling my body as it moves, when in fact that is an illusion created by an arrangement in which the body moves on its own at the precise time the soul wishes it to move. Leibniz likened the situation to two clocks or watches that keep the time perfectly, seemingly working together when they are actually two independent devices with no influence on each other.[88] He went so far as to describe the soul as a "spiritual automaton," which may not operate in a strictly mechanical manner but "contains in the highest degree all that is beautiful in mechanism."[89]

The Body-Machine: The Mechanistic Physiologists

In the decades immediately after Descartes's death in 1650, his physiological ideas made a significant impact in the medical field, inspiring a veritable revolution in the study of the natural body. Doctors and physiologists of the period sought to replicate in their field the great advances made by natural philosophers in physics and astronomy. To throw off the yoke of Aristotelian and Galenic ideas as well as more recent animistic notions from

Hermetic philosophy, the most obvious path to take was the mechanistic one that aimed at gaining a quantitative and mathematical grip on the biological subject through a rigorous empirical approach.[90] As Giorgio Baglivi, one of the leading figures in Italy of the movement that was sometimes called "iatromechanism," put it in 1696:

> Since doctors have begun to examine the structures and actions of the animate body on the basis of geometrical and mechanical principles, as well as of physical, mechanical, and chemical experiments, they have not only discovered innumerable phenomena unknown to preceding centuries, but have also realized that as far as its natural actions are concerned the human body is nothing more than a complex system of mechanical and chemical movements that obey mathematical laws.[91]

As Cartesian philosophy spread in the larger intellectual scene of western Europe, Descartes's medical ideas were continuously evoked and analyzed in texts of mechanistic physiology. That does not mean, however, that every mechanist accepted all of his views; many took issue with both specific findings and general methodology in his works. For instance, Thomas Willis, one of the most important doctors in England, who was at Oxford in the 1650s with other major mechanistic thinkers like Robert Boyle and Robert Hooke,[92] laid out his physiological ideas in his *Two Discourses Concerning the Soul of Brutes* (1672), in which the natural forces of animal spirits play a more significant role in the harmonious function of the body-machine than in the Cartesian model.[93] Likewise, Giovanni Borelli, the central figure of the new medicine in Italy, disproved some of Descartes's ideas including the cause of fevers and the mechanics of brain-to-nerves-to-muscle action, seriously considered the role of the soul in the cause of motion, and criticized his overly deductive methodology with little empirical work.[94] Despite their differences, what these figures had in common was their commitment to achieving a comprehensive understanding of the natural body through a mechanistic methodology that precluded the role of any forces or entities whose existence could not be quantified or proved through observation or experimentation. This agenda is revealed in the frequent appearance of the machine-body analogy and the automaton in their works.

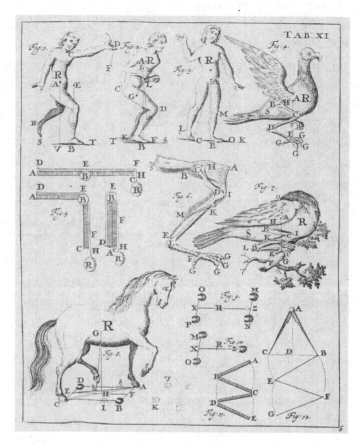

Giovanni Borelli, illustration from *On the Movement of Animals* (1680)—trans. Paul Maquet (Berlin: Springer-Verlag, 1989). With kind permission of Springer Science+Business Media.

Borelli's posthumously published book *On the Movement of Animals* (1680) was a central text of mechanistic medicine. Early in the work he asserts that

> Geometry and Mechanics are the ladder by which we climb the wonderful knowledge of the movements of animals . . . however, since we were not more used to mechanics, we missed the truth so far, not because we tolerated error and what has been said hitherto must be rejected but because, although some of it is true, we did not follow more exact science.[95]

Like Descartes, Borelli finds the automaton a useful conceptual device for illustrating his ideas. In a section on the involuntary movement of the

THE MAN-MACHINE IN THE WORLD-MACHINE 127

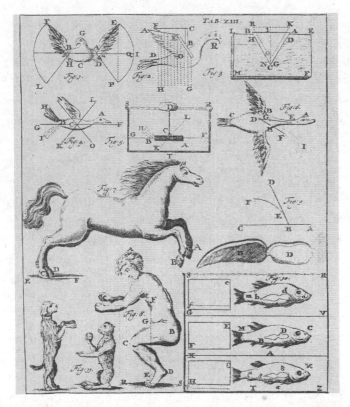

Giovanni Borelli, illustration from *On the Movement of Animals* (1680)—trans. Paul Maquet (Berlin: Springer-Verlag, 1989). With kind permission of Springer Science+Business Media.

heart, he phrases his proposition this way: "The movement of the heart can result from an organic necessity as an automaton is moved."[96]

> Above all a clock seems to represent the heart of life. The clock is made of cog-wheels by which the force of a weight acting permanently moves a hand which indicates time. The pendulum oscillates with repetitive falls downward like waves, in the same manner as the heart moves.[97]

He goes on to question whether automatic movement is sufficient to explain the actions of the heart and whether there is also a nonmechanical force involved, a point on which he remains ambivalent.[98] In another passage he asserts: "An automaton seems to present some resemblance with animals since both are mobile organic bodies which comply with the laws of mechanics and both are moved by natural faculties."[99] Just as a clock contains a weight or an oscillatory pendulum to prevent its wheels from spinning too fast, causing it to spend itself too soon, natural bodies have a regulating principle in the flow of blood that carries air throughout the body, he calls this principle an "aerial machine" that is the "automaton of Nature."[100] Finally, on a section on the fertilization of the egg, he writes:

> Among the technical achievements, we do not see anything more comparable to animals and plants than an automaton or a clock made of cog-wheels. In such devices we must consider three points above all: (1) the shape of the wheels which can carry out different movements; (2) the addition of a motive force such as a hanging weight or the force of wind or a flow of water; (3) the way of action of the automaton.[101]

After Borelli, the movement and its mechanistic methodology was carried forward in Italy by Lorenzo Bellini and Giorgio Baglivi.

In England, the doctors of the College of Physicians, including Willis, Charles Goodall, Walter Charleton, and William Cole, tried to increase respectability for their profession by adopting the "modern" medicine and experimental method of William Harvey (i.e., his model of the pumping action of the heart and the circulation of blood throughout the body) but rejecting his adherence to such older notions as epigenesis.[102] George Castle in his *The Chymical Galenist* (1667) comments on the works of Willis, affirming:

> It is not, I think, to be question'd, that a man is as Mechanically made as a Watch, or any other Automaton; and that his motions, (the regularity of which we call Health) are perform'd by Springs, Wheels, and their Work from those pieces of Clock-work, which are to be seen at every Puppet-play.[103]

Robert Hooke best summarizes the views of English mechanistic natural philosophers in his pioneering work of microscopic observation *Micrographia* (1665), in which he hopes that their work will produce many admirable advantages, towards the increase of the Operative, and the Mechanick Knowledge, to which this Age seems so much inclined, because we may perhaps be inabled to discern all the secret workings of Nature, almost in the same manner as we do those that are the productions of Art, and are manag'd by Wheels, and Engines, and Springs, that were devised by humane Wit.[104]

In the following generation, the mechanistic project in medicine was carried on most notably by Stephen Hales and the Scottish doctors Archibald Pitcairne and James Keill, who worked under the powerful influence of Isaac Newton.

In France, ideas of mechanist physiology are found in the works of Claude Perrault, Louis de la Forge, François Bayle, Jacques Rohault, Pierre Sylvain Régis, François Lamy, and in the following generations those of Pierre Chirac, chief physician to Louis XV, and Claude-Nicolas Le Cat, who competed with Vaucanson in the attempt to create an anatomically correct automaton.[105] After La Forge collaborated with Claude Clerselier in editing the first French edition of Descartes's *Treatise on Man*, he published his own *Traité de l'espirit de l'homme* (1664), a work of Cartesian philosophy.[106] In the nineteenth chapter, on memory, La Forge begins with a discussion of Descartes's dualism, including an explanation of mechanistic physiology in which he praises the philosopher for his admirable comparison of natural bodies with hydraulic machines that move diverse figures, a reference to Descartes's descriptions of the Saint-Germain-en-Lay automata.[107] Pierre Sylvain Régis, who emerged in the last decades of the seventeenth century as a leading proponent of Cartesian philosophy in France, uses the same analogy in his *Cours entier de philosophie* (1691), describing the hydraulic statues that can be seen in the grottoes of great gardens.[108] Unlike La Forge, however, he does not give credit to Descartes for the comparison, so an unfortunate confusion resulted later, as the analogy became a well-quoted one in the philosophical and medical discourse of the eighteenth century and was attributed to Régis himself. David Renaud Boullier, a Dutch philosopher who was a major figure of opposition to animal-automatism, wrote in his *Essai philosophique sur l'ame des bêtes* (1737) of "the comparison made by Mr. Régis, of certain hydraulic machines, which one can see in the grottoes and fountains of certain houses of the great."[109] In the article "L'Ame des bêtes" (The soul of animals) in the *Encyclopédie* that Boullier cowrote, he made the wrong attribution again, contributing to its spread.[110]

Given the title of Julien de La Mettrie's enormously controversial treatise *Man a Machine*, it would be easy to regard it as the ultimate culmination of the mechanistic movement in physiology started by Descartes, the radical materialist taking the final logical step of jettisoning the notion of an immaterial, transcendent soul in man and turning him into an organic automaton and *nothing more*. In the tradition of using the automaton analogy—now a century old—La Mettrie describes the brain as "a well-enlightened machine" and asks "why would a man's having a share of the natural law make him any less a machine? [With] a few [more] cog wheels and springs than in the most perfect animals, the brain [is] proportionately nearer the heart so it receives more blood."[111] He further asserts that the "human body is an immense clock, constructed with so much artifice and skill."[112] But recent scholars of La Mettrie's works have pointed to significant ways his ideas differ from those of traditional mechanists. Aram Vartanian has claimed that for La Mettrie the concept of the man-machine is "no more than a mechanical model, or a picturable analogy, of the mind—not its essence."[113] In addition, the exact type of machine La Mettrie pictures man to be is a different one from that envisioned by mechanists, namely an organic perpetual-motion machine that is better described as a dynamic living machine: "the human body is a self-winding machine, a living representation of perpetual motion. Food sustains what fever excites. Without food, the soul languishes, goes into a frenzy, and, exhausted, dies."[114] Given the unique nature of such a device, and the presence of vitalistic ideas in La Mettrie's work, Vartanian sees him as a transitional figure who utilized the concepts and language of mechanicism but went beyond some of its fundamental tenants in explicating the nature of life in materialist terms, an act that amounts to "an attempt to combine two traditionally opposed attitudes into a unified standpoint."[115] Kathleen Wellman also points out that La Mettrie was less interested in carrying on the project of mechanistic physiology than in showing the material nature of the soul and, consequently, eliminating all essential differences between man and animal as biological subjects.[116] Equating the man-machine with the animal-machine was simply the most provocative way of demonstrating his materialism.

La Mettrie's materialist tract marks the end point of the classical mechanist movement that commenced in the 1630s. His departure from some of its aspects allows us to also clarify its major characteristics, by way of contrast. As mentioned, significant disagreements existed among thinkers like Descartes, Willis, and Borelli on issues such as the role of the soul in the movement of bodies, the nature and function of animal spirits, and the existence of souls in animals. We can, however, distill three elements central to mainstream mechanistic physiology from 1637 to 1748 that also shed light on the role of the automaton as a conceptual object in the period.

First, most classical mechanists were dualists who affirmed that the soul and the body were entities of different essences, the former being immaterial and transcendent, the latter material and mechanical.[117] Second, classical mechanists, as dualists, tended to restrict mechanistic descriptions to the body, excluding the soul. Third, the use of the automaton and other machines for an analogical purpose was generally positive in the sense that the description of the body-machine was often expressed in laudatory language. From Descartes on, the classical mechanists did not utilize the analogy in a neutral manner but celebrated it in a veritable refrain of praising God for having created such a beautiful, intricate, and well-functioning machine as the natural body. Just as the automaton was a wonder of natural magic for the Hermetic philosopher, it was an equally astounding and admirable work of natural mechanics for the seventeenth-century natural philosopher, as the automaton-body comparison invited readers to appreciate what a piece of mechanical work is man. The positive nature of such descriptions is evident in the flattering adjectives featured in them. For example, Descartes in the *Discourse* says that the body-machine, "having made by the hands of God, is incomparably better ordered than any machine that can be devised by man"; Leibiniz asserts that the divine automata that are natural bodies "infinitely surpasses all artificial automata"; and Willis extols the "Great Workman" for the construction of souls and bodies, "the Excellent structure of the Organs, most Exquisitely labored," giving evidence of skills far "beyond the Workmanship and artificialness of any other Machine."[118] Even in colonial America, the famed Puritan minister Cotton Mather in his 1721 work of natural theology *The Christian Philosopher* marvels at the human body: "a *Machine* of a most astonishing Workmanship and Contrivance! *My God, I will praise Thee, for I am strangely and wonderfully made!*"[119]

What is apparent here is that the wonder and awe that Renaissance philosophers felt in response to the automaton was still in full effect in the seventeenth century, and the aura of the marvelous around the self-moving machine remained as radiant as ever through the transition from natural magic to natural mechanics. Intellectuals also made good use of that aura in political theory where the machine analogy appears with some frequency in the period.

The State-Machine

The Roman historian Livy recounts an episode in the early history of the Republic when an upsurge of popular unrest in the city was quelled by Senator Menenius Agrippa, who appeared before the people and told them the Aesop tale of the revolt of all the body parts against the belly.[120] When the belly is accused of living off of all the good works provided by the rest of the body while doing nothing itself, it answers by explaining all the essential functions it performs in governing and nourishing the whole. The Roman Senate, then, was the belly of the city without which the state would wither and die.

The use of a body as a representation for an entire society has a long history that goes back to Plato and Aristotle and continues today in the common usage of such terms as "the body politic," "the social body," and "the government body." Mary Douglas writes that the comparison is an easy one to make because the "body is a model which can represent any bounded system.... The body is a complex structure. The functions of its different parts and their relation afford a source of symbols for other complex structures."[121] What makes the history of the analogy so rich is that throughout the centuries it was used to describe and advocate all types of political system, from the most centralized and autocratic to the more open and democratic, depending on which model of physiology was used.[122] Advocates of authoritarian forms of government emphasized the innate superiority of the ruling organ (the belly, the brain, or the heart) and the rest of the body's dependence on it, while those of more liberal forms wrote either of the reverse (i.e., the central organ's reliance on the rest of the body to sustain it) or of the body as a cooperative system of mutually dependent parts. The state-body comparison was also greatly informed by the prevalent medical ideas of a given period on such issues as the relative importance of various organs, the nature of life in the body, and what an "ideal" body looked like. To give one significant example of this interaction of medical and political thought, when William Harvey dedicated his *De Motu Cordis* (1628), the pioneering work on the pumping action of the heart, to Charles I of England, he called the monarch "the *Heart* of the Commonwealth" and advised him to study his work on the organ since it was "the Principle of Man's Body, and the Image of your Kingly power."[123]

In the period that followed, as mechanistic physiology became the dominant movement in medical thought, the social and political body became mechanized as well. Michel Foucault has noted that the man-the-machine concept of the period was written in two registers: the "anatomico-metaphysical," in the manner of Cartesian physiology, and the "technico-political."[124] The most explicit illustration of such a linkage of medical and political ideas can be found in Hobbes's *Leviathan* (1651). Hobbes begins the work with the description (quoted earlier) of automata, or "engines that move themselves by springs and wheels as doth a watch," before comparing them to animals that work by the same means, the heart, the nerves, and the joints functioning like the spring, the string, and the wheel.[125] He then makes the analogical leap of asserting that since the state is like a gigantic man, composed of all the citizens who belong to it, it is also like a great automaton made of up many small automata.

> *Art* goes yet further, imitating that Rationall and most excellent worke of Nature, *Man*. For by Art is created that great LEVIATHAN called a COMMON-WEALTH, or STATE (in latine CIVITAS) which is but an Artificiall Man; though of greater stature and strength than the Naturall, for whose protection and defence it was intended.

And just as different parts of a body or an automaton fulfill different functions within the larger system, the different parts of the state are assigned their roles in the great mechanical man, with the ruler in the controlling role of the soul:

> the *Soveraignty* is an Artificiall *Soul*, as giving life and motion to the whole body; The *Magistrates*, and other *Officers* of Judicature and Exe-

cution, artificiall *Joynts; Reward* and *Punishment* (by which fastned to the seate of the Soveraignty, every joynt and member is moved to performe his duty) are the *Nerves*, that do the same in the Body Naturall.[126]

These analogies form a logical sequence that runs: the natural body is an automaton; the state is a great body; therefore, the state is a great automaton.

Leviathan became one of the most controversial and vilified books of the period not only for its pragmatic argument for absolutism that precluded any mention of divine right but also for its materialism. Yet his portrayal of the state as a great machine and its sovereign as an engineer of the state-machine had a decisive impact on political theory. Almost a century after the book's publication, Frederick II of Prussia wrote:

> As an able mechanic is not satisfied with looking at the outside of a watch, but opens it, and examines its springs and wheels, so an able politician exerts himself to understand the permanent principles of courts, the engines of the politics of each prince, and the source of future events.[127]

Likewise, Condillac, in his 1749 *Treatise on Systems*, asserts in the chapter on political systems that

> [the] people is an artificial body. It is up to the magistrate watching out for its survival to maintain harmony and strength in all its members. He is the mechanic who must restore the springs and wind up the whole machine as often as circumstances require.[128]

Condillac is most famous for his work of sensationalist psychology *A Treatise on Sensations* (1754), in which he borrows from Descartes the device of postulating an artificial construct (i.e., a statue) for the purpose of detailing his ideas on the mind-body problem: "I forewarn the reader that it is very important to put himself exactly in the place of the statue we are going to observe."[129]

Otto Mayr has compiled a large collection of such quotations from all over western Europe in the second half of the seventeenth century and throughout the eighteenth on "the clockwork state."[130] He also makes a broad claim that the mechanistic analogy in political writings was used in support of absolutism, in direct opposition to more liberal forms also advocated in the period.[131] In a similar vein, Lionel Rothkrug quotes from a 1698 work by the political writer Jean Pottier de la Hestroye, who asserts:

> All subjects must work; in the state everyone must be occupied. The state is, properly speaking, a machine, the movements of which altogether different must be regulated without interruption; we cannot interfere with the movement without running the risk of destroying the state. Similarly subjects must act and work in a state to support it and to render it flourishing.[132]

Rothkrug analyzes this and other French political writings of the period in their support of Louis XIV and his finance minister Colbert's efforts to transform France's economy into a streamlined mechanical system.[133] Margaret Jacob has also shown that many of the major mechanistic thinkers, including Descartes, Leibniz, and Newton, were aware of the political implications of their ideas and allowed them to be used to support the established order.[134] The mechanistic conception of the state reflected one of the primary concerns of intellectuals in the era, namely the establishment of order and stability after the chaos of the mid-seventeenth century. For that reason, the organization-machine idea was also utilized as a method of operating many other institutions in society. As Foucault has detailed, it was applied in the military, the school, the hospital, the prison, and the factory, with increasing emphasis on regular and uniform movements, specialization of tasks and group obedience, reducing individuals in such systems into what he calls "docile bodies."[135]

In this period, the word "automaton" denoted a device run by clockwork mechanism, often of the kind that was self-moving once it was set in motion through the use of a spring. Given the fact that the automaton was repeatedly evoked when the mechanistic analogy was used in both medical and political works, it is not difficult to see how it suggested an autocratic form of government. A self-moving machine usually features a single controlling mechanism and power source, with all the other parts obeying its dictates in a regular and controlled manner. It is a technical marvel precisely because of the way all of its intricate components are interconnected and move in conjunction with one another. If a defective part fails to fulfill its function, like a recalcitrant subject disobeying his sovereign, it can cause the entire network to break down. This translates into the view of the state as a vast machine, all of its sectors and subjects functioning under the orders of the central authority of the king, who is both the spring and the mechanic of the great device. As I will show in the next chapter, this association of mechanistic politics with absolutism became a major theme in the political discourse of the eighteenth century, especially for radical thinkers who denigrated the automaton idea for its association with the status quo.

It would be a mistake, however, to assume that all political writings that made use of automatic machinery for an analogical purpose did so in the service of absolutism. Kathleen Wellman points out that by La Mettrie's time the use of machine imagery in medical writings became so commonplace that its appearance in any given text of the period should not lead one to automatically assume that the writer was a mechanist in the classical sense.[136] The same can be said about machine imagery in political writings, such notions as the "machinery of government," "body of the state," and "manipulating the gears of political power" had become clichés by the early eighteenth century.[137] For instance, Henry Fielding in his great picaresque novel *Tom Jones* (1749) refers to the "judgment which can penetrate into the cabinets of princes and discover the secret springs which move the great state wheels in all the political machines of Europe."[138]

The casual description of politics as mechanics in the novel takes us to another realm in which the impact of mechanistic ideas on the larger literate culture can be discerned. In the literature of the late seventeenth and early eighteenth centuries, many writers made complex and fruitful use of cutting-edge philosophical and scientific ideas of their day. Notions from mechanistic physiology in particular captured the imagination of poets and novelists in significant ways.

The Man-machine and the Animal-machine in Literature

A number of scholars have examined early modern works in which poets, essayists, and novelists reacted to prevalent mechanistic ideas, especially that of animal-automatism.[139] Jean de La Fontaine, famous for his moral *Fables* featuring animals, felt compelled to defend the honor, rationality, and intelligence of the creatures in several of his stories, in direct response to Descartes.[140] But there was by no means a universal rejection of mechanistic ideas in the literary world. Jean de La Bruyère, author of cynical portraits of his contemporaries in *Les Caractères* (1688), was a royalist who was generally critical of new ideas from "free-thinkers." But when it came to the Cartesian theory that the "actions in a dog are not in the effect of either passion or sentiment, but proceed naturally and necessarily from a mechanical disposition caused by the multiple organization of the material parts of his body, I may, perhaps acquiesce in this doctrine."[141] Other writers made imaginative uses of mechanistic notions, participating in the ongoing meditations on the nature of humanity.

An entertaining example of an early appearance of animal-automatism in literature can be found in Cyrano de Bergerac's fantastic novel *The Voyage to the Moon and the Sun* (1657), which is filled with advanced scientific ideas of the time. Cyrano the fictional character travels to the Moon, only to be captured by its four-legged inhabitants, who assume that he is an exotic animal and put him in a cage.[142] He tries to prove that he is a sentient creature, but even after he learns their language and speaks to them, they consider him to be a kind of parrot capable of imitating intelligent speech. This is clearly a critique of Descartes's assertion of animals' soulless nature. Yet Cyrano also utilizes the Cartesian notion that the true test of one's sentience is the ability to master a language and to carry on a meaningful conversation. After he persuades some of the Lunarians that he is indeed a rational creature, they help him take his case to a court where he is ultimately successful in attaining his freedom through the demonstration of his reason.

This work was an inspiration for Jonathan Swift's novel *Gulliver's Travels* (1726), in which a similar story is told in a different setting. When Gulliver is captured by the gigantic inhabitants of Brobdingnag, he is initially treated not as a soulless animal but an automaton, which amounts to the same thing in Cartesian philosophy. He is given the name Grildrig, which means "what the *Latins* call *Nanunculus*, the *Italians Homunceletino* and the *English Mannikin*."[143] When he is taken to the king of the realm, who has been "educated in the Study of Philosophy, and particularly Mathematicks," he initially considers Gulliver to be "a piece of Clockwork ... contrived by some ingenious Artist."[144] As does Cyrano, Gulliver demonstrates his humanity by delivering a speech in his defense before some learned men. Although they remain skeptical, suspecting him to be a clever animal that has been trained to speak, the king accepts the Cartesian test of intelligence and becomes convinced of Gulliver's rational nature.

The theme is taken up yet again in Voltaire's story "Micromégas" (1752), in which colossal travelers from the star Sirius and planet Saturn visit Earth on a comet but are so big that they can barely see whales in the ocean through a microscope. They finally discover humans on a ship (Maupertuis and his cohorts returning from their 1737 expedition to Lapland to measure the meridian within the Arctic circle) who are so small in their eyes that one considers them to be "tiny machines moving about" and the other finds it unlikely that such miniscule creatures could talk since in "order to speak one must be able to think, more or less. But if they could think, they must have the equivalent of a soul. Now to attribute the equivalent of a soul to this species seemed to him absurd."[145] Once the alien from Sirius fashions a speaking-trumpet, using his thumbnails, he is able to establish contact with the humans, and their speech, once again, proves that they are creatures of reason, however feeble.

In those works, the idea of animal-automatism is criticized by placing a human being in the position of being the "other" to dominant beings who doubt his sentience because of his radically different appearance, in shape or size. Yet the stories seem to agree with Descartes that the definitive demonstration of sentience is the mastering of language and the carrying on of rational conversation, which animals and automata cannot do. So in the literary reaction to ideas like animal-automatism, we find a series of rich and imaginative meditation on them rather than wholesale rejection or uncritical acceptance.

The most provocative use of mechanistic ideas in literature can be found in John Cleland's pornographic novel *Fanny Hill, or the Memoirs of a Woman of Pleasure* (1748–1749), in which the characters' sexual bodies are constantly described as machinery.[146] Throughout the novel, the male genital is referred to as a machine or an engine (mighty, wonderful, spit-fire, oversized, etc.), and in one description of a sexual encounter, a boy nicknamed Good-natur'd Dick turns into a "man-machine, strongly work'd upon by the sensual passion" while his partner becomes "mere a machine, as much wrought on, and has her motions as little at her own command."[147] One long passage describing copulation is worth quoting in full for its particularly vivid use of such imagery, to such an extreme that makes it liable to be confused with the description of actual machines in action.

> I not only then tightened the pleasure-girth round my restless inmate, by a secret spring of suction and compression, that obeys the will in those parts, but stole my hand softly to that store-bag of nature's prime sweets, which is so pleasingly attach'd to its conduit-pipe, from which we receive them; there feeling, and most gently indeed squeezing those tender globular reservoirs, the magic touch took instant effect, quickn'd and brought on upon the spur, the symptoms of sweet agony, the melting moment of dissolution, when pleasure dies by pleasure, and the mysterious engine of it overcomes the titillation it has rais'd in those parts by plying them with the stream of warm liquid, that is itself the highest of all titillations, and which they thirstily express, and draw in like the hot-natured leach, who, to cool itself, tenaciously attracts all the moisture within its sphere of exsuction.[148]

That even the most intimate of physical acts was imagined in terms of mechanical motions by organic automata attests to the powerful hold of the man-machine idea on the imagination of the period.

Vaucanson Redux

An evidence of the central importance of the Vaucanson works in the history of the automaton is the shift in the very definition of the word "automaton." After Cornelius Agrippa and François Rabelais revived this obscure Greek word from Homer and Aristotle, it was taken up by the mechanistic philosophers of the seventeenth century, for whom it meant a self-moving machine. In the numerous usages of the word quoted in this chapter, it encompassed any such device, including a clock or a watch. When Robert Boyle praised God for having created such a great automaton as the world, he had in mind the astronomical clock of Strasbourg, which featured the moving figures of the Three Magi and a cock.[149] After the appearance of the Vaucanson automata, however, the word was increasingly used in the more specific sense of a self-moving machine built for the explicit purpose of imitating a living creature. The more general sense can be found in the late eighteenth century and through the nineteenth, but the more specific definition became more dominant.

In the *Encyclopédie* article "Automate," the definition is given as an "engine which moves by itself, or machine which contains the principles of its motion."[150] After a discussion of the Greek roots of the word, the very first example that is provided of an automaton is the legendary flying dove of Archytas. The article then briefly mentions that some authors also consider mechanical instruments like clocks and watches to be automata, but the vast majority of the piece is taken up with the description of Vaucanson's duck. In the piece itself, then, the more general definition, which was the dominant one before the period, is made out to be a marginal and even quaint one, as the article cites such usage in the works of the Renaissance figures Giambattista della Porta and Julius Caesar Scaliger (Giulio Cesare della Scala) when it could have just as well referred to Boyle and Borelli. Similarly, Charles Hutton writes in his *Mathematical and Philosophical Dictionary* (1796), in the article "Automaton," that "clocks, watches, and all machines of that kind" are automata, but then gives examples only of life-imitating machines, including the dove of Archytas, the wooden eagle and iron fly of Regimontanus, Dr. Hook's model of a flying chariot,[151] the Vaucanson automata, the Jaquet-Droz automata, and the chess-playing automaton of Wolfgang von Kempelen.[152] Later dictionaries, encyclopedias, and other reference books continued to include the general definition of any self-moving machine and to mention clocks and watches but usually privileged the life-imitating self-mover as the best example of an automaton.

For a full appreciation of the significance of the Vaucanson automata in the context of the Enlightenment, we must also take into account the very manner in which they were presented, as part of the culture of public demonstration of scientific knowledge. As detailed above, the natural philosophers of the seventeenth century inherited from the Renaissance Hermetics the experimental method in investigating nature and the interest in the technological arts. But there was an important difference between the alchemists and the natural philosophers in the practice of the presentation of knowledge. In direct opposition to the Hermetic obsession with secrecy, embodying the ethos of keeping vital knowledge within a restricted circle of trustworthy adepts, Francis Bacon and Robert Boyle advocated an open approach, especially through the public demonstration of experiments.[153]

In Bacon's utopian work *New Atlantis* (1627) that new approach is exemplified by "Salomon's House," an institution "dedicated to the study of the Works and Creatures of God."[154] Among the many wonders described there, the visitors are shown to "engine-houses, where are prepared engines and instruments for all sorts of motions," including those that imitate "motions of living creatures, by images of men, beasts, birds, fishes, and serpents"—automata.[155]

As mechanistic ideas began to take hold in the intellectual scene of western Europe, this effort toward public discourse on new knowledge about the world was greatly facilitated by the formation of real-life versions of Salomon House—the Royal Society in Restoration England in 1660, the Académie Royale des Sciences in France in 1666, and comparable organizations in other countries in later years. They provided regular forums for the demonstrations of scientific experiments and discussion of new ideas that allowed natural philosophers to learn about and build on new findings. Such activities in those settings drew some controversy, as Steven Shapin and Simon Schaffer have detailed in their classic account of the debate between Hobbes and Boyle over air-pump experiments.[156] Nevertheless, the institutions proved to be powerful means of disseminating knowledge, further solidifying the hold of mechanistic ideas in the intellectual culture. Various scholars have also shown how in the rest of the seventeenth century and throughout the eighteenth, such public discourse about scientific ideas moved beyond the restricted organizations of learned men to the larger public through popular lectures and experimental demonstrations in both private homes and public venues.[157] In the generation following that of Boyle, certain natural philosophers became famous for such activities, including, in England, Francis Hauksbee, Stephen Gray, William Whiston, and most notably John Theophilus Desaguliers, a Huguenot émigré from France who became an important disciple of Newton and an effective advocate of his ideas; and in France, Pierre Polinière and Jean-Antoine Nollet, who were encouraged by Fontenelle, the permanent secretary of the Académie Royale des Sciences and a supporter of scientific education of the public. Their activities played a crucial role in Enlightenment culture as tools of propaganda in spreading the ideas of the mechanistic vision of the world. From earlier demonstrations of the air-pump and anatomical dissections to later experiments on gravity and electricity, people were simultaneously entertained and edified by these shows, which were sometimes of spectacular nature, as in the presentations of Gray and Nollet on electricity. It is in this context of Enlightenment science and the culture of public demonstration that we can properly understand the phenomenon of Vaucanson's automata in the 1730s.

As already discussed, one aspect of the success of Vaucanson's automata that makes it such an interesting case for a historical analysis is the fact that it was both a popular and an intellectual phenomenon. In Paris, London, and elsewhere, the common people flocked to see his works to indulge in the sense of wonder that was aroused by the mechanical marvels. Intellectuals also found the works of great interest for their demonstration of mechanical ingenuity and mechanistic ideas, especially those of physiology. Vaucanson himself, in his effort to gain intellectual respectability for his works so as to advance in society beyond the status of a successful showman, deliberately presented his inventions as a public demonstration of scientific knowledge in his *Mémoire descriptif*, including the fraudulent claim that he had created an actual digestive system in the duck. It makes sense, then, that Fontenelle approved of the work and praised the author in an attached note for its submission to the Académie Royale des Sciences, and that it was Desaguliers who translated the text into English after the automata were demonstrated in London.

While advanced scientific ideas were disseminated through public lectures and demonstrations, both intellectuals and the larger public found in the works of the mechanical genius the very embodiment of the most prevalent ideas about the world, the state, and the body. When Roland Barthes described the automaton as the craze of the eighteenth century, he was referring to not just the popularity of the actual object but also the concept as a means of organizing and understanding the world.[158] Of particular importance, then, is Vaucanson's revelation of the inner workings of his automata, both in the written form of the *Mémoire descriptif* and in the physical act of exposing the machinery inside the flute-player to show that its movement was the result of pure mechanical arrangement, just as in the God-made world-machine, the manmade state-machine, and the body-machine. In a sense, the success of the automata as pure machines, with no doubt as to whether it was done "by artifice or by necromancy," marks the fulfillment of the promise Descartes made in the previous century in his incomplete dialogue *The Search for Truth*:

> After causing you to wonder at the most powerful machine, the most unusual automatons, the most impressive illusions and the most subtle tricks that human ingenuity can devise, I shall reveal to you the secrets behind them, which are so simple and straightforward that you will no longer have reason to wonder at anything made by the hands of men. I shall then pass to the works of nature, and after showing you the causes of all her changes, the variety of all her qualities, and how the souls of plants and of animals are different from ours, I shall present for your consideration the entire edifice of the things that are perceivable by the senses.[159]

In a world in which nature was a machine and the machine a representation of the divine mechanistic order, the automaton emerged as the central emblem of the worldview that manifested itself physically in 1738, playing the flute and the drum and defecating for the pleasure and edification of the machine-citizens of the machine-state in the machine-universe.

Given such a significant impact of the Vaucanson works on both the history of the automaton and eighteenth-century culture, it must be pointed out that their success marked not only the culmination but also one of the last celebrations of the worldview it represented. The success of the automata did set off a popular craze that resulted in the construction and presentation of increasingly sophisticated figures throughout the rest of the century. At the same time, however, starting from the 1740s, the ideas they embodied went into decline in the intellectual scene, as an increasing number of thinkers began to criticize the mechanistic conception of the world, the state, and the body. In other words, just as the popular interest in the mechanical marvel blossomed, its significance as a conceptual object underwent a radical change. So despite the fact that many automaton-makers created works of greater technical sophistication than those of Vaucanson, none in the second half of the century received the kind of intellectual approbation he had, as their works were regarded as little more than entertaining diversions.

The novelist Frances Burney, who was acquainted with John Joseph Merlin, the chief mechanic at Cox's museum in London, wrote an interesting scene in her epistolary novel *Evelina* (1778) that takes place in front of the automata that were featured at the museum. When some of the characters visit the museum, the protagonist finds the whole place "very astonishing, and very superb; yet, it afford me but little pleasure, for it is a mere show, though a wonderful one."[160] While viewing some of the mechanical marvels, including a pineapple that opens up to reveal a nest of chirping birds, the xenophobic Captain Mirvan associates them with "French taste" (perhaps in reference to Vaucanson or Merlin, a French-speaking Belgian) and confronts the group's guide by demanding an explanation of their use. The guide's answer is an appeal to their aesthetic value, which the captain summarily dismisses.

> "Why, Sir, as to that, Sir," said our conductor, "the ingenuity of the mechanicsm,—the beauty of the workmanship,—the—undoubtedly, Sir, any person of taste may easily discern the utility of such extraordinary performances." "Why then, Sir," answered the Captain, "your person of taste must be either a coxcomb, or a Frenchman; though for the matter of that, 'tis the same thing."[161]

When Joseph Spence saw Vaucanson's duck in 1741, it naturally evoked the idea of animal-automatism, but not so in this encounter with the Merlin automata in which the beautiful object is judged trivial and useless. In fact, such a contemptuous attitude toward the automaton became commonplace in philosophical, scientific, medical, and literary writings of the second half of the eighteenth century. Only in the following century did scientists like Charles Babbage, David Brewster, and Hermann von Helmholtz come to appreciate the practical potential of automaton technology.

I have outlined three common elements essential to understanding mechanistic description of humanity in the period 1637–1748: its dualist vision of both man and the world, its restriction to the description of the body and not the soul, and its generally celebratory nature. A glaring and crucial exception to this can be found in La Bruyère's work, *Les Caractères*, in which he provides a definition of a fool:

> an automaton, a piece of machinery moved by springs and weights, always turning him about in one direction; he always displays the same equanimity, is uniform, and never alters; if you have seen him once you have seen him as he ever was, and will be; he is at best but like a

lowing ox or a whistling blackbird; I may say, he acts according to the persistence and doggedness of his nature and species. What you see least is his torpid soul, which is never stirring, but always dormant.[162]

While a general kind of dualism is maintained in this passage in the contrast between the inactive soul and the automatically moving body, this analogy departs significantly from the period's mainstream mechanistic description of humanity in two ways. It is obvious here that when a fool is likened to an automaton, it is his personality and behavior that are being described, not the physiological functioning of his body. And the description is far from celebratory, as this automaton-man is regarded as low and contemptuous. Such a negative use of the machine-man analogy was rare until the middle of the eighteenth century, when it became so prevalent in intellectual and literary works that it was a veritable cliché in the period.

It is the purpose of the following chapter to describe how this shift from the celebratory man-machine to the contemptuous automaton-man occurred in the larger culture and to explore the reasons behind the transformation.

4
From the Man-machine to the Automaton-man, 1748–1793

From the Mechanist to the Vitalist Enlightenment

In 1741, Jean-Jacques Rousseau found himself at a gathering also attended by the famed mechanic Vaucanson—an interesting moment in the history of Enlightenment automata as the encounter between the great advocate of the natural man and the celebrated maker of artificial men. In a letter written in the following year, Rousseau notes Vaucanson's flute-player and Voltaire comparing him to Prometheus but then claims to have used the automaton for a metaphoric purpose on the occasion.

> As for myself, I said at the time, my admiration must be so much less suspect to the extent that I am accustomed to sights which I dare say are more marvelous. I was looked upon with astonishment. I come, I added, from a land filled with quite well made machines, which know how to dance the Quadrille and play Faro, that swear, drink Champagne, and spend the day reciting lies to other quite wonderful ma-

chines that pay them back in kind. People began to laugh; and what's funny is that two or three machines which were there laughed even more than the others.[1]

Given Rousseau's well-known awkwardness in large company, this passage has the air of an after-the-fact witticism, but its meaning is clear. People in polite society are nothing but automata going through the motions of their programmed activities. Rousseau makes the point again in his novel *The New Heloise* (1761) where the character St. Preux describes Parisians.

> You would think that isolated individuals who are independent would at least have a mind of their own; not at all: just more machines that do not think, and are made to think with springs. You have only to inquire into their assemblies, their coteries, their friends, the women they frequent, the authors they know: on this basis you can determine in advance their future sentiment on a book about to appear and which they have not read, on a play about to be performed and which they have not seen, on such and such an author they do not know, on such and such a system of which they have no notion. And just as the clock is ordinarily wound for twenty-four hours only, all those people go out every evening to learn in their assemblies what they will think the next day.[2]

In his political philosophy, Rousseau thought that there were aspects of community formation that were ennobling for humanity, especially at a primitive stage when people gathered together for mutual aid and comfort. After his break with Denis Diderot and other Encylopedists in the 1750s, however, he portrayed the urban life of his time in an increasingly negative manner.[3] He saw modern society with its conventions of behavior and thought as a corrupting environment that debilitates all that is natural, humane, and passionate in man. And in his mind, this process was imagined as that of the natural man turning into a machine-man.

This use of the man-as-automaton imagery for the purpose of describing problematic aspects of humanity was not idiosyncratic to Rousseau's writings. In fact, the negative description of the automaton-man was ubiquitous in the literary and philosophical works of the second half of the eighteenth century. A survey of the numerous instances of a person being either derided or pitied as a machine reveals this association to be a convention that was applied to a wide variety of people. In general, it was applied to four different types—the stupid, the oppressed (e.g., peasants), the conformist (especially aristocrats and other members of high society), and the tyrannical. But what did a mentally handicapped person, a peasant toiling in the field, a fashion-obsessed nobleman, and an absolute monarch have in common with one another that they were all called automata? In the view of Enlightenment philosophes they were all machine-like beings who led lives devoid of the principle of freedom, whether due to mental deficiency, external oppression, or inner conformism. So the automaton concept came to denote the lack of autonomy.

The irony of this development lies in the fact that the word "automaton" means "self-mover" and so implies autonomy, and it was precisely in that sense that the natural philosophers of the seventeenth and early eighteenth centuries utilized the automaton as a conceptual object. For instance, when Boyle describes the world as an automaton, he does so in direct opposition to the Aristotelian model of the world, which he likens to a puppet that requires "the peculiar interposing of the artificer or any intelligent agent employed by him."[4] In contrast, the automaton-world of mechanistic philosophy does not require such micromanaging on the part of the creator, as it operates according to the "virtue of the general and primitive contrivance of the whole engine."[5] But less than a century later, the idea of the automaton became virtually synonymous with the puppet, in the sense of a thing that is moved by a force beyond it with no free will of its own.

So how did something that is defined by its independence come to signify something that lacks independence? To pursue this question, we must examine how the mechanistic vision of humanity changed from referring only to the corporeal body and having a generally positive connotation in the 1637–1748 period to the description of certain types of personality and carrying an overwhelmingly negative sense in many writings from the 1740s on. For a full understanding of this development three interrelated contexts must be analyzed: first, the eclipse of mechanistic medicine; second, the revolution of sentimentality that swept across European culture in midcentury; and third, the greater prominence and mainstream acceptance of radical political ideas that associated mechanistic philosophy with the status quo.

Even during the period when the mechanistic movement was dominant in medical discourse, its adherents were not without major detractors who raised significant objections to the body-machine idea. In England, Francis Glisson (1598–1677) described a nonmechanical living force present in all substances that manifested itself as "irritability" in animal fiber.[6] Later on in Germany, Georg-Ernst Stahl (1660–1734) gave credit to the achievements of the mechanists in advancing medical science but also pointed to the limitations of studying the body in their terms, discerning in its function the presence of an animating force.[7] And the Swiss physiologist Albrecht von Haller (1708–1777), a student of the Dutch mechanist Herman Boerhaave, remained committed to the mechanistic project throughout his career, but his works on irritability as an essential phenomenon in animate bodies became a major source of vitalist ideas for medical thinkers of the following generation.[8]

In the mid-eighteenth century there was a discernible shift in the focus of medical scholarship, as physiologists became less interested in the further elaboration of the mechanistic functioning of the body and more concerned with understanding the operation of the vital force theorized by Glisson and Stahl. In their vision the corporeal entity was no longer a composite of discrete parts fulfilling their functions under the dictates of a governing organ but an organic, holistic vessel animated by a living force that allowed the entire body to act, change, and grow in harmonious unity. Consequently, the central emphasis of medical study shifted from mechanical actions to sensations and feelings, and the body began to be viewed as an entity of sensibility.[9]

In England, doctors like William Cockburn, George Cheyne, and Richard Mead, all students of the mechanist Archibald Pitcairne, began their career as mechanists but departed from strict adherence to the methodology starting in the 1720s. Cheyne, one of the most celebrated doctors in England of his time, once described the animal body in the quintessentially mechanistic terms as "nothing but a Compages or Contextures of Pipes, an *hydraulic Machin*," but he later adopted vitalist ideas, especially after his extensive studies on nerves, whose operation he could not explain solely on mechanistic principles.[10] In midcentury, figures such as Robert Whytt, William Cullen, Richard Brocklesby, and John Hunter led the vitalist movement in Britain.[11] An important element of this development was the intellectual as well as personal influence of Isaac Newton. The dominance of his ideas at the turn of the century was such that they played an essential role in the development of both mechanicism and vitalism.[12] Earlier physiologists like Pitcairne, James Keill, and Stephen Hales were inspired by Newton's *Principia* (1687), with its vision of the cosmos-machine, whose movements and actions could be measured, formulated, and predicted, something the doctors sought to achieve for the body. Vitalists from Cheyne to Hunter, on the other hand, admired the Newton of the *Opticks* (1704) with its descriptions of his celebrated experiments, as one of their chief complaints against the mechanists was that their machine-body analogy had become such an idée fixe that they seemed no longer able to observe the organic subject at hand in an objective and holistic manner. Furthermore, in the revised version of the *Opticks* of 1718, Newton revived his notion of the "etherial medium" that encompassed all of nature and played a role even in bodily movement: "Is not animal motion performed by the vibrations of this medium, excited in the brain by the power of the will and propagated from thence through the solid, pellucid, and uniform capillamenta of the nerves into the muscles, for contracting and dilating them?"[13] This allowed the vitalist doctors to interpret the medium as a living force at work in the organic body.

In France, the most significant challenge to mechanicism came from the vitalist doctors of the Montpellier University of Medicine, including François Boissier de Sauvages, Théophile de Bordeu, Gabriel-François Venel, and Paul Joseph Barthez, who considered the elucidation of the vital force as the essential subject of study in the new medicine.[14] And in 1749 the natural historian Georges-Louis Leclerc Buffon published the first three volumes of his monumental *Historie naturelle*, which not only brought together all the knowledge available at the time about the natural world but also narrated a grand epic of development, diversification, and generation of life on Earth that was replete with vitalist ideas.[15] In its "Initial Discourse," Buffon rails against mechanistic methodology as useful in such fields as physics, astronomy, and optics but completely inappropriate for the study of the living world, which requires careful observation of individual and particular phenomena:

> when one wishes to apply geometry and arithmetic to quite complicated subjects of natural philosophy. . . . One is obliged in all cases to

make suppositions which are always contrary to nature, to strip the subject of most of its qualities, and to make of it an abstract entity which has no resemblance to the real being.... When the subjects are too complicated to allow the advantageous application of calculation and measurement, as it almost always is the case with natural history and the physics of the particular, it seems to me that the true method of guiding one's mind in such research is to have recourse to observations, to gather these together, and from them to make new observations in sufficient number to assure the truth of the principal facts.[16]

Medical thought has at this point reached the polar opposite of that of the days when the sixteenth-century mechanist Giorgio Baglivi placed his hopes in the application of "geometrical and mathematical principles" to the study of the body to reveal its essential secrets.[17]

Yet just as there was significant room for disagreements among mechanists like Descartes, Borelli, and Wills, the vitalists were hardly united on all major issues of physiology. Montpellier doctors like Bordeu, Venel, and Barthez frequented the circles of Parisian philosophes and contributed important articles on medicine to the *Encylopédie*, but there were major differences between their ideas and those La Mettrie, Paul-Henri Thiry d'Holbach, and Diderot, despite the fact that Diderot turned Bordeu into his mouthpiece in his 1769 philosophical dialogue *D'Alembert's Dream*.[18] Diderot and d'Holbach were vitalistic materialists who believed that all matter was suffused with an animate or sensitive force of differing degrees (inherent in matter like Newtonian gravity) and that its operation made the physical world self-sufficient in its deterministic course.[19] The Montpellier doctors stayed away from such philosophical assertions, affirming the autonomy of medical science and the distinction between living creatures, which were endowed with vital force, and inorganic material, which was devoid of it. Peter Reill has claimed that this was a deliberate strategy of "epistemological modesty" that the Enlightenment vitalists pursued—carefully delineate both the extent and limit of human knowledge in navigating a course between the Scylla of mechanicism, with its unwarranted metaphysics, and the Charybdis of retrograde animism of the Hermetic type, with its pantheistic excesses.[20]

Furthermore, individual vitalists rejected or utilized the mechanistic analogy of the body in varying degrees. There were those who took a hardline position in completely rejecting the usefulness of the body-machine idea, like the Scottish doctor Robert Whytt, who in his work *An Essay on the Vital and Other Involuntary Motions of Animals* (1751) thought it evident that animals possess degrees of sentient qualities like memory and reason and found it fantastic that Descartes and his followers would believe them to be "mere machines formed entirely of matter, and, as it were, so many pieces of clock-work wound up and set a-going."[21] Likewise, Diderot ridiculed the idea in his unfinished *Elements of Physiology*.

> The animal is a hydraulic machine. What idiotic things can be said following this one supposition. The laws of motion of hard bodies are unknown, for there are no perfectly hard bodies. The laws of motion of elastic bodies are no more certain, for there is no perfectly elastic body. The laws of motion of fluid bodies are quite uncertain, and the laws of motion of bodies which are sensitive, animated, organized, living, are not even outlined. Anyone who omits from the calculation of this last kind of motion, sensitiveness, irritability, life, spontaneity, does not know what he is doing.[22]

Others, however, saw their vitalism not as an outright rejection of mechanist medicine but as a continued elaboration of it and used the machine-body analogy when appropriate. Boissier de Sauvages is regarded as the first of the major vitalist doctors of Montpellier in the midcentury period, yet Elizabeth Williams has shown the mechanist-vitalist connection in his works, especially in his frequent use of mechanistic imagery.[23] While he opposed Cartesian physiology, he expressed admiration for Borelli and Baglivi and sought in his own vitalist work to establish a "genuinely mechanical" method in medicine, as opposed to the "pseudomechanical" one of the Cartesians. In his final work, *Nosologia methodica* (posthumously published in 1768), he claimed that "man is an aggregate or a being composed of a living soul and a mobile body or a hydraulic machine united together."[24] This sounds close to the Cartesian idea of the automaton-body controlled by the soul, but the crucial difference is that while in the Cartesian man-machine the motive power is inherent in the body and the soul provides only rational guidance, in Sauvages's hydraulic man-machine it is the soul that is the animating force in the body.[25] The majority of vitalists, in fact, did not reject the machine-body analogy completely, as they found it useful for the description of certain aspects of physiology like the actions of muscles and bones. They were careful to assert, however, that the mechanistic perspective was appropriate *but only up to a point* and lost its usefulness when it came to such essential matters as sensitivity and vitality. This is an important point to keep in mind when I examine the automaton-man imagery in the literary works of the late Enlightenment. In an intellectual culture dominated by ideas of vitality, sensibility, and sentimentality, the machine-man idea did not disappear but was made much fruitful use of as an image of an incomplete human being (in the way the mechanist model of the body came to be seen as an incomplete picture)—a person with something essential missing in him that causes him to be stupid, bumbling, callous, cruel, vapid, or just plain tedious.

Anne Vila has demonstrated the cultural importance of medical ideas of sensibility, as evidenced by their impact on French literature, first in mid-century novels by writers like Prévost, Pierre de Marivaux, Prosper de Crébillon, and Françoise de Graffigny.[26] There was a growing interest starting in this period in individual sentiments throughout European culture, as evidenced by the emergence of the man or, with increasing importance, woman of sensibility as the central character in the development of the modern novel, represented in English literature by such successful works as Samuel Richardson's *Pamela* (1740) and *Clarissa* (1747–1748), Oliver Goldsmith's *Vicar of Wakefield* (1764), and Henry Mackenzie's *Man of Feeling* (1771).[27] In France, Rousseau, the greatest prophet of sentiment, became a cult figure, and his sentimental novel *The New Heloise* (1761) was possibly the best-selling work of the century.[28] An important aspect of these novels is the representation of what the culture of sentimentality viewed as the ideal man or woman.

In many of these works, two types of personality are contrasted. The first is the ideal man of the mechanistic age, a rational individual who avoids enthusiasms and other forms of extreme passion, which can lead to everything from religious fanaticism to libertine debauchery, favoring a calm, organized, and decorous mode of living based on moderate, reasonable principles. His home is the urban environment, where he finds happiness in the pursuit of enlightened self-interest by respecting the community's laws and the social hierarchy, which guarantees stability. In direct contrast, the protagonist of the sentimental age is a passionate and sensitive nonconformist constantly struggling with the powerful emotions that overtake him, compelling him to commit actions of a grand but sometimes destructive nature. He is an idealist, an ardent lover, and a loner who is constantly misunderstood by people and consequently shunned, ostracized, and persecuted by them. But he is also a genius whose brilliance is fueled by his soul's undying fire and whose true home is in nature, among living fauna and flora, away from the madding crowd of the cities. In the literature of sentimentality, the person of the first type is described as an incomplete being and often derided as an automaton, in direct contrast to the man or woman of passion. The success of works that featured the tension between the sentimental and the mechanical occurred at the same time that the vitalistic challenge to the mechanistic in medical discourse was arising—leading to the question why such new views became popular at this particular time. For a full understanding of this issue, one must take into account the larger historical and political context of this period.

After the death of the militaristic Louis XIV in 1715, western Europe enjoyed a period of relative peace and prosperity, especially in France and England during the 1720s and 1730s under the leadership of Cardinal Fleury in France, regent to the young Louis XV, and prime minister Robert Walpole in England. The constitutional monarchy that was established in England in the bloodless revolution of 1688 was stabilized with the Hanoverian succession of 1714, creating an environment conducive to economic growth and innovation. It was in this period that the vision of the state as a vast machine, attended to by the learned mechanics of politics and economics who saw to its smooth and efficient functioning, was thought by establishment thinkers to have come to fruition.

The War of Austrian Succession began in 1740 and concluded eight years later, only to flare up again in 1756 as the Seven Years War. These conflicts, along with endless fighting between French and English forces in foreign colonies for imperial dominance, took their toll on the stability of both nations. In France, the situation gave rise to a series of political and economic crises that eventually led to revolution; in England, representatives of the lower orders began to make serious challenges to establishment Whigs with demands for reform, especially in the 1760s, with John Wilkes as the central radical figure, supported by such intellectuals as Richard Price and Joseph Priestly. In this context, the expanding middle class of both countries, at the center of the significant rise in literacy in the first half of the century, felt alienated from the culture of the ruling elites. Consequently, they sought an alternative to the worldview dominated by mechanistic ideas, whose powerful adherents not only had failed to sustain the peace and prosperity of the previous decades but were regarded as directly responsible for the troubles of the midcentury.

As Margaret Jacob and Jonathan Israel have shown in their works on the radical Enlightenment, ideas that did not conform to the dominantly rationalist, mechanist, and dualist views of the classical Enlightenment were present in the intellectual scene of western Europe throughout the seventeenth and early eighteenth centuries.[29] These ideas ranged from the pantheism of Spinoza to the materialism of Hobbes and the deism of John Toland, as well as pantheistic, materialistic, and deistic versions of the ideas of Descartes, Leibniz, Locke, and Newton, who would have rejected these versions as distortions of their original thought. Such views, professed by marginalized radicals, found greater mainstream acceptance in the crisis period of the midcentury, gaining increasing numbers of adherents from the disgruntled middle class.

Whereas mechanistic philosophers tended to think of the world, the state, and the body in terms of machinery, of parts and devices fulfilling their individual functions according to the dictates of a central authority (i.e., the laws of God, the enlightened despot, or the soul), thinkers of the radical Enlightenment regarded the subject of their study in a holistic and organic manner. In political discourse, the vital force was often used as an antiauthoritarian concept, an analog for the "General Will" of the people and the true essence of a just state that bound the community together in the same way the true essence of the natural body animated and harmonized.[30] Unlike the clockwork model of the state and the body in which power emanated from one superior organ, the living force manifested itself in the mutual cooperation of all. It was from this perspective that the machine became an idea associated with tyranny, inequity, and static hierarchy. Vitality and sensibility, in contrast, became central ideas in the advocacy of reform and agitation for revolution.

This association of clockwork imagery with centralized and authoritarian forms of government compelled advocates of more balanced or egalitarian models of the state to favor more organic metaphors in their descriptions of the state. Immanuel Kant put it most aptly in his *Critique of Judgment* (1790) when he asserted that "a monarchical state is represented as a living body when it is governed by constitutional laws, but as a mere machine (like a hand-mill) when it is governed by an individual absolute will."[31] There was, in fact, a general air of skepticism toward mechanistic ideas among major philosophers in the second half of the eighteenth century. Hume, in *Dialogues Concerning Natural Religion* (1779), criticized the classic proof of divinity through the argument by design, denying the validity of the mechanist analogy at its heart and opting for an organic alternative.

> The world plainly resembles more an animal or a vegetable than it does a watch or a knitting loom. . . . The cause of the former is generation or vegetation. The cause therefore of the world, we may infer to be something similar or analogous to generation of vegetation. . . . In like manner as a tree sheds its seeds into the neighbouring fields and produces other trees; so the great vegetable, the world, or this planetary system, produces within itself certain seeds, which, being scattered into the surrounding chaos, vegetates into new worlds.[32]

In France, the materialist d'Holbach in his atheistic tract *Le Bon sens* (1772) ridiculed in the same breath the argument by design, mechanistic physiology, and religion in its entirety by denigrating the Cartesian idea of the wondrous body-machine of divine manufacture:

> One pretends that in forming the universe God had no object but to render man happy. But, in a world made expressly for him and governed by an all powerful God, is man after all very happy? . . . This human machine, which is shown to us as the masterpiece of the Creator's industry, has it not a thousand ways of deranging itself? Would we admire the skill of a mechanic, who should show us a complicated machine, liable to be out of order at any moment, and which after a while destroys itself?[33]

A significant passage in Louis-Sebastien Mercier's utopian novel *Memoires of the Year 2440* (1771) provides evidence of the use of medical vitalism for a political purpose.[34] The Montpellier doctors generally avoided philosophical and political discussions, but others freely appropriated their ideas in discourses outside the medical. Bordeu, who regarded life in the animate body as defined by the faculty of sensitivity, made a point of emphasizing the autonomous nature of individual parts that work in harmony with one another, constituting a "federation of organs" in the body, in direct contrast to the monarchical supremacy of the brain-intellect-soul in the mechanist model.[35] In Mercier's novel, the protagonist awakes in a Paris of the distant future in which the ideals of the progressive philosophes have been realized, including the establishment of a constitutional state. The passage on how the decentralized political system was established by a "philosophic prince" makes use of the vital body analogy, including its federal nature:

> in an extensive kingdom there should be an union of the different provinces in order to its being well governed; as in the human body beside the general circulation, each part has one that is peculiarly adapted to itself; so each province, while it obeys the general laws, modifies those that are peculiar to it, agreeable to its soil, its position, its commerce and respective interests. Hence all lives, all flourishes. The provinces are no longer devoted to serve the court and ornament the capital.[36]

But to avoid painting a simplistic picture of the intellectual milieu of western Europe in the second half of the eighteenth century, I must note that the prominence of vitalist ideas in the period did not signify a total collapse of mechanist ideas or of the positive use of the machine analogy. In physics there was a vigorous neomechanistic movement, represented by such figures as Jean Le Rond d'Alembert, the coeditor of the *Encyclopédie*, Joseph-Louis Lagrange, and Pierre-Simon Laplace, who refined Newton's ideas with sophisticated mathematical measurements and calculations, a development concomitant with the pioneering chemical works of Antoine Laurent Lavoisier.[37] In addition, Simon Schaffer has pointed to the description of workers as automata in the discourse of early Industrial Revolution in which the human machine is envisioned as part of actual machinery in the factory.[38] Adam Ferguson asserts in his *Essay on the History of Civil Society* (1767):

> Many mechanical arts require no capacity. They succeed best under a total suppression of sentiment and reason, and ignorance is the mother

of industry as well as of superstition. Reflection and fancy are subject to err, but a habit of moving the hand, or the foot, is independent of either. Manufactures, accordingly prosper most, where the mind is least consulted, and where the workshop may, without any effort of imagination, be considered as an engine, the parts of which are men.[39]

Similarly, Diderot in the *Encyclopédie* article "Bas" describes the operation of an automated silk loom, marveling at how the machine "makes hundreds of stitches at once . . . and all without the worker who moves the machine understanding anything, knowing anything or even dreaming of it."[40] The case of Diderot illustrates the complexity of the usage of the machine-man imagery in the late Enlightenment. He celebrates the budding industrial system that features the reduction of human workers to parts of machinery; but in his literary works he also makes frequent use of the idea to express pity for those who have been deprived of their free will and to ridicule dull conformists (as will be shown below).

The proper way to understand the intellectual scene of the late Enlightenment is not as a transition from the mechanist worldview to the vitalist one but as an ever increasing tension between the two, with radical ideas on the fringes of intellectual and political life gaining greater mainstream acceptance at midcentury, especially among middle-class readers attracted to these ideas' antiestablishment tendency. And sometimes that tension can be discerned in the writings of a single writer like Diderot.

The Automaton-man in Late Enlightenment Literature

The demise of mechanist medicine in the mid-eighteenth century provided intellectuals with a new vision of the vital body and with the view of the automaton-man as that of an incomplete person lacking something essential to humanity. Such ideas appeared in the sentimental literature of the period, in which the passionate and independent character of sensibility was often contrasted with machine-like authoritarians and conformists. This cultural development was a part of a general rejection of mechanist ideas among advanced intellectuals who were reacting against the establishment that they held responsible for the crisis of the midcentury. In their writings the machine as a conceptual object signified tyranny, ignorance, and the lack of autonomy, and the vision of the automaton-man represented a person devoid of freedom either through external oppression or inner conformism. This is the historical origin of the paradox noted by Mark Seltzer of the word "automaton" denoting both an object capable of independent motion and a person incapable of independent thought or action.[41]

In this section, I present an impressionistic survey of many instances in which this formulation of the automaton-man as an image of a person without freedom was deployed in late Enlightenment writings by such major literary and intellectual figures as Samuel Richardson, Rousseau, Voltaire, Diderot, Mercier, Choderlos de Laclos, Marquis de Sade, and Thomas Paine. It would take a full volume to analyze in detail how each of these writers made use of the idea for individual purposes in his works, so within this chapter I can only demonstrate that the imagery was indeed a widespread one in the writings of the mid- to late eighteenth century. I will also show that the automaton-man was used to characterize such disparate people as the mentally handicapped, the ignorant, the oppressed, the conformist, and the tyrant, all those the Enlightenment philosophes alternatively despised, pitied, and ridiculed for their lack of autonomy and the deficiency of character.

Significant instances of the imagery appear in the novels of Samuel Richardson, who was a friend of the mechanist-turned-vitalist doctor George Cheyne.[42] In *Pamela*, the reformed rake Mr. B lectures his wife on the importance of his keeping regular hours in organizing the activities of the day, even though some may ridicule him for his strictness, saying, "It signifies nothing to ask *him:* he will have his own way. There is no putting him out of his course. He is a regular piece of clock-work."[43] The machine analogy is employed here in a negative manner, to point to his inflexibility, with intimations also of an authoritarian nature. Mr. B, however, does not reject the imagery but adopts it to demonstrate the need for such strictness—"man is as frail a piece of machinery, and, by irregularity, is as subject to be disordered as a clock." The mechanist philosophers of the previous century marveled at the intricacy, efficiency, and beauty of the body-machine of divine manufacture, yet here the man-machine is portrayed as a fragile thing that requires an active and disciplined will of the individual to maintain its order. This view is echoed in a much more trenchant manner by d'Holbach in the already quoted passage from *Le Bon sense* in which he rails against the notion of the wonderful human machine by pointing out that this supposed masterpiece has "a thousand ways of deranging itself" and asks what respect one should pay to a mechanic who builds a machine that is "liable to be out of order at any moment, and which after a while destroys itself?"[44] The libertine Robert Lovelace in Richardson's *Clarissa* questions his own mechanical nature, as an entity who might be manipulated by those he has nothing but contempt for. As he meditates on possible marriage to Miss Clarissa Howe, the victim of his sexual machinations, he imagines himself becoming an implement in her hands, debasing himself to the extent that he must make peace with her family that despises him. He thinks ". . . 'tis poor . . . to think myself a machine—I am *no* machine—Lovelace, thou art base to thyself, but to *suppose* thyself a machine."[45] For both Mr. B and Lovelace the human machine is far from being the autonomous or admirable entity it was for the philosophers and physiologists of the classical Enlightenment. Whether as an object that needs constant regulating or as a thing that is used and manipulated by others, it represented a person who was deficient in some vital way.

As shown in the last chapter, one of the earliest instances of this pejorative description of a personality type occurs in La Bruyère's 1688 work *Les Caractères,* in which a fool is defined as "an automaton, a piece of machinery moved by springs and weights."[46] Voltaire makes use of the idea in "The Story of a Good Brahmin," (1761) in which a traveler meets a wise Brahmin philosopher who is miserable because the accumulation of knowledge had only made him more aware of his ignorance. The former, on his way to see the wise man, encounters an old woman whom he asks if "she had ever been afflicted by the thought that she was ignorant of the nature of her soul. She did not even understand the question." As he then confronts the Brahmin about her:

> "Are you not ashamed to be unhappy," I said, "when outside your garden there is an old automaton who thinks about nothing and yet lives happily?" "You are right," he replied: "I have told myself a hundred times that I would be happy were I as brainless as my neighbour, and yet I would not want such happiness."
>
> This answer from my Brahmin impressed me more than all the rest. I set to examining myself, and saw that in truth I would not care to be happy at the price of being an imbecile.[47]

As does La Bruyère, Voltaire equates an ignorant person without the capacity for self-meditation with an automaton.

Rousseau in *Emile* (1762) uses the idea of mechanical imbecility in his discussion of the helpless nature of newborn infants whose actions and cries are "purely mechanical effects." He imagines such a creature born with the body of an adult.

> This man-child would be a perfect imbecile, an automaton, an immobile and almost insensible statue. He would see nothing, hear nothing, know no one, would not be able to turn his eyes toward what he needed to see. Not only would he perceive no object outside of himself, he would not even relate any object to the sense organ which made him perceive it: the colors would not be in his eyes; the sounds would not be in his ears; the bodies he touched would not be on his body; he would not even know he had one. . . . However little one may have reflected on the order and the progress of our knowledge, it cannot be denied that such was pretty nearly the primitive state of ignorance and stupidity natural to man before he learned anything from experience of his fellows.[48]

Rousseau's attitude toward women has been roundly criticized by Mary Wollstonecraft and many others, but later in the book he stresses the importance of giving young girls a proper education, precisely to prevent them from turning into mechanical imbeciles. If they are to become stimulating companions for boys and men (the main function of educated women for Rousseau, revealing the limited nature of his sympathy for them) they should not be "raised in ignorance of everything and limited to the housekeeping functions alone."

> Will man turn his companion into his servant? Will he deprive himself of the greatest charm of society with her? In order to make her more subject, will he prevent her from feeling anything, knowing anything? Will he make her into a veritable automaton? Surely not. It is not this that nature has spoken in giving women such agreeable and nimble minds.[49]

In yet another passage of the book, he shifts the automaton-man image to that of the oppressed peasant whose existence is as predictable and mindless as that of La Bruyère's fool, but through no fault of his own. What Rousseau elicits here is not contempt but pity as he contrasts the peasant to a savage (i.e., a man living in harmony with nature). Rousseau points out that both of them lead highly active lives, but their essential natures could not be more different:

> nothing is duller than the peasant and nothing sharper than a savage. What is the source of this difference? It is that the former, doing always what he is ordered or what he saw his father do or what he has himself done since his youth, works only by routine; and his life, almost an automaton's, constantly busy with the same labors, habit and obedience take the place of reason for him.[50]

As he elaborates on the nature of the savage, he asserts that what makes him so different from the peasant, so unlike an automaton, is that he is free in the open world of nature: "Attached to no place, without prescribed task, obeying no one, with no other law than his will."[51]

This convention of describing the state of being forced to do things against one's will as that of being turned into a machine or an automaton was utilized by many other writers. D'Holbach, again, in *Le Bon sens* points to the collusion of political and religious leaders in maintaining their power through oppression and imposed ignorance. In such a state, "man was a pure machine in the hands of tyrants and his priests, who alone had the right to regulate his movements; always treated as a slave, he had in all times and in all the places, the vices and character of the slave."[52]

In 1760 Diderot wrote the novel *La Religieuse* to decry the abusive practice of forcing powerless women into convents against their will. As the pitiful character of Suzanne is made to go into a religious house and take the habit, she goes into a trance-like state, losing all self-will and understanding of the world around her.

> I heard nothing of what was being said around me, I had almost reached the state of an automaton. I saw nothing, but occasionally little convulsive shudders ran over me. I was told what to do, and often it had to be repeated because I did not understand the first time, and I did it. It was not that I was thinking of anything else, but I was absorbed.[53]

Diderot also makes use of the automaton-man image to comic effect in his metafictional novel *Jacques the Fatalist*.[54] The character he refers to as an automaton is neither an imbecile nor an oppressed person but the master of the title character, an unnamed aristocrat who is ridiculous because of his boring personality. In one episode of the duo's journey through the novel, the master sends Jacques back to the inn they stayed in to retrieve some belongings they left behind. After the servant's departure, the narrator provides us with a description of the master's shortcomings.

> He has very few ideas in his head at all. If he happens to say something sensible, it is from memory or inspiration. He has got eyes like you and me but most of the time you cannot be sure he is actually seeing anything. He does not exactly sleep, but he is never really awake either. He just carries on existing simply because it is what he usually does. Our automaton carried straight on ahead.[55]

Later on, when the narrator goes off on one of many digressions in the novel, he anticipates the criticism of readers who may want a conventional narrative by affirming his independence in choosing to write the story in an unorthodox manner.

> Reader, you're treating me like an automaton. That's not polite. "Tell the story of Jacques' love life," "Don't tell the story of Jacques' love life," "I want you to tell me about Gousse," "I've had enough . . ."
>
> It is no doubt necessary that I follow your wishes, but it is also necessary that I sometimes follow my own. And that is without considering the fact that anyone who allows me to begin a story commits himself to hearing it through to the end.[56]

The narrator asserts his freedom to take the narrative in any direction he wants to, so he objects to the reader's demand that he unfurl the plot in a conventional manner, which amounts to the writer acting like a mere storytelling machine.

Toward the end of the novel, as Jacques explains to his master the nature of his fatalistic philosophy of life, he describes actions performed without or against one's will as that of a machine.

MASTER: What are you thinking?

JACQUES: I am thinking that, although you were speaking to me and I was answering you, you were speaking without wanting to and I was answering without wanting to.

MASTER: And?

JACQUES: And? That we are nothing but two living and thinking machines.

MASTER: Well, at this moment what do you want?

JACQUES: My God, that doesn't make any difference. That only brings one more function of the two machines into play.[57]

Many philosophes expressed their greatest contempt for conformists as automaton-men, those who parroted the most fashionable ideas with no original thought of their own, especially those of high society. Mercier, in the course of describing the ideal society of the future in *Memoires of the Year 2440*, ridicules every major figure of authority in his own time, with constant reference to them as automata, all caught up in the trappings and mannerisms of their station in society. As the time traveler explains to his host, in his own time, "a puerile and destructive luxury" had turned the brains of the nobility to the extent that "a body without a soul was covered with lace; and the automaton then resembled a man."[58] That is why in the future republic "ancient livery of pride" is despised, since when "a man is known to excel in his art, he had no need of a rich habit, nor of magnificent apartments, to recommend him . . . his actions speak, and each citizen is desirous that he should receive recompence of his merit."[59] Monks are not to be found in that society since, as the time traveler's host explains, "We no longer fatten, in our state . . . a set of automatons, as troublesome to themselves as to others, who make a foolish vow never to be men, and hold no connection with those that are."[60] In the academia of the future, scholars are judged on their merits alone, as exemplified by the titles of their works, which are displayed for public viewing, unlike in the old days when undeserving "bishops, marshals and preceptors" held positions of authority. As the author explains in a footnote, "We have seen on the Boulevards . . . an automaton that articulated sounds, and the people flock to admire it. How many automata, with human faces, do we see at court, at the bar, in the academies, who owe their speech to the breath of invisible agents; when they cease, the machines remain dumb."[61]

Images of the automaton-man and automaton-woman appear in the libertine works of Choderlos de Laclos and Marquis de Sade in similar terms.[62] Laclos's 1782 epistolary novel *Liaisons dangereuses* portrays the high society of the ancien régime as a gaming ground on which the master manipulators move individuals about like chess pieces. In the letters exchanged between the two players of the drama, the Marquise de Merteuil and the Vicomte de Valmont, who begin as allies but end up mortal enemies, Valmont describes the people he has been using as "automata, near whom I have vegetated since this morning," while Merteuil advises against further using the silly and easily manipulated Cécile de Volanges, since women like her are "absolutely nothing but pleasure machines" and "very soon everybody gets to know the springs and contrivances of these machines."[63]

Roland Barthes has analyzed the prevalence in the pornography of de Sade of the machine image, which appears in two forms—human bodies as sexual devices and the frequent use of actual machines to alternatively arouse and torment people.[64] Susan Sontag, in her reading of de Sade's *120 Days of Sodom*, has also noted the work's depiction of a person "as a 'thing' or an object, of the body as machine, and of the orgy as an inventory of the hopefully indefinite possibilities of several machines in collaboration with each other."[65] Given the nature of the interminable sex scenes in such works as *Justine, ou les malheurs de la vertu* (1791), *La Philosophie dans le boudoir* (1795), and *L'Histoire de Juliette* (1797), one might interpret them as extreme fantasies of sexual mechanics, extended narratives of mechanistic pornography like those in Cleland's *Fanny Hill*. There is, however, a crucial difference in the physiological ideas represented in the works of Cleland and de Sade, highlighted by de Sade's obsession with arousal that traverses the boundaries of pain and pleasure. In *Fanny Hill*, the sexual bodies are indeed machines that derive pleasure out of their mechanical intercourse, but in de Sade's imagination human bodies are vital entities of sensibility. While the bodies of mechanistic medicine would feel no discomfort at being used like the machines that they are, vitalist bodies would protest when subjected to such treatment as being tightly restrained, made to perform endlessly regular movements, and abused with blows and sudden penetrations. And that protest of sensitivity manifests itself in the most acute and exquisite sensation of pain-pleasure. In other words, the sadomasochistic enterprise works precisely because the vital body is treated like the machine that it is *not*. In fact, for the Sadean orgy-master, such as the horrendous Noirceuil of *Juliette*, people who act like they are nothing more than automata in the elaborate sex acts he organizes spoil the fun for him with their dull and monotonous behavior. Noirceuil complains during a break about some of the participants in an orgy: "Feeble-minded creatures," he murmured; "pleasure-machines, sufficient to our purposes, but, truly, their appalling insensibility depresses me."[66]

Finally, while political radicals described subjugated people as automata in the sense of vital beings who have been reduced to machinery through oppression, they also saw tyrants as automata, beings trapped in their own system of despotism and moved by its dictates. In a social and political situation without liberty, the perpetrators of inequities are just as lacking in the principle of freedom as their victims. It is in this sense that Thomas Paine asserts in *Rights of Man* (1791):

> Hereditary succession is a burlesque upon monarchy. It puts it in the most ridiculous light, by presenting it as an office which any child or idiot may fill. It requires some talents to be a common mechanic; but to be a king requires only the animal figure of man—a sort of breathing automaton. This sort of superstition may last a few years more, but it cannot long resist the awakened reason and interest of man.[67]

Two years after the publication of the work, in the bloody year 1793, Maximilien Robespierre, in a report to the Convention, had occasion to refer to the recently beheaded king as "that crowned automaton called Louis XVI."[68]

The Post-Enlightenment Legacy of the Automaton-man

At this point I would like to jump ahead of the historical narrative with a representative sampling of the appearance of automaton-man imagery in psychological, scientific, and literary works in the periods following the Enlightenment. This is for the important purpose of demonstrating that the eighteenth-century notion of the mechanical being as a representation of humanity lacking in freedom, self-will, and originality is one of the most powerful and enduring legacies of the late Enlightenment to the modern imagination, something that is very much present today.

One of the more beguiling phenomena of late eighteenth-century France was the popularity of mesmerism, a medical practice based on the mystical ideas of the German doctor Franz Anton Mesmer that combined elements of vitalist medicine, animist philosophy, and the science of magnetism and electricity.[69] It featured a practice that involved putting an ailing subject into a trance that made him or her more receptive to the healing influence of what Mesmer called animal magnetism, an all-encompassing vital force in nature. Its popularity prompted the formation of an official scientific commission, which included Antoine Lavoisier and Benjamin Franklin, to investigate Mesmer's ideas and claims. In the commission's debunking report that attributed the experiences of the mesmerized subjects to their imagination rather than actual physical causes, it noted the loss of autonomy on the part of the patient, putting him "entirely under the government of the person who distributes the magnetic virtue."[70] In the same vein, the physician Jean-Jacques Paulet in his 1782 antimesmerist pamphlet *L'Antimagnétisme* claimed that in the world of mesmerism "we are all like real puppets, ignorant and utterly blind slaves."[71] Despite the difference between the actual objects of automaton and puppet—the self-moving nature of the former— the two became virtually synonymous as descriptions of people without free will. In other words, Paulet would have lost none of his meaning if he had written "we are all like real automata." Consequently, in the mesmerist discourse of the following century, the automaton was frequently evoked in the description of the hypnotized subject, as Alison Winter demonstrates in her study of the subject in Victorian Britain.[72]

More than a century later, Sigmund Freud makes an interesting use of this idea in his *Psychopathology of Everyday Life* (1901) in the chapter "Bungled Actions," in which he examines unconscious reasons behind mistaken actions. Freud explains that he has been treating an old lady as a patient with daily visits to provide her with medicine. All he has to do is put some drops in her eyes and give her an injection of morphine, tasks so simple that he can accomplish them without conscious thought. "During the two operations my thoughts are no doubt busy with something else; by now I have performed them so often that my attention behaves as if it were at liberty."[73] But then he realizes one day that he has made a mistake.

> One morning I noticed that the automaton had worked wrong. I had put the drop into the white bottle instead of the blue one and had put morphine into the eye instead of collyrium. I was greatly frightened and then reassured myself by reflecting that a few drops of a two percent solution or morphine could not do any harm even in the conjuctival sac. The feeling of fright must obviously have come from another source.[74]

A number of elements in this story are significant in connection with the Enlightenment notion of the automaton-man. First, Freud refers to himself as an automaton because he acted without conscious thought; second, he is also an imperfect automaton that is liable to commit "bungled actions" if left to its own devices (just like the clockwork man of Mr. B in Richardson's *Pamela*); and third, he is an automaton because he is controlled by an unknown force, just as in the mesmerized subject. In the following passage, Freud analyzes his own actions and concludes that he was "under the influence of a dream which had been told me by a young man the previous evening and the content of which could only point to sexual intercourse with his own mother."[75] So his preoccupation with the notion of doing violence to an old woman led him to commit an unconscious action that could have endangered his elderly female patient. His "fright" at what he had done, even with his medical knowledge that he could not have caused harm, came from the feeling of having been manipulated by an external force, another sign of the automaton's lack of autonomy. In other major psychological works, Erich Fromm in his classic study of totalitarianism *Escape from Freedom* (1941) characterizes the blind obedience of followers as "automaton conformism"; and the poststructuralist psychoanalyst Jacques Lacan also sets in opposition what he calls "tuché," defined as "the encounter with the real," against the "automaton," which refers to the repetitive compulsion of a person acting in a disassociated state.[76]

In the scientific realm, the problem of automatism and free will became a central issue of contention when the debate over mechanistic physiology was revived in 1874 by T. H. Huxley in his controversial lecture "On the Hypothesis That Animals Are Automata, and Its History." He begins with a discussion of past ideas on mechanistic physiology, particularly those of Descartes, and assesses them in light of contemporary discoveries in biology. While he notes errors in classical mechanicism, he affirms the validity of the machine analogy as extended even to the human mind. In his particular brand of evolutionary psychology, consciousness is regarded not as a transcendent feature of humanity but the sum of concomitant reaction in the brain to activities that are occurring in the nerves, "thoughts" as such being linked to humans' biological function in a causal fashion. But on the difficult question of free will, he makes a rather vague claim that there is apparent freedom of action in people's everyday function, but that such actions can also be accounted for in a larger scheme of cause and effect.

> We are conscious automata, endowed with free will in the only intelligible sense of that much-abused term—inasmuch as in many respects we are able to do as we like—but none the less parts of the great series of causes and effects which, in unbroken continuity, composes that which is, and has been, and shall be—the sum of existence.[77]

While there is no place for supernatural agency in his worldview, this idea is reminiscent of Leibniz's notion of the local efficacy of free will within the larger system of God's pre-established harmony (which is replaced by a secular form of universal causality in Huxley).[78]

When William James responded to Huxley's lecture in his "Are We Automata?" (1879) he used arguments similar to Kant's in his *Critique of Practical Reason* where he criticized Leibniz precisely on the point of free will as demonstrated through the nonmechanical functioning of the mind, even evoking Vaucanson's automata for a negative comparison.[79] James points explicitly to the Kantian view of the mind as a naturally organizing instrument to show that

> there are a great many things which consciousness *is* in a passive and receptive way by its cognitive and registrative powers. But there is one thing which it does ... which seems an original peculiarity of its own; and that is, always to choose out of the manifold experiences present to it at a given time some one for particular accentuation, and to ignore the rest.[80]

After elaborating on this selective function of the mind, James concludes:

> The utility of reasoned thought is too enormous to need demonstration. A reasoning animal can reach its ends by paths on which the light of previous experience has never shone. . . . I had found it impossible to symbolize by any mechanical or chemical peculiarity that tendency of the human brain to focalize its activity on small points which seems to constitute its reasoning power.[81]

And it is on this basis of the human ability to make "autonomous" decisions and discoveries from reason, which cannot be explained through external causal factors alone, that James rejects Huxley's characterization of people as conscious automata.[82]

The Enlightenment origin of this controversy lies not just in the revival of the age-old debate on animal automatism and mechanistic physiology, though recast in the language and concepts of nineteenth-century biology and psychology. The deployment of the automaton as an idea denoting the lack of true free will and the state of being moved by larger forces (again, the irony of a thing defined as a self-mover representing beings that are moved by others) points to the clear impact of late Enlightenment use of the object.

The automaton-man image also makes significant and imaginative appearances in literary works of the nineteenth century. The political radicals of the early Romantic period took up the association of the mechanical with the tyrannical and the enslaved, employing it frequently in their writings. A powerful example of this is Percy Bysshe Shelley's early work *Queen Mab* (1810), in which he meditates on the corrupting effect of power by describing the opposition of oppression and freedom in terms of the natural versus the artificial.

> Nature rejects the monarch, not the man;
> The subject, not the citizen: for kings
> And subjects, mutual foes, for ever play
> A losing game into each other's hands,
> Whose stakes are vice and misery. The man
> Of virtuous soul commands not, nor obeys.
> Power, like a desolating pestilence,
> Pollutes whate'er it touches; and obedience,

Bane of all genius, virtue, freedom, truth,
Makes slaves of men, and the human frame,
A mechanized automaton.[83]

In Stendhal's masterpiece *The Red and the Black* (1830), the Romantic hero Julien Sorel finds the stultifying society of Restoration France to be filled with human automata with no sign of the originality or ambition of the Napoleonic era. What is peculiar is that people in that society keep referring to one another as machines and automata, aware of the dullness and pettiness of their existence in their conservative age. Monsieur de Rênal, the mayor of the fictional town of Verrières where the story begins, unaware of the budding romance between his neglected wife and Julien, the tutor to his children, keeps commenting on the fragility of his wife by referring to women in mechanical terms: "Typical woman . . . there's always something to put right with those particular machines"; "That's women all over . . . there's always something ailing those complicated machines."[84] At one point, however, Julien pays him back in kind as he considers his employer's inability to properly care for his wife: "That automaton of a husband might as well do her more harm than good."[85] Toward the end of the novel, as Julien sits in prison waiting to go on trial for the attempted murder of his once beloved, the Abbé de Frilair tries to save him by manipulating the jury. He is confident that he can achieve his purpose because five members are beholden to him as members of his congregation and three out-of-towners can be controlled as well: "The first five are machines. Valenod is an agent of mine, Moirod owes everything to me, de Cholin is an imbecile scared of everything."[86]

In Alexandre Dumas's historical novel *Twenty Years After* (1845), which takes place in the France and England of 1648–1649, d'Artagnan, the young hero of *The Three Musketeers* (1844), who has become a middle-aged officer, is summoned by Cardinal Mazarin to be sent on a secret mission. As Dumas describes d'Artagnan's appearance before the powerful minister, "D'Artagnan, erect, imperturbable, waited without impatience and without curiosity. He had become a military automaton, acting, or rather obeying, at the touch of a spring."[87] According to Foucault, it was precisely at the period in which the story takes place that a mechanistic approach to military discipline was being systematically implemented in western Europe, the view of the ideal soldier as the embodiment of individual strength, courage, and pride shifting to that of a perfectly obedient and regulated machine.[88] D'Artagnan hopes to become the captain of the Musketeers, so he has deliberately turned himself into a military automaton for that purpose. As he goes about obeying the orders of contemptible and conniving superiors, the reader is made to feel that he is debasing himself and has become something less than the dashing swashbuckler of twenty years before. Dumas describes the sorry state he is in, due to his separation from his great companions of the past:

> While his friends surrounded him, d'Artagnan remained in his youth and his poetry. His was one of those delicate and ingenious natures which easily assimilates to itself the qualities of others. Athos imparted to him of his greatness of soul, Porthos of his enthusiasm, Aramis his elegance. If d'Artagnan had continued to live with these three men, he would have become a superior man.[89]

In another scene, Cardinal Mazarin complains to Anne of Austria, the mother of the boy king, that

> every day I suffer affronts from your princes and titled servants—all automata who do not see that I hold the string that moves them, and who under my patient gravity have not divined the laugh of the irritated man who has sworn to himself to become one day their master.[90]

As the plot thickens, d'Artagnan becomes increasingly embroiled in the nefarious plans of the cardinal and ultimately has to decide whether he will remain his "machine" or follow a more honorable course with the companions of his past. As do many of the images of the automaton-man in nineteenth-century literature, d'Artagnan as a military automaton yet again embodies the ambivalence at the heart of the metaphor: the mechanical officer is the ideal in the modern army yet is far from an ideal human being with his lack of self-will, sentiments, and independence.

A wonderful example of the use of the automaton-man image for political purposes can be found in American literature in Henry Adams's anonymously published *Democracy, An American Novel* (1880), a biting satire of the state of U.S. politics in the late nineteenth century. Madeline Lee, a society widow from New York, takes up an interest in the workings of the nation's democracy and decides to move to Washington, D.C. Her desire to see for herself the process of government at work is described as

> the feeling as a passenger on an ocean steamer whose mind will not give him rest until he has been in the engine-room and talked with the engineer. She wanted to see with her own eyes the action of the primary forces; to touch with her own hand the massive machinery of society; to measure with her own mind the capacity of the motive power. She was bent upon getting to the heart of the great American mystery of democracy and government.[91]

This mechanistic description of the state is particularly interesting in the American context since it is democracy that is being described, not the absolutist monarchy that many of the late Enlightenment philosophes decried as a monstrous, oppressive machine. As the story of Madeline progresses in the nation's capital, however, it turns out that the above passage is a setup for a ridiculous sight that awaits her at a White House reception. What she witnesses is none other than a pair of automata. As Madeline and her escort move up a line to meet the president of the United States and the first lady,

> Madeline found herself before two seemingly mechanical figures, which might be wood or wax, for any sign they showed of life. These two figures were the President and his wife; they stood stiff and awkward by the door, both their faces stripped of every sign of intelligence, while the right hands of both extended themselves to the column of visitors with the mechanical action of toy dolls. Mrs. Lee for a moment began to laugh, but the laugh died on her lips. To the President and his wife this was no laughing matter. There they stood, automata, representatives of the society which streamed past them.[92]

As Madeline considers why she finds the sight so ridiculous and horrifying at the same time, she realizes that all the artificial ceremonies there, all so formal and mechanical, are really those inherited from the courts of European royalty. She thinks them wholly inappropriate to the center of what is supposed to be a democratic state but then realizes that no one else seems to be scandalized by it. She then makes a leap from the mechanical show before her to a vision of what all of American society is becoming, namely an entire nation of automata.

> To all the others this task was a regular part of the President's duty, and there was nothing ridiculous about it. They thought it a democratic institution, this droll aping of monarchical forms. To them the deadly dullness of the show was as natural and proper as ever to the courtiers of the Philips and Charleses seemed the ceremonies of the Escurial. To her it had the effect of a nightmare, or of an opium-eater's vision. She felt a sudden conviction that this was to be the end of American society; its realization and dream at once. She groaned in spirit.
>
> "Yes! at last I have reached the end! We shall grow to be wax images, and our talk will be like the squeaking of toy dolls. We shall all wander round and round the earth and shake hands. No one will have any object in this world, and there will be no other. It is worse than anything in the 'Inferno.' What an awful vision of eternity!"[93]

The Enlightenment legacy can be discerned here in the association of mechanical behavior with that of monarchical absolutism, the description of a state without liberty, and the vision of a society of unoriginal conformists going through the motions of tedious social rituals.

A False Automaton at the Twilight of the Enlightenment

In returning to the late Enlightenment, I want to clearly distinguish at this point two historical eras that have been the subject of the last and current chapters. The first is *the golden age of the automaton*, the time of the object's intellectual primacy from 1637 to 1748 (Descartes to La Metrrie). The second is the period of *the automaton craze*, or the public interest in the performance of actual automata that began with the Vaucanson automata, climaxed with the Jaquet-Droz creations, and, as I show below, ended with Wolfgang von Kempelen's chessplayer. There is an overlap between the two, around the time of the initial success of the Vaucanson works in the 1730s, but for the reasons detailed in this chapter the intellectual celebration of the automaton concept went into decline in the following decade.

In the golden age of the automaton, the concept signified much more than an ingenious piece of mechanical craft. It was used as a heuristic device to illustrate the nature of the body, the state, and even the entire universe constructed by an engineer-God, functioning as the central emblem of the mechanistic cosmos of the classical Enlightenment.

As for the automaton craze, the public continued to be awed and amused by objects that moved, spoke, and played music even after the worldview they represented was challenged by vitalist thinkers and writers. Even in the nineteenth century, new works by ingenious mechanics were produced with varying degrees of success. In England, a jeweler named Thomas Weeks displayed mechanical works from the Cox collection and his own inventions at his "Mechanical Museum" in the first decade of the new century; Friedrich Kaufmann, a German mechanic from Dresden, presented a trumpet-playing automaton in 1810; and starting in 1846, Joseph Faber from Freiburg gave performances in London of a remarkable speaking machine called "Euphonia" that garnered some public and media attention.[94] Christian Bailley has also identified the period of 1848–1914 as one in which there was a renewed vogue for automata in France, as he demonstrates in his beautifully produced book on the subject.[95]

None of those objects, however, achieved the kind of cultural and intellectual impact that the works of the previous century had, as they were regarded largely as objects of entertainment, with only a few exceptions.[96] The enduring legacy of Enlightenment automata is further evidenced by the fact that in the new century when Carlyle, Helmholtz, Huxley, and others referred to the automaton it was Vaucanson's name they evoked, not any of their contemporaries'. In fact, despite having produced only three working automata, he is the only automaton-maker in history to have attained fame that lasted well beyond his lifetime.

The success of Vaucanson's works began the eighteenth-century automaton craze, and its end was marked by the highly controversial celebrity of a false automaton, one that embodied ideas and attitudes toward the mechanical mimicry of life that were at odds with those of the early works. The object was Kempelen's so-called chess-player, also known as the Turk, and to understand why this astounding device of trickery represented the closing of the Enlightenment's love affair with the automaton, one must examine its story as well as its nature in direct contrast to those of the earlier machines.

If the chess-player of Baron Wolfgang von Kempelen (1734–1804) had been an actual automaton, it would have been the rival of Vaucanson's defecating duck as the most famous in the history of the object. The significance of its fame, however, points not to the celebration of the mechanical, as in the works of the previous decades, but precisely the disillusionment with it in the larger culture. A substantial amount has been written on the device, both during the time it was performed, some by major intellectual and literary figures like the philosophe Friedrich von Grimm, the British scientist-inventor Charles Babbage, and the American writer Edgar Allen Poe, and recently, with renewed interest after the defeat of the chess master Gary Kasparov by the IBM computer Deep Blue in 1996 and 1997.[97] Despite the fact that the Turk is still commonly referred to as an automaton, Philip Thicknesse, an eighteenth-century debunker of the device, was absolutely right when he asserted in his 1784 article "The Speaking Figure, and the Automaton Chess-Player, Exposed and Detected" that "to call it AN AUTOMATON, is an imposition," pointing out in a footnote that if the definition of an automaton is "a self-moving Engine, with the principle of motion within itself" (giving the examples of the dove of Archytas, and the eagle and fly of Regiomontanus), "the modern Chess-Player is no such thing."[98] Instead, Thicknesse refers to it as a doll, though puppet would have been more appropriate.

Wolfgang Von Kempelen, a Hungarian nobleman from Pressburg (today's Bratislava in Slovakia), was an extraordinary polymath who served as a privy councilor at the Viennese court of Maria Theresa and was noted for his multilingual fluency and his translation of the revised legal code. Although his name became associated most with the chess-player, he used his considerable mechanical talents during his lifetime to also design bridges, a hydraulic system for Schönbrunn Palace, and machines for canal construction. He also built a speaking automaton and a writing machine for the blind.[99]

According to the account by Karl Gottlieb von Windisch, Kempelen attended a magic performance that was given at the royal court in 1769 by a Frenchman named Pelletier who used magnets to create his illusions. After the act Kempelen told the empress that he could construct a machine "whose effects should be more suprizing, and the deception more complete, than any thing her Majesty had then seen."[100] A year later, he fulfilled his

Wolfgang von Kempelen, the Turk chess-player. Illustration from Karl Gottlieb von Windisch, *Briefe über den Schachspieler des Herrn von Kempelen*, reproduced in Charles Michael Carroll, *The Great Chess Automaton* (New York: Dover 1975).

promise by presenting to her his chess-player, a life-sized figure in Turkish costume sitting behind a large cabinet with a chess board on top. Before the demonstration got under way, the inventor performed an act that is essential to understanding the device's problematic place in the history of automata. Kempelen aped Vaucanson by opening up various small doors on the cabinet to reveal the wheels and gears within to demonstrate that, like the flute player, his invention had nothing but machinery inside. After the device beat a series of opponents in several courtly performances, news of the astounding object spread to England and France through letters written to *Gentleman's Magazine* and *Le Mercure de France* by Louis Dutens, an English clergyman who witnessed a performance in Vienna. Like Vaucanson after the success of his automata, Kempelen wanted to move on to more important tasks, so he dismantled the chess-player and began his work on the hydraulic system at Schönbrunn. He was, unfortunately, not as apt at capitalizing on his work as Vaucanson had been, and his position at the royal court diminished after the death of his patron Maria Theresa in 1780.

In the following year, he was ordered by the successor Joseph II to prepare the machine for a revived performance for a delegation from Russia. After a successful demonstration, Kempelen took up the emperor's suggestion to take the device on a tour throughout Europe. In 1783, the chess-player embarked on a two-year journey across western Europe, capturing the imagination of regular spectators and intellectuals alike in many nations. It beat most opponents, though not François-André Danican Philodor, the greatest grand master of his time—in a match in Paris, where it also played against Benjamin Franklin, with an unknown outcome.[101] After Kempelen returned to Vienna in 1785, he dismantled the device again.

After his death in 1804 Johann Nepomuk Maelzel, a Bavarian inventor, bought the chess-player from Kempelen's son and put it back into operation. By 1809, it was ready to be demonstrated for the pleasure of the conquering Napoleon Bonaparte when he occupied Schönbrunn Palace after crushing the Austrian army at the battle of Wagram. The encounter with Napoleon apparently did take place (the emperor kept making illegal moves that made the Turk sweep all the pieces off the board, much to his amusement),[102] but legends concerning its encounter with other royalty, including Frederick the Great, Catherine the Great, George III, and Louis XV, are all spurious.

After the Napoleonic Wars, Maelzel took the chess-player and his own mechanical works on another tour starting in 1818, at one point adding a speaking device to the Turk allowing the figure to say "check" or "échec," depending on the country it was in. In 1826, he travelled with the device to the United States, starting in New York and moving on to Boston, Philadelphia, and Washington, D.C., playing against some of the best American masters and igniting the imagination of people in that nation as well. When the Turk came to Richmond, Virginia, in 1835, Edgar Allen Poe witnessed a performance, inspiring him to write two essays: "Maelzel's Chess-Player" (1836) and "Von Kempelen and his Discovery" (1849). After a few more years of touring across America, Maelzel was returning from an unsuccessful trip to Cuba when he died aboard the ship in 1838.

After some performances in Philadelphia under new ownership, the device was destroyed in a fire in 1854.

The chess-player was no doubt a remarkable simulacrum of a thinking automaton that may have fooled some people, but a careful reading of the numerous writings on the device reveals that with the exception a few gullible commentators, none were actually fooled into thinking that it was actually playing the game. The vast majority of contemporaries, even those who praised it effusively, explicitly mention that it was an ingenious piece of trickery, inspiring many to try to solve its mystery. Louis Dutens, in his 1771 letter to *Gentleman's Magazine*, calls the device an automaton and praises its skill in playing but is also certain that "the contriver influences the direction of almost every stroke played by the Automaton," though he is unable to account for how exactly this is done.[103] Similarly, Friedrich von Grimm was impressed by the device when he saw it during its first tour in Paris but asserted that a machine "would not know how to execute so many different movements, which could not be determined in advance, unless it was under the continual control of an intelligent being."[104] In the nineteenth century, Charles Babbage, the mathematician and pioneering designer of the Difference and the Analytic Engines, had an opportunity to play against it in 1820 and seriously considered the possibility of constructing a real game-playing machine.[105] But this is not to say that he thought Kempelen had succeeded in creating such a thing. In his description of his two encounters with the device he was apparently trying to figure out how the trickery was done, noting that a small boy could fit into the cabinet and that there was a trap door just behind the Turk figure.[106]

It would be unfair to call the device a fraudulent one since Kempelen himself never claimed to have built an actual chess-playing machine. In preparation for the first tour, the baron's friend Windisch published a pamphlet, *Brife über den Schachspieler des von Kempelen* (Letters on the chess-player of von Kempelen), in 1783 that was quickly translated into French and English as an advertisement, in the latter case under the highly suggestive title *Inanimate Reason*. In this pamphlet, which describes the workings of the device without revealing its inner secrets, Windisch affirms that

> it can only be a *deception*, in this the inventor, and every other reasonable being will readily agree with you. But in what this deception con-

sists, is a gordian knot more difficult to untie, than that in the days of yore, which Alexander cut asunder.

'Tis a *deception!* Granted; but such an one as does honor to human nature; a deception more beautiful, more surprising, more astonishing, than any to be met with, in the different account of mathematical recreations.[107]

In response to those who thought that a master chess-player must be manipulating the Turk figure using magnets, Kempelen invited people to bring to the performance a loadstone or any other device that would disrupt such an operation. Others thought a human controller was hidden inside the cabinet, like a dwarf or a small boy. As it was unlikely that Kempelen had found such a diminutive person who also happened to be a chess prodigy, there had to be a way a master player was communicating with the person inside to direct the movements of the Turk. The correct solution was arrived at by a doctor named Robert Willis, who published a piece entitled *An Attempt to Analyse the Automaton Chess Player of Mr. de Kempelen* in 1821 in which he related how at a performance he had managed to stealthily measure the dimensions of the cabinet using his umbrella.[108] He showed that the object was designed to appear much smaller than it really was and that, in actuality, it had enough room to fit an adult. Considering the fact that numerous expert chess-players were hired as hidden operators by Kempelen, Maelzel, and others during countless performances from 1770 to 1854, it is rather astonishing that none of them let the cat out of the bag until quite late in the day. It was only in 1834 that one of the players, Jacques-François Mouret, revealed its secrets in an anonymous article in the French journal *Pittoresque*.[109]

Long after its demise, the beguiling device continued to captivate the imagination of both Europeans and Americans throughout the modern period. The American writer Ambrose Bierce wrote a horror story entitled "Moxon's Master" (1893) in which an inventor builds a chess-playing machine. But when he is able to beat it in a match, it reaches over and chokes him to death—a significant episode in the history of automaton literature as possibly the first time a machine *intentionally* murders a human being.[110] Walter Benjamin famously made conceptual use of the chess-player in his essay "Theses on the Philosophy of History" (written in 1940, published posthumously in 1950) in which he described the relationship between

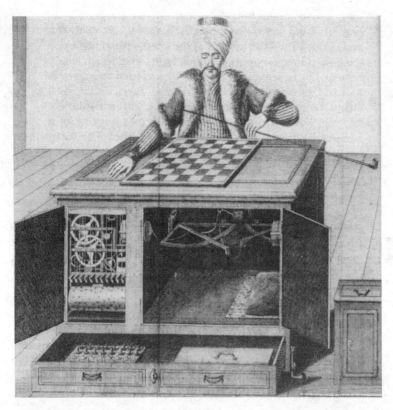

Wolfgang von Kempelen, the Turk chess-player exposed. Illustration from Karl Gottlieb von Windisch, *Briefe über den Schachspieler des Herrn von Kempelen*, reproduced in Charles Michael Carroll, *The Great Chess Automaton* (New York: Dover 1975), pg. 6, fig.2.

modern historical materialism and traditional theology by comparing the former to the "puppet" (i.e., the automaton figure of the Turk) and the latter to the controller inside (Benjamin assumes that it was a dwarf).[111] Recently, the Slovenian philosopher Slavoj Žižek, in his Lacanian analysis of contemporary religious phenomena cleverly revised this formula in his book *The Puppet and the Dwarf* (2003), asserting that it is historical materialism that is the dwarf that is controlling the puppet of theology.[112] In contemporary literature, Arturo Pérez-Reverte mentions the chess-player to good effect in his mystery novel *The Flanders Panel* (1990), and the German writer Robert Löhr has written *The Chess Machine* (2007), a clever and entertaining narrative of the device from the point of view of an Italian dwarf who is recruited to control the chess-player.[113]

Vaucanson and Kempelen were the two figures who marked the beginning and the end of the automaton craze of the Enlightenment, as their works achieved the greatest popular and intellectual success in their respective periods. What is crucial to understand is that their inventions embodied very different ideas and attractions for their viewers. And it is that difference that provides essential insight into the nature of the rise, the decline, and then the transformation of the automaton idea in the course of the eighteenth and early nineteenth centuries.

The success of the Vaucanson automata was the culmination of the already century-old obsession with the idea of a self-moving machine that began with Descartes. The defecating duck and the two music players appeared as the embodiment of the entire worldview of the mechanistic age, when the language of machinery dominated the intellectual discourse of the West. Even the act of opening up the flute-player to show the gears and wheels within was a quintessential gesture of rational revelation that marked the period's scientific enterprise. In the decades following Vaucanson's initial success, however, confidence in the mechanical worldview was undermined by the deteriorating political situation and the challenge by radical thinkers of the vitalist persuasion, resulting in the automaton coming to signify something that was ridiculous, pitiful, and tyrannical.

The intellectual significance of Kempelen's chess-player was of a completely different order, one that was emblematic of late Enlightenment culture. While Vaucanson went out of his way in his *Mémoire descriptif* to claim that in constructing his duck he had replicated natural physiology as far as possible, Kempelen never claimed that he had created an actual chess-player. In other words, just as vitalist doctors were denying that natural bodies operated like machines or more moderately asserted that they were like machines but only to a certain extent, Kempelen presented what appeared to be an automaton that did things that most people agreed a clockwork machine could not possibly do. So the chess-player was both a vitalist illustration of the *impossibility* of man as machine and a challenge to the viewer to figure out how exactly human guidance and reason invisibly controlled its mechanism. This is analogous to the effort by physiologists to elucidate how exactly vital force animated and brought about the growth and transformation of living bodies. There is an ironic element, then, in Kempelen's aping of Vaucanson in opening the panels of the chess-player's cabinet to reveal the machinery within, which was there just for show, as was his pretense at winding up the device every once in a while as if its motive power depended on it. Despite its nature as an illusion in appearing to be what its creator never claimed it was (i.e., a true automaton and a chess player), it was a more honest construct than that of Vaucanson's duck, which perpetrated a real fraud in his claim that it had an actual digestive system when it did not. Since the vast majority of writings on the chess-player took up the challenge of trying to uncover its trickery, the fascination with the object signified not the obsession with and celebration of the mechanical, as in the reception of Vaucanson's works in the 1730s, but rather the exact opposite, namely the disillusionment with and skepticism toward the grand claims of mechanistic philosophy. The chess-player was not an automaton in the sense of a self-moving machine but an articulate puppet that did an excellent job of pretending to be an automaton. As radical political thinkers of the period also derided the state-machine notion that was related to the world-machine and body-machine ideas, one can interpret the French Revolution as an act of breaking open the false Leviathan-automaton of the monarchy and exposing the mere human players hiding within. So when Robespierre referred to the executed Louis XVI as a "crowned automaton," he was speaking not just about a single individual who had been beheaded but the entire absolutist-mechanistic order that the revolution had taken apart.

In the Middle Ages and the Renaissance, even as people were captivated by the automaton image, they felt uneasy in the face of its liminal nature (animate/inanimate?; natural/preternatural?; living/dead?). They articulated that uneasiness in the language of magic, through which they worried over whether the self-moving object was the work of legitimate or diabolical art. Once the mechanistic philosophers of the seventeenth century reasoned away its magical aura, they adopted the automaton as an essential concept in their intellectual works, completely assimilating it into their worldview. This is evidenced by the fact that in the century and a half from the end of the Thirty Years War to the ascendance of Napoleon, there was remarkably little expression of a sense of uneasiness and the uncanny in the face of the automaton. The predominant emotional reactions toward it were awe and wonder during the classical Enlightenment of 1637–1748 and contempt, ridicule, and pity in the late Enlightenment. While political radicals hated the power of the state-automaton ruled by absolutist automata, there is not much expression of horror toward the self-moving machine as something inherently disturbing. There is no doubt that the common people who flocked to see demonstrations of automata did so to indulge in the sense of wonder aroused by the strange and the marvelous. Yet they were also learning through these presentations that there was indeed nothing diabolical or supernatural about such machines, that inside were just so many gears, wheels, and springs. To borrow the language of Mary Douglas, the object that used to be a categorical anomaly was, in the course of the seventeenth century, reinterpreted, redefined, and then assimilated into the emerging mechanistic worldview.

Zakiya Hanafi in her book *The Monster in the Machine* asserts that in the course of the seventeenth century, as explanations of monsters (i.e., deformed or unusual humans, animals, and plants) lost their sacred dimension and became strictly secular, all the powerful emotions and ambiguities attached to them were transferred to the era's machines, including automata.[114] I have endeavored to demonstrate, on the contrary, that the automaton could be elevated and celebrated as the central emblem of the mechanistic worldview precisely because the mechanistic thinkers of the period stripped away every aspect of the object that gave it the aura of the preternatural and the monstrous, turning it into a representation of pure rationality. There was a revival of the view of the magical automaton but not until after the end of the Enlightenment. From the magical automaton of the Middle Ages and the Renaissance, the rational automaton of the mechanistic age, and the ridiculous and pitiful automaton of the late Enlightenment, a new image of the uncanny automaton emerged in the fevered imagination of the Romantics. It will be the purpose of the next chapter to explain this development in the historical, intellectual, and cultural contexts of postrevolutionary Europe.

5
The Uncanny Automaton, 1789–1833

The Return of the Magical Automaton

Christopher Clark, in his history of Prussia, recounts how in the summer of 1796 Berliners flocked to see a theatrical show by Karl Enslen, a Swabian entertainer.[1] The performance began with a series of automata—mechanical figures of musicians that played the flute, the glass-organ, and the trumpet (the latter could also speak), which was followed by floating animal figures and yet another automaton that enacted gymnastic movements. So far, this seems like a late instance of the automaton craze that began with Vaucanson, yet another demonstration of mechanical wonder. But in the final act of the event, the show took a different turn that highlights its distance from the Enlightenment entertainments of the past. The light in the theater was extinguished, and the sound of a thunderclap was heard before a ghostly apparition appeared. A small star on the stage enlarged itself until it produced the image of the deceased Frederick II, which also grew until it became life-sized.

A decade after the death of the monarch who had been dubbed "the Great," the reign of his nephew Frederick William II was deemed a failure, marked by repressive political and cultural policies at home, through the edicts of his minister Johann Christof Wöllner, and by military and diplomatic bungling abroad that left a precarious international situation after commencement of the French Revolutionary Wars. It seemed like the age of enlightened monarchy, lauded by Immanuel Kant in "What Is Enlightenment?" (1784), was over and had been replaced by an era of political extremes: the republican revolution in France degenerating into the Terror and conservative reaction in Prussia and elsewhere. It is no wonder, then, that in this illusionist show, as the image of Frederick II faded back into the star, the people begged so earnestly for the dead king to stay with them that he was returned twice before the performance was brought to a close.

Clark interprets this episode as one of anxiety and nostalgia in a time of uncertainty for the nation that was to face even greater trials and tribulations in the following decades.[2] For the purpose of this study, it also illustrates a dramatic shift in the automaton motif that occurred in the late eighteenth century—namely, the resurgence of its magical aura. The mechanistic philosophers of the Enlightenment used the automaton as a representation of the world-machine, the state-machine, and the body-machine, while dismissing the idea of mechanical devices embodying magical knowledge as the result of ignorance and superstition. It is striking, then, that at the end of the century the demonstration of automata was coupled with an act of the supernatural, something unthinkable in Vaucanson's time. And this renewed association of the object with the magical became a long-term trend that went beyond this period.

To give a significant example from a later time, Jean-Eugène Robert-Houdin (1805–1871) was the most famous European magician of the nineteenth century, often referred to as the father of modern magic.[3] In his memoir, he recounts that as a son of a clock-maker, he exhibited from an early age a talent for creating mechanical devices, including automata.[4] But his father wanted him to enter a more lucrative profession than his own and persuaded him to become a clerk at a notary's office. Robert-Houdin describes the monotonous work there as an "automaton's labor" unsuited to his talents—an interesting statement in itself, in that he desired to leave this automaton-like work so he could follow his true passion of making automata.[5] He eventually convinced his father to let him return to the watch-making shop, where he constructed more clever devices. In a moment he describes as the most important in his life, a mix-up at a local bookstore caused him to take home a book of magical tricks thinking it was a book on clock-making—sending him on the path to fame and fortune as a celebrated entertainer.[6] Throughout his career he continued to make mechanical figures and even incorporated them into his acts, utilizing them in his illusionist tricks.[7] Wolfgang von Kempelen had already set a precedent for combining mechanics with mechanical trickery, but Robert-Houdin went a step further in using his devices for overtly magical purposes. Later in the century, the Englishman John Nevil Maskelyne made similar use of his automata "Psycho" and "Zoe" in mentalist acts.[8] This does not mean, of course, that people like Enslen, Robert-Houdin, and Maskelyne were regarded by their audience as true incarnations of the medieval sorcerer or the Hermetic magus who wielded magical powers in the literal sense. As they appeared to the public as entertainers, they created illusions that titillated the people with intimations of the supernatural but with the implicit understanding that their performances were mere trickery.

The working automata that were produced following the automaton craze of the eighteenth century were of limited cultural significance, as most of them were regarded as objects of entertainment, with only a few major exceptions.[9] Even the inventors of those constructed for more serious purposes failed to receive the intellectual respect and fame Vaucanson garnered. So the most important development in the history of the automaton in this period is not in its appearance as an actual object but the conspicuous frequency with which it was featured in fictional works, especially in German Romantic literature.[10] In fantastic stories the self-moving machine no longer represented rational or mechanistic ideas but appeared as an uncanny entity of mysterious nature that unsettled and frightened people, sometimes to the point of madness. Stories involving automata belong to the period's fantastic genre, which positively teems with ghosts, vampires, reanimated corpses, and other creatures that cross the boundaries of the living and the dead, the animate and the inanimate, and the natural and the artificial.

This shift in significance raises certain essential questions that need to be pursued by means of a historical analysis. Given the fact that the automaton was a symbol of rational order in the classical Enlightenment, how did it become an uncanny thing in the Romantic imagination? Even during the late Enlightenment, when radical thinkers used the automaton-man idea in a negative fashion, the emotions it aroused were contempt, ridicule, and pity, not horror. What was the source of this fear in response to the mechanical that is apparent in the literature? And why was the uncanny automaton associated with the magical and the supernatural, in direct opposition to the naturalizing efforts of the Enlightenment? In the same vein, why was the automaton-maker so frequently depicted as a modern-day sorcerer, again in contrast to the previous view of the mechanic as a colleague of the philosophe? Why so in an age when actual magicians were understood to be entertainers offering tricks and illusions? And finally, given the strongly negative emotions aroused by the uncanny automaton, what fears and anxieties exactly of the larger culture were being expressed toward the object?

To answer these questions, I must begin witht a general overview of two subjects, in order to provide the background for the analysis of fictional works that feature the automaton: first, the intellectual development from the worldview of the Enlightenment to that of Romanticism, especially in the scientific ideas of *Naturphilosophie;* and second, the rise of the fantastic genre in European fiction and its significance in the cultural history of the period. I will then examine key works by writers like Jean Paul (Friedrich Richter), Georg Büchner, E. T. A. Hoffmann, and Ludwig Achim von Arnim.

From the World-machine to the World Soul

During the Enlightenment, Germany produced two important mechanistic philosophers in Gottfried Leibniz and Christian Wolff. In the last decades of the eighteenth century, however, German thought tended to follow the larger trend of the late Enlightenment in moving away from classical mechanicism in favor of more vitalist and organic views on the world, the body as well as the state. I have already shown how the greatest philosopher of the period, Immanuel Kant, criticized the mechanistic view of the state and of human nature. In his major work on natural philosophy *Metaphysical Foundations of Natural Science* (1786) he also rejected the notion of the world as consisting of dead matter that moves like parts of a machine, opting for a more dynamic model in which the very essence of matter consists of attractive and repulsive forces, the apparently solid and inert state of things being the result of their perfect balance.[11] In the medical field, the vitalist challenge to mechanicism was carried out most prominently in Germany by the famed physiologist and pioneering anthropologist Johann Friedrich Blumenbach. In his vision of the vital body in which not one but a number of distinct forces were at work to provide formation, motion, and sensation to the corporeal entity, he introduced the notion of the *Bildungstrieb* (the "formative drive" that guided the growth and the development of an organism).[12]

As thinkers of the late Enlightenment who were reacting against the metaphysical excesses of the previous era, both Kant and Blumenbach were acutely aware of the limitations not only of mechanistic ideas but of human thought in general. Kant's monumental achievement in philosophy is the elaboration of the full extent of reason as well as the boundaries beyond which it cannot go, setting up an impenetrable barrier between the knowable world of phenomena and the unknowable realm of noumena. Likewise, Blumenbach adopted what Peter Reill calls the strategy of epistemological modesty of late Enlightenment vitalist medicine in his refusal to foray into the metaphysical with speculations on the source and essence of the *Bildungstrieb*.[13]

The new Romantic worldview that emerged in the following period, that of *Naturphilosophie*, associated with figures like Johann Wolfgang von Goethe, Alexander von Humboldt, Friedrich Schelling, Hans Christian Oersted, and Lorenz Oken, owes a great deal to Enlightenment vitalist thought.[14] Kant's force theory of matter became one of the fundamental ideas of Romantic physical science, especially in the works of Schelling.[15] So did the notions of vital force and formative drive that came out of the works of Buffon, Herder, Kant, Blumenbach, and others, in which a non-mechanical force acts on living entities through time to create variations and improvements in an orderly but not predetermined manner.[16] This connection between late Enlightenment science and *Naturphilosophie* complicates the still persistent notion of Romanticism as an anti-Enlightenment, antirationalist, and antiscientific movement that drew its inspiration primarily from premodern Christian theology and Hermetic pantheism.[17] What the Romantics rejected was not rationality and the Enlightenment in their entirety but ideas associated with mechanistic philosophy of the classical Enlightenment, with its conception of reason in terms of the mathematical, the logical, and the quantifiable and its static models of the world, the state, and the body. In addition to such ideas as the vital force and formative drive, the Romantics also inherited from the vitalists an expanded notion of reason that included the use of imagination, intuition, and sensation.[18] This does not mean, however, that *Naturphilosophie* was simply a continuation of Enlightenment vitalism into the nineteenth century. The essential difference in worldviews that is at the heart of the Enlightenment-Romanticism split must be seen in the context of the tumultuous historical events of the time.

Just as the crises of the mid-eighteenth century significantly weakened the hold of the mechanistic position of the classical Enlightenment, the destabilization of the political situation in Europe at the end of the century caused many of the younger generation to lose faith in the predominant ideas of the day. The degeneration of the French Revolution into the Terror, its final demise at the hands of the Corsican tyrant who plunged the entire continent into a devastating war, and the stultifying conservative reaction after his fall all contributed to this disillusionment with the hopes and promises of the age of reason. The case was especially acute in Germany, where liberals and radicals failed to bring about a revolution of their own and conservatives and reactionaries proved equally inept at mounting a defense against the shocks coming from the West, with even the mighty Prussian army finding itself shattered at Jena and Auerstedt in 1806.[19]

Many intellectuals of the period were drawn to late Enlightenment vitalism for its critique of mechanistic ideas that they found ridiculous and contemptible in the era of ongoing crisis. But they were also dissatisfied with vitalism's epistemological modesty, its delineation of the limits of reason that precluded metaphysical certainty and a clear vision of an eventual goal at the end of the transformative process for both human history and natural development. The new worldview that emerged out of post-Kantian idealism in the early nineteenth century held out to these intellectuals the possibility of absolute knowledge of all things in motion and transformation, as well as certain hope for an ultimate resolution of all the contingencies of history in a time of anxiety and uncertainty.

The Romantic *Naturphilosophen* adopted the basic vitalist model of the world as animate, dynamic, and transformative, of matter undergoing a temporal process of change, variation, and improvement through the agency of a living force. But in direct reaction to the intellectual circumspection of the vitalists, the Romantics superimposed on their vision a grand idealist narrative that was both metaphysical and teleological, a comprehensive scheme of development that encompassed everything from biology and history to geology and cosmology. The central story they told was of a primordial unity that was broken at the birth of the cosmos, resulting in the splitting of all things into countless sets of opposing forces. In the manner of the dialectic, the antipathetic elements in every pair come into constant conflict with each other but ultimately resolve themselves into a new unity that, once again, comes into conflict with its opposite to eventually form another union. At the heart of the dynamic is the binary pair Spirit and Matter—contradictory in nature but endowed with an essential desire for reunion, a longing to a return to the original oneness. Spirit is trying to realize itself by taking on characteristics of matter and becoming an actual presence in the natural world, while Matter yearns to transcend itself from its earthly nature. This tumultuous interaction of opposing entities is the central source of power that moves all things in the world, from the revolution of the stars to the formation of mountain ranges, the transformative development of animals and vegetables, the rise and fall of states and systems of government, even the yearning unto death of a lover in the throes of passion. No matter how chaotic, violent, and maddening this process of opposition and resolution through unification may seem at any given moment in time, the narrative carries the promise of an eventual reunification of all, as Spirit and Matter will become one at the end of cosmic history. For the early nineteenth-century thinkers, this vision provided not only an explanation of why events in the world can be so violent and disorienting but also hope for an ultimate unity in which all things will eventually be healed and made whole. A century and a half after the mechanistic philosophers ousted animate Nature from the world picture and turned the cosmos into a machine made of dead matter, the *Naturphilosophen* attempted to reanimate it by using the vitalists' ideas of *Lebenskraft* (vital force) and *Bildungstrieb* in their idealist cosmology.

In the elaboration of this grand narrative, one of the central ideas is that of polarity, or the opposition of antithetical forces. This is another notion inherited from the late Enlightenment, especially in the chemical ideas of Buffon and Lavoisier.[20] As Goethe explains its function and importance to the development of all things in the world:

> Whatever appears in the world must divide if it is to appear at all. What has been divided seeks itself again, can return to itself and reunite. This

happens in a lower sense when it merely intermingles with its opposite, combines with it; here the phenomenon is nullified or at least neutralized. However, the union may occur in a higher sense if what has been divided is first intensified; then in the union of the intensified halves it will produce a third thing, something new, higher, unexpected.[21]

In the description of polarity's operation, the world in its entirety is often described in animistic terms as one colossal organism or metaphorically a great animal, featuring all of its biological functions.[22] In direct opposition to the world-machine of mechanistic philosophy, Schelling presented the notion of the world soul *(Weltseele)*, the representation of living nature that guides everything toward the final mergence of Spirit and Matter. As he elaborates on the entity in a 1799 poem—

> Don't know how I could be in terror of the world,
> Because I know it inside and out,
> It is just an inert and tame animal,
> Which threatens neither you nor me,
> Which must submit itself to laws,
> And lie quietly at my feet.
> Yet there is a gigantic spirit trapped within,
> But it is petrified with all its senses,
> It cannot get out of its tight shell,
> Nor break its iron prison,
> Although it often stirs its wings,
> Stretches itself forcibly and moves,
> In dead and living things
> Strives mightily after consciousness.[23]

He goes on to describe how the imprisoned animal's efforts to know and free itself is the cause of all actions in the world, its eventual attainment of consciousness allowing it to understand that "It is One force, One interplay and weaving, / One drive and impulsion to ever higher life."[24]

Other major *Naturphilosophen* utilized animistic imagery to illustrate the living nature of all things, even of seemingly inanimate objects and matter. Hans Christian Oersted, Danish physicist and pioneer of electromagnetic studies, in his 1824 dialogue "The Spiritual in the Material," points to the activities of inorganic matter over time:

> Subterranean forces are constantly striving to raise or sink the rock, which, when apparently in a condition of repose themselves, are by no means inactive. When they really effect any elevations or depressions, it takes place so slowly, that it would escape observation, if one century did not assist another. Amidst all these formations and transformations of the earth, the materials, out of which it is compounded, are also formed and transformed, for these materials are not distinct from the earth, but belong to it as much as bones, flesh, and blood, belong to the bodies of animals.[25]

So the deadness of inorganic matter is only apparent or temporary, as it is in the process of gradual transformation through the actions of the same vital force that is at work in living beings.

Among the Enlightenment vitalists there was some disagreement about the essential distinction between animate beings and inanimate matter. Montpellier doctors like Bordeu and Barthez asserted a clear difference between entities endowed with the vital force and therefore alive and those devoid of it and so dead.[26] A materialist like Diderot, on the other hand, saw the quality of "sensitivity" in all things, organic and inorganic.[27] In Romantic *Naturphilosophie*, the vital force of the world soul is everywhere, manifesting itself in differing degrees as necessitated by the universal process. In view of the interaction of Spirit and Matter, the boundary between the organic and the inorganic became blurred to a much greater degree than in the worldview of the Enlightenment vitalists. According to Romantic cosmology, the final union of all things will result in the complete erasure of the distinction between the spirit and the matter of the world. But this leads to a perplexing tension within Romantic culture as a whole, specifically in the relationship between scientific-philosophical ideas and literary imagination.

The cosmic narrative of *Naturphilosophie* both explained the chaos of the world and provided hope in a time of instability and anxiety. Given this idea, we are faced with a confounding cultural situation when we read the fantastic literature of the period, by figures like Ludwig Tieck, E. T. A. Hoffmann, Clemens Brentano, and Ludwig Achim von Arnim, who were compatriots of the major *Naturphilosophen* and frequented the salons organized by the brothers August Wilhelm and Friedrich Schlegel in Jena and Berlin, the centers of the early Romantic movement in Germany. In stark contrast to the natural philosophers' effort to establish certainty and absolute knowledge, their literary friends wrote imaginative stories full of anxiety and uncanny horror, often in the face of supernatural forces and entities. In fact, there was a proliferation of such fantastic works all over western Europe featuring ghosts, lunatics, and religious fanatics. What is the meaning of this discrepancy between the philosophical and scientific search for certainty and the literary expressions of uncertainty? How do these two seemingly contradictory aspects of Romantic culture go together? Before we can look at individual works for answers to those questions, we must first consider the genre of the fantastic as a whole and its place in late eighteenth- and early nineteenth-century literature.

The World of Fantastic Literature

Various literary scholars have examined the rise of fantastic literature in the eighteenth century, starting with Horace Walpole's gothic novel *The Castle of Otranto* (1764), and the proliferation of works dealing with the supernatural, the horrific, and the uncanny in the 1780s and 1790s, including those by Ann Radcliffe, William Beckford, Matthew Lewis, and William Godwin.[28] In terms of their historical context, the popularity of such works has been linked to the perception of the crisis of reason in the larger culture. As I have shown, many mechanistic thinkers of the scientific revolution denigrated ideas, beliefs, and symbols that were associated with the magical, the miraculous, and the superstitious, denying them legitimacy as subjects of scholarly study. In the mid-eighteenth century, however, when the political situation in Europe destabilized, there was a concomitant disillusionment with the very ideas and institutions that sought to quash all that was associated with unreason. As a result a kind of return of the repressed in literature occurred, as writers of fantastic fiction produced stories involving sorcerers, ghosts, and lunatics with popular success. This does not mean, of course, that because there was a general skepticism toward the promises of the Enlightenment people returned to a premodern mindset to believe literally in the reality of otherworldly entities. The uncanny creatures and their eerie effects were, rather, powerful devices with which to express and meditate on the fears aroused by the chaotic state of the world, of things going out of control and posing terrifying dangers to people and their sanity. For example, scholars of Mary Shelley's 1818 novel *Frankenstein, Or the Modern Prometheus* have noted the connection between the story of the reanimated corpse and the French Revolution, an entity that is created as a product of reason and science but turns into the murderous monstrosity of the Terror.[29] Many stories that featured unnatural or supernatural beings and forces also employed uncertainty as a narrative strategy, making the reader wonder what is "truly" going on, whether there is a mundane explanation for the strange events or if there is something otherworldly afoot. This strategy is well explained in Tzvetan Todorov's explication of the genre of the fantastic.

Todorov describes three types of nonrealist literature: "fantastic," "marvelous," and "strange."[30] A story of the "marvelous" is one in which, due to the convention of the narrative or explicit signals within it, the reader understands that the occurrence of the anomalous is supposed to be taken as literally supernatural—when reading a horror novel by Stephen King, for instance, one knows from the start that a vampire or a ghost is to be accepted as an actual being within the narrative. In a story of the "strange," in direct contrast, one understands that a natural explanation is forthcoming no matter how weird and eerie an event appears to be. In Arthur Conan Doyle's *Hound of the Baskervilles*, a supernatural monster seems to be stalking humans in the countryside, but since it is a story about the detective Sherlock Holmes, one knows from the outset that an ordinary solution will be presented in the end. The purpose of the category of the "fantastic" is to deliberately make it unclear whether an anomalous occurrence or entity is indeed supernatural or natural, delaying the revelation as long as possible, sometimes indefinitely. Much of the fantastic literature of the early nineteenth century employs this narrative strategy, including stories involving automata.

The *Naturphilosophen* blurred the distinction between the natures of living beings and inert objects, seeing in the latter the latent presence of the world soul that was striving to realize itself in matter. This was part of the ultimately hopeful narrative of the cosmic dance of Spirit and Matter but also suggested the possibility of disturbing scenarios in which dead things come alive, wrecking havoc on the reader's conventional view of reality, which naturally divides things into the categories animate and inanimate. Ironically, then, the worldview of *Naturphilosophie* inspired dark fantasies of *unnatural* events and entities, of corpses becoming reanimated, objects turning into animals, and automata taking on characteristics of living beings. As a result, the mechanical object became uncanny, that is, a thing that causes psychological insecurity through its transcategorical nature. Premodern people wondered at the automaton by questioning whether the thing was made through "artifice or necromancy," but writers of the Romantic period, for whom the medieval notion of the preternatural was no longer available, expressed their unease in more uncertain terms, prompting their readers to wonder whether an automaton was just a mechanical device or some unknown force was at work. For this reason the automaton also appeared in performances of magical and supernatural trickery, as part of the sublime and cathartic entertainment described by Edmund Burke. A century and a half after mechanistic philosophers turned the automaton into a wholly mechanical thing representing the rational order of the world, the Romantics put it back among the wonders of the marvelous, sometimes in the debased context of public spectacle.

In William Wordsworth's autobiographical epic *The Prelude* (1805), he recounts a visit to St. Bartholomew's fair, an annual London event held at Smithfield, where he saw displays of automata, including "The bust that speaks and moves its goggling eyes, / The waxwork, clockwork, all the marvelous craft / Of modern Merlins." But these things are spoken of in the same breath as the "albinos, painted Indians, dwarfs, / the horse of knowledge, and the learned pig, / The stone-eater, the man that swallows fire, / Giants, ventriloquists, the invisible girl." And far from the objects suggesting to him the image of an ordered world, he sees them in a low, uncanny light as things that point to human hubris and irrationality.

> All out-o'-th'-way, far-fetched, perverted things,
> All freaks of Nature, all Promethean thoughts
> Of man—his dulness, madness, and their feats,
> All jumbled up together to make up
> This parliament of monsters.[31]

While Romantic thinkers sought comfort in the face of their anxieties over the state of the world around them in their idealist cosmology, fiction writers gave expression to those very anxieties in stories that recounted the disturbing consequences of the reanimation of the world picture—as if it had engendered a parliament of monsters. Since the same historical concerns informed their scientific-philosophical ideas and their fantastic stories, these seemingly contradictory aspects of the culture actually formed a cultural whole (in the manner of a polarity), like Janus faces that express hope on the one side and horror on the other.

A representative case that embodies many of the themes discussed above can be found in E. T. A. Hoffmann's novel *The Life and Opinions of the Tomcat Murr* (1820) in the figure of Master Abraham Liscov. A confidant and advisor of Iraenaeus, a prince, Liscov is supposed by many people to be a sorcerer who is "obviously and ominously in league with many strange, uncanny powers."[32] Yet throughout the story, it is deliberately made unclear whether this alternatively mysterious and comic figure is a real sorcerer wielding actual magic or just a clever trickster whose reputation comes from his proficiency at creating illusions and mechanical objects, including marvelous automata:

> the Master went running about the room with the speed and liveliness of a young man, wound up the machinery and arranged the magic mirrors. In every corner devices came to life: automatons strutted about, while screeching parrots flew circling around them....
> "Let these devices here, alive yet dead, go through their tricks."[33]

His story begins with an account of a disastrous display he has prepared for the prince on the occasion of the festivities for his wife's name day. He has arranged for a chubby manikin representing Genius to hover in the air by a system of ropes and float before the royal family as they proceed to their castle, illuminating their way with a pair of torches. Unfortunately, the figure turns out to be too heavy to move and ends up spilling drops of hot wax on the luminaries gathered for the event. In several such episodes, what are initially perceived as examples of his "strange, uncanny powers" turn out to be mechanical tricks, revealing him to be a figure like Enslen or Robert-Houdin. At one point in the story he is confronted by his irate friend, the court composer Johannes Kreisler, who is skeptical of his magical reputation from the start. On the way to visit Liscov, Kreisler is terrorized by strange sounds in the air and what appears to be a horrific vision of Kreisler's doppelgänger, only to be told by him that a wind-operated harp and a trick mirror were responsible for his fright. Kreisler berates him and his false magic:

really, I can't understand your strange liking for such silly tricks. Like a skilled cook, you concoct wonders from all kinds of spicy ingredients, and you think that people whose fancy has become dulled, like the glutton's palate, must be stimulated by such practices. Nothing is more absurd than to discover that those accursed tricks which constrict the breast have all happened naturally.

Master Abraham's response to this is significant, since it points to both Romantic *Naturphilosophie* and its connection to mechanical wonders with the revived magical aura.

Naturally! Naturally! . . . As a man of some sense, you should see that nothing in the world happens naturally, nothing at all! Do you think, my dear Kapellmeister, that because we can produce a certain effect with the means at our disposal, we have a clear view of the cause of that effect, which proceeds from the mysterious organism? . . .

The most ordinary and easily calculable mechanisms are often connected to the most mysterious wonders of nature, and can then produce effects which must remain inexplicable.[34]

Since the supernatural (spirit) is at work in the natural (matter), one who creates wonders through the manipulation of worldly forces can justifiably be regarded as a wielder of supernatural powers. In that sense he is not perpetrating a fraud by keeping the mechanics of his tricks hidden and mysterious, as they were made possible by his knowledge of the world's forces in the first place. In other works from this period that deal with similar themes, this blurring of the distinction between the natural and the supernatural, the mechanical and the magical, the animate and the inanimate is presented in a less comic and more fearful fashion.

Before I look at the fantastic works that feature automata, I would like to correct from the outset a persistent misreading of such works from this period. In the discussion of fantastic literature that feature machines and other artificial entities that take on animate characteristics, some scholars have lumped stories by Hoffmann, Mary Shelley, and others with works from later in the century and analyzed them in terms of people's anxieties about the Industrial Revolution. Turning the machine into a monster is interpreted as a way of depicting the negative consequences of industrialization, a device with which to interrogate modernity itself. Peter Gendolla notes that in such works produced between 1790 and 1820, the automaton poses a danger to humans not physically as bodily threats but conceptually as representations of people who act like machines.[35] He attributes this to what he calls "technological development of human nature" that occurred at the turn of the century, with the automatization of human workers in the modern factory.[36] The problem with this view is that it not only ignores the proliferation of mechanistic descriptions of humanity in the Enlightenment but wrongly associates these late eighteenth- and early nineteenth-century automaton stories with modern industry. The technological revolution commenced in Britain in the second half of the eighteenth century, prompting an early critic like William Blake to condemn modern factories as "dark Satanic Mills" in his 1804–1808 *Milton*.[37] But on the Continent, large-scale industrialization did not begin until decades later, and few literary figures there produced extensive critiques of it until the 1830s.[38] Furthermore, none of the automata in the literature of the period was based on the new steam technology. They were mainly the type of clockwork device of the previous era, some with explicit references to the works of Vaucanson, Jaquet-Droz, and Kempelen. And none of their creators were portrayed as industrialists seeking to make automatic laborers, but rather as sorcerers and entertainers wielding seemingly supernatural powers. Even in the case of the nonmechanical reanimated corpse in Shelley's *Frankenstein*, Chris Baldick has pointed out that interpretations of the work as a critique of industrial technology and the monster as a representation of the working class (i.e., a neglected and exploited mass rising up to wreak vengeance on the masters) are informed more by later versions of the story, from numerous dramatizations for the stage and film.[39] The works of this period must be analyzed in the context of the events and ideas specific to the era, namely the political instability of the period and the development of Romantic natural philosophy and literature. Later on, anxieties over industrialization not only played a role in the transformation of the automaton idea but increasingly became the central issue in its literary representation.

In the automaton stories of the late eighteenth and early nineteenth centuries, the object appears in the two modes of the satirical and the uncanny. As shown in the last chapter, many late Enlightenment writers used the automaton-man as a figure of pity, ridicule, and contempt. The German Romantics inherited this image and utilized it in fiction in which the mechanical being was regarded as an image of debased humanity. The uncanny automaton, on the other hand, was an entity with which they expressed their anxiety over the loss of faith in the principles of the Enlightenment and explored the more disturbing consequences of *Naturphilosophie*.

The Satirical Automaton: Jean Paul, Friedrich Schlegel, and Georg Büchner

Jean Paul (1763–1825, born Johann Paul Friedrich Richter; adopted pen name in admiration of Jean-Jacques Rousseau) is obscure in the Anglophone world but occupies an important place in the history of German literature both as an idiosyncratic figure of the first generation of Romantic writers and as the premier satirist among them.[40] In the short works he wrote in the late eighteenth century, the automaton-man image appears a number of times. As a literary object it functions in line with the late Enlightenment theme of debased or manipulated people, which Jean Paul enlivens through the ironic and comic tone of his tales.

In an early fragmentary work entitled "Menschen sind Maschinen der Engel" (Men are machines of angels, 1785), he ridicules the notion of the primacy of humanity in the universal order, in the manner of Swift and Voltaire, by asserting that we are nothing but so many objects built for the use and amusement of superior creatures.

> If it is necessary to be very enlightened to give up the proud conviction that the entire world exists only for us and that the stars are in fact nothing but brass buttons shining on the surface of the chariot or world that carries us, then our heads demand even more enlightenment to convince us that we live here merely for the purposes of higher beings we call angels, and they are the true inhabitants of this earth, and we are simply their household utensils.[41]

The specific utensil he has in mind is the machine.

> When the angels first set foot on this earth, it did not yet have the numberless human machines that they would wish; over time they invented first this machine and then that machine or men, as we like to

say, until their number grew so great that there are now the most splendid machines (or people) for all purposes.⁴²

Jean Paul elaborates on this idea by pointing to Kempelen's chess-player, but then, in words of caution to the reader about taking this metaphor too literally, he asserts that he is not reverting to mechanistic physiology.

> It is overlooked that there is a great difference between the [machines made by men and machines made by angels], and the angels' work is vastly superior to that of men. The angels' work consists of flesh and blood—no chemist can formulate blood—and the creation of men consists of mere wood and some metal.⁴³

As was the automaton-man of the late Enlightenment, the image is deployed to deflate humanity's self-importance in the world, but there is also a denial of a literal identification of man with machine in biological terms.⁴⁴

In his 1789 collection *Auswahl aus des Teufels Papieren* (Selections from the Devil's papers), Jean Paul included a particularly biting satire, "Unterthängiste Vorstellung unser, der sämtlicher Spieler und redenden Damen in Europa entgegen und wider die Einführung der Kempelischen Spiel- und Sprechmaschinen" (Most humble remonstrance of all players and talking ladies in Europe against the introduction of von Kempelen's playing and talking automata) in which a protest against Kempelen's speaking and chess-playing automata begins with the concern that machines may take jobs away from people but concludes with the clear implication that the machines are such a threat because the vapid and conformist society ladies are such mechanical creatures themselves that they fear being replaced by them.⁴⁵

The most significant and interesting automaton story by Jean Paul can be found in the same collection, in a piece entitled "Machinenmann nebst seinen Eigenschaften" (The Machine-Man and his characteristics).⁴⁶ The narrator of the tale imagines his soul's transcendence to an idyllic world on Saturn, which he would make his home and whose inhabitants he would entertain with stories of Earth, especially the follies of his fellow humans. He describes to the Saturnians a particularly horrendous specimen of humanity he is glad to have left behind on his home world: the Machine-Man, a creature he once visited in his machine realm. The lord of the mechanical world is a man who is so enamored of machines that he surrounds himself with them, as they fulfill all of his necessary actions: a writing machine writes his letters for him, a calculating machine does his accounts, a chewing machine breaks up his meals, and a speaking machine utters his prayers in church. In the winter, the Machine-Man gives a concert of music composed of notes determined by a dice-throwing machine, written down by an inscription machine, and played by the automaton musicians of Vaucanson and Jaquet-Droz. The spectacle gives the narrator a horrifying vision of future humanity's utter dependence on machinery, which turns people into machines themselves, with artificial limbs, sense organs, and even stomachs from which waste material is excreted by hydraulic mechanism. They would also live in a world where all animals have been replaced by automata, made, again, by Vaucanson and other mechanics.

When Jean Paul rewrote the story as "Personalien vom Bedienten- und Maschinenmann" (Particulars of the Servant- and Machine-Man) for his 1798 collection *Palingenesien*, he made the conspicuous alteration of turning the Machine-Man into the Machine-King (of an island called Barataria). With this change, he turned the story into a satire of absolute monarchy, utilizing the late Enlightenment theme of portraying tyrants and autocracy in mechanical terms.[47] The Machine-King is served by myriad machines, but also by an army of human servants who are described as kinds of machines as well, since they have been reduced to mere devices. The narrator points out, however, that the Machine-King is as dependent on his servants as they are on him, and in that awful state of mechanical rule, no one is free, not even the master of both human and real machines.

> Now he lets his people set him up like a ninepin, in order to promenade with me.... All of his porters came running at once... apparatus carriers, hat carriers, canned food carriers, eyeglass carriers, and an instructor with a book under his arm.... He is condemned to the porters; who else carries his necktie, his boots, stockings, summer clothes, pantaloon and everything else? And who then gives him a machine, a clockwork, that would set him in motion?[48]

The narrator then meditates on the fact that human progress is often measured by the extent to which people use machinery, that the more mechanical people are, the closer they are to perfection. In such a view, animals would be at the lowest state, peasants at a higher, artisans yet higher, and the rich and the powerful, with all the machines and devices at their disposal, at the highest. Yet what kind of a world do the latter occupy—the apathetic and suffocating society of rich pensioners, vapid court ladies, and know-it-alls leading empty lives.[49] Isn't that indeed the most depraved existence on Earth? The obvious implication here is that the opposite is true. The more mechanical people become, the further removed they are from natural perfection.

At the end of the earlier version of the story, the Saturnians ask the narrator about the identity of the Machine-Man.

"Which, then, was the living century of your Machine-Man?"
"The eighteenth!" I say.
"But what is he actually called?" they say.
"Just that, namely the eighteenth century, or the genius of the eighteenth century."[50]

To the Romantic writer working at a time when traditional monarchy came into serious crisis, there was no better representation of the Enlightenment's rational man and the absolute monarch of the classical age than the Machine-Man and the Machine-King who fashion a machine-world around them, depriving all, including themselves, of vital freedom.[51]

A brief but significant instance of the contrast of the automaton-man with the vital-man can be found in the 1799 novel *Lucinde*, by Friedrich Schlegel, who was the chief organizer of the Romantic movement in Germany along with his brother Augustus Wilhem. A roman à clef based on Schlegel's love affair with his future wife, Dorothea Veit (daughter of the Jewish philosopher Moses Mendelssohn), it was decried at the time as an indecent work for its defense of free love.[52] In one of its many philosophical interludes, the character of Julius (Schlegel's alter ego) dreams up an "Idyll of Idleness" in which he praises a kind of passive, unconscious state of being that allows one to be inspired by the forces of nature without the obstructions of rational thought and material compulsion. He goes so far as to claim that "the more divine a man or a work of man is, the more it resembles a plant; of all the forms of nature, this form is the most moral and the most beautiful. And so the highest, and the most perfect mode of life would actually be nothing more than *pure vegetating*."[53] To elaborate, Julius narrates a fantastic scene in a theater in which the audience is watching two central figures representing the contrasting principles of mechanistic rationality and vitality. One is the creator Titan Prometheus, described as "the inventor of education and enlightenment," who is bound up in chains and tormented by "several monstrous creatures who were continually driving and whipping him" while he creates human beings through a mechanical process.[54] As soon as he completes one, he throws the person out of his workshop into the audience, "where he immediately became indistinguishable from the others; they were all so much alike."[55] The other figure is that of Hercules, who has Hebe, the goddess of youth, on his lap and is surrounded by cupids and satyrs who are contrasted to the mechanically made people in the audience in that "every one of them had his own peculiar manner, a striking originality of facial expression."[56] Hercules is also a prolific creator of living beings, but through natural procreation, as he can "keep fifty girls busy during a night for the good of humanity, and heroic girls to boot."[57]

In the description of the mechanically constructed audience, the artificial represents the conformist, the unoriginal, and the uniform, deployed for the purpose of social critique, something already familiar from the automaton-man writings of the late Enlightenment. The device was used again to comic effect in the 1838 play *Leonce and Lena*, by the short-lived but significant dramatist of the era Georg Büchner. Its contrived plot is borrowed from traditional Italian comedies in which a series of misunderstandings are ultimately resolved through fortuitous coincidences. Prince Leonce of the Kingdom of Popo is due to marry Princess Lena of the Kingdom of Pipi through an arrangement between their royal parents, although they have never even met. Since neither of them wants to enter into such a union, they both run away but then meet by chance and fall in love without knowing each other's identity. When Leonce's fool, Valerio, realizes what has happened, he arranges for his master's return home with his beloved through a clever spectacle. To the king of Popo, grieving the loss of his son, the fool presents himself as an impresario who has come to display two marvels for the pleasure of the monarch: automata that are actually Leonce and Lena in disguise. The speech he gives in presenting these wonderworks begins with a description of them.

> Here, ladies and gentlemen, you see two individuals of either sex, a man and a woman, a gentleman and a lady! They are nothing more than artifices and mechanical ingenuity, pasteboard and watch springs! Each is equipped with delicate, delicate ruby spring under the nail of the small toe of the right foot. Press this lever ever so gently and the mechanism will be set for fully fifty years. These individuals are so consummately constructed that they cannot be distinguished from other human beings, unless one knows that they are merely pasteboard; they might even be accepted as members of human society.[58]

At this point, the presentation takes a rather unexpected turn, as Valerio uses the very idea of the automaton to deliver what is obviously a satire of the conformist society to which his audience belongs. The mechanical objects are identified with the members of the upper class, whose lives are dictated by rules, routine, and fashion, even in matters of love and courtship. He goes on to describe the qualities of his automata:

> They are very well born, because they speak High German; they are very moral, because they rise punctually at the stroke of a bell, because they eat punctually at the stroke of midday, and because they retire punctually at the stroke of another bell; and then, too, they have a good digestion, which attests to a clear conscience. They possess a delicately ethical sense of feelings, because the lady never had leave to speak of the concept of women's drawers, and it is utterly impossible for the gentleman to precede the lady, much as a single step. They are very well educated, because the lady sings all the new operas and the gentleman wears cuffs. Pay attention now, ladies and gentlemen, for they are now in an interesting stage of their development: the mechanism of love is beginning to set itself in motion. The gentleman had already given the lady her scarf a number of times, and several times now the lady has rolled her eyes and turned them toward heaven. Both of them have whispered together a number of times: about faith, and love, and hope! They already seem very much in accord, all that's lacking is that paltry word: Amen.[59]

The king gets the idea to marry his lost son to Lena in effigy, with the two automata serving as their substitutes. It is only when the ceremony is done and the two are betrothed that Leonce reveals himself to his father and Valerio explains his ruse. The implication here is that these royalty and their loyal subjects leading their unoriginal lives can be replaced with machines with no real difference in their natures.

This satirical use of the automaton in ridiculing the ruling class also points to their tyranny, a topic of particular concern for Büchner, whose most famous work is his unfinished play *Woyzeck*, in which a humble, impoverished soldier is driven to madness and murder by the abuses of his superiors, who treat him with contempt because of his low class status. Büchner was a political radical and revolutionary agitator who had to flee his native Hesse after an arrest warrant for treason was issued following the publication of his 1834 pamphlet *The Hessian Courier*, which detailed political and economic abuses in the Grand Duchy. In the call for the overthrow of the status quo at the end of the work, he presents a provocative image of the people rising in revolutionary fervor.

> Germany is now a field of battle, soon it will be a paradise. The German people is a *single body*, you are a member of it. It is immaterial where this seeming corpse begins to move and show signs of life. When the Lord gives you His sign through the men through whom He will lead the people from servitude into freedom, then raise yourself up and the whole body will rise up with you.[60]

This organic view of the people and the idea of their uprising as that of a reanimated corpse is a highly significant one, since scholars have pointed to the same political dimension in Shelley's *Frankenstein*.[61] In the latter novel, however, the image is employed in a manner that is both cautionary and uncanny—taking us to other works of the era in which the automaton appears in a much more sinister and frightening light.

The Uncanny Automaton: E. T. A. Hoffmann and Ludwig Achim von Arnim

The writer who made use of the uncanny automaton to the greatest and the most imaginative extent was E. T. A. Hoffmann, a master of fantastic literature in the period. As an active participant in the German Romantic movement, especially during his last years in Berlin, he was well versed in the ideas of his philosophical and scientific compatriots, especially those of the doctor and naturalist Gotthilf Heinrich von Schubert, a disciple of Schelling. Schubert is an interesting figure in that while he adhered to the central worldview of Romantic *Naturphilosophie*, he explored the darker implications of its ideas, as is indicated by the title of his major book on the subject, *Ansichten von der Nachtseite der Naturwissenschaft* (Views of the night side of natural science [1808]).[62] In this work he elaborated on the role of degeneration and decay in the unfolding cosmic story, on the necessity of the material world becoming progressively degraded before it can be regenerated through its interaction with the spiritual. He also wrote on inorganic matter and the uncanny process by which it merges with the organic, but only after a period of significant decay.[63]

In his automaton stories, Hoffmann explored the psychological consequences of the breakdown of the distinction between the animate and the inanimate. For this purpose, the narrative device he used to great effect was that of uncertainty, centering on a series of questions that crop up repeatedly: whether a particular automaton is nothing but a mechanical construct or some unknown force is at work in it; whether the perception that there is indeed something otherworldly about an automaton is based on reality or a misunderstanding or, in extreme cases, the madness of the viewer; and whether the feeling of the uncanny that is aroused by the automaton emanates from the object itself or from the mind of the perceiver who finds its operation difficult to assimilate into a worldview based on the categories of ordinary reality. Hoffmann also employed the narrative strategy of the fantastic, as described by Todorov, delaying the revelation of the answers to those questions as long as possible, sometimes ending the story with only possible explanations rather than a definitive one to further beguile the reader. This method of alternating between the natural and the supernatural can be clearly seen in "The Automaton" (1814).

Hoffmann presents an automaton that is a combination of Kempelen's chess-player and his speaking machine—a "Talking Turk" that appears explicitly in the context of the magical as it functions as an oracle. Two young men, Ludwig and Ferdinand, attend a public performance of the device that is the talk of the town, though Ludwig is uneasy at the very idea of the device since such an object always gave him "a horrible, eerie, shuddery feeling . . . the oppressive sense of being in the presence of something unnatural and gruesome; and what I detest most of all is the mechanical imitation of human motions."[64] The less squeamish Ferdinand decides to try it out and takes his turn in asking a question.

It is he who ends up being seriously affected by the few words the automaton utters, which convince him that there is something more than mere mechanical trickery at work. The possibility of the supernatural enters as Ferdinand seriously considers that the machine is a medium through which an otherworldly being has communicated with him. To explain why he believes this, he tells Ludwig about a trip he took several years earlier, during which he heard a woman singing in the middle of the night and fell in love with her voice. The next day he managed to catch only a glimpse of her as she was leaving in a carriage, but he was so taken with her image that he drew a miniature portrait of her and has kept it with him at all times in a locket hanging next to his heart. The question he asked the automaton was, "Will there be a time for me again like that which was the happiest in my life?" to which it answered, "I am looking into your breast; but the glitter of the gold, which is toward me, distracts me. Turn the picture around." And then, "Unhappy man! At the very moment when you see her next, you will be lost to her forever!"[65]

Ludwig asserts that it has all been a clever illusion perpetrated through some contrivance of acoustics that allows someone to speak through the automaton, but he is willing to consider that whoever controls it may possess psychic powers that have allowed him to read Ferdinand's mind. That person may be Professor X———, the inventor of the automaton, whom they visit to inquire about the nature of his creation. The professor welcomes them into his house and puts on a show of three automata playing musical instruments, but Ludwig and Ferdinand find no answer to their questions. Later on, they discuss the necessary inferiority of a machine playing music to a human doing so since only the latter can tap into spiritual impulses that guides one's feelings through the sounds. Ludwig explicitly mentions the ideas of Gotthilf von Schubert in telling the story of how people were in tune with the natural harmonies of the world in the pure, primeval state of the human race but then degenerated in the course of historical development.

At the conclusion of the story, what appears so far to be a fairly simple moral tale of the superiority of the natural and spiritual to the artificial and the mechanical takes a very strange turn. The last conversation between

Ludwig and Ferdinand takes place just after they leave Professor X——'s workshop and walk to a garden outside the town (from the artificial environment to the natural). There they not only hear a woman singing from inside a house, which turns out to be the very voice Ferdinand fell in love with years ago but also see Professor X—— in the garden, having mysteriously been transported there. After the professor goes into the house and the music stops, the two friends find themselves disturbed by the encounter. (How did the professor get there so fast? Who is that woman singing in the house? What is the connection between her and the professor?) The story ends with one further twist. Ferdinand leaves town but then sends Ludwig a letter in which he reveals that the automaton's prophecy has come true. While visiting a small village, he came across a wedding in which the bride was the very singer he loves, and she was being given away by Professor X——. Ludwig finds this both puzzling and worrisome, since he knows for certain that the professor has not left town.

So the story ends not with a revelation that answers the questions presented by the narrative but with a series of beguiling mysteries that are never resolved. Is the talking automaton just a clever mechanical device or is it a conduit for a supernatural power? Is Professor X—— a mere mechanic or does he possess psychic abilities that allow him to read the minds of those who interact with his invention? Is Ferdinand's beloved, the mysterious singer, a real woman, a figment of his imagination, or perhaps even an automaton built by Professor X——? And can the whole thing be explained by the fact that Ferdinand may be a liar or a lunatic, all the stories of his encounter with the singing woman, the automaton's prophecy, and the wedding that fulfilled it either fabrications or delusions of his disturbed mind?

For the purpose of this study, the possibility that the singer may have been an automaton is an intriguing one since it threatens to invalidate Ludwig and Ferdinand's affirmation that artificial music can never rival human beings playing in accordance with natural and spiritual impulses. If the singer was indeed an artificial construct, like the other automata in the professor's workshop, then it was the singing of a machine that Ferdinand fell in love with and considered the highest expression of beauty and spirit. This points to an interesting paradox within *Naturphilosophie* itself. The Romantic attitude is one that privileges the natural, the organic, and the instinctual over the artificial, the mechanical, and the rational. But if one also believes that the world is undergoing a process in which Spirit and

Matter are merging and the distinction between the organic and the inorganic is gradually breaking down, one could imagine an artificial construct that is designed to facilitate the arousal of the latent spirit in its inorganic matter, becoming a device through which the supernatural manifests itself in the world. And so the artificial becomes the natural through the agency of the supernatural. Just as the speaking automaton may be a machine with which Professor X——— exercises his psychic powers, if Ferdinand's beloved is a singing automaton, it may also be performing its music through the actions of a higher power, just like the Aeolian harp (a musical instrument designed to be played by the wind), which Ludwig mentions and praises for its tempting of "Nature to give forth her tones."[66] The blurring of the distinction between the organic and the inorganic, then, not only arouses the feeling of the uncanny before objects and entities that seem to cross the boundary line between the two but also leads to this ironic situation in which the privileging of the natural over the artificial becomes problematic because the two are revealed to be not absolute opposites but the two sides of a polarity that is ultimately seeking unity.[67] In fact, a purely mechanical object that is under total control of a higher power may surpass humans in its performance.

This is exactly the point made in Heinrich von Kleist's celebrated dialogue "On the Marionette Theater," which was published four years before the Hoffmann story. In the exchange between the narrator and a celebrated dancer called Herr C———, the latter expresses his admiration for a performance of dancing puppets, implying that their movements are superior to those of humans.[68] He explains that the lack of artifice and self-consciousness on the part of the inanimate objects puts their actions better in tune with nature, transcending constraining forces of the material world like gravity. In fact, he believes that the natural beauty of a puppet's dance could even be enhanced by eliminating the role of the puppeteer altogether, replacing the human controller with a mechanical crank (i.e., the complete automatization of the puppet), and so making the object even more attuned to natural forces. This idea, found in both Hoffmann and Kleist, creates an interesting dilemma within the Romantic imagination. The Romantics inherited from the late Enlightenment the notion of the automaton-man as an image of debased humanity, but the automaton becomes a much more complicated entity within their worldview as something that is the very opposite of the natural but at the same time embodying the potential to come alive as the ultimate instrument of the supernatural. This is the source of the anxious uncertainty in response to the uncanny nature of the Romantic automaton.

Hoffmann's better known automaton story "The Sandman" (1816) was turned into musical form as one of the episodes in Offenbach's opera *Les Contes d'Hoffmann*. In the face of a series of bizarre and unsettling events in the narrative, various characters present explanations of the episodes that alternate, once again, the natural with the supernatural. As S. S. Prawer puts it, in this story "the wondrous, the transcendent, the demonic are playing a game of hide-and-seek—or, more accurately, of cat-and-mouse—with the characters."[69] A young man, Nathaniel, is tormented by a sense of impending doom that he attributes to a strange childhood memory. On certain nights, he and his siblings were sent to bed early because, their mother told them, "The Sandman is coming." In answer to the curious boy's question, his mother explained that it was a mere convention for describing the onset of sleepiness, but an old woman gave him an alternate explanation, a description of a wicked man who blinds and abducts recalcitrant children. Nathan wanted to see the Sandman, so he stayed awake and stole into his father's room, where he found him in the company of an old family friend named Coppelius, engaged in some mysterious activities. The boy became frightened and cried out, whereupon Coppelius expressed his desire to take his eyes. Nathaniel's father begged him not to do so, and the visitor agreed, but he then used the boy in a horrific and uncanny manner.

> Coppelius laughed shrilly and cried: "The boy can have his eyes then, and the use of them, But now let us observe the mechanism of the hands and feet."
> And with that he seized me so violently that my joints cracked, unscrewed my hands and feet, and fixed them on again now in this way, now in that.
> "They don't look right anywhere! Better where they were! The Old One knew what he was doing!" Coppelius lisped and hissed. But everything went dark around me, a sudden spasm shot through my frame—I felt nothing more.[70]

In this passage, which Freud interprets as the source of Nathaniel's castration complex, the boy's feeling of utter helplessness is expressed as his reduction to a machine, an automaton with detachable parts that can be easily manipulated.[71]

Clara, Nathaniel's levelheaded beloved, tries to assuage his disturbance with a natural explanation of the memory, that his father and Coppelius were probably engaged in alchemical experiments, and that his fear of the visitor and the trauma of getting caught while snooping combined to create this bizarre episode of dismemberment in his mind. What has triggered the whole thing in Nathaniel's memory is the appearance of a barometer dealer named Giuseppe Coppola, whom he suspects is Coppelius himself, though an acquaintance of his, an Italian scientist named Spalanzani, informs him that he knows Coppola and that he is Piedmontese, not German, as Coppelius was.

The central plot begins with Nathaniel attending Spalanzani's lectures and noticing his beautiful daughter Olympia. Despite the fact that she never speaks, he falls in love with her because of her perfect grace, posture, and, later, skill at singing and playing the piano. Nathaniel dances with her at a ball and finds her stiff in movement and ice-cold to the touch, but he is still moved by her eyes, which shine with love. When his feelings for her become apparent to his friends, one of them, Siegmund, is mystified.

> Do not take it ill, brother, but she has appeared to us in a strange way rigid and soulless. Her figure is well proportioned; so is her face—that is true! She might be called beautiful if her eyes were not so completely lifeless. I could even say sightless. She walks with a curiously measured gait; every movement seems as if controlled by clockwork. When she plays and sings it is with the unpleasant soulless regularity of machines, and she dances in the same way. We have come to find this Olympia quite uncanny . . . it seems to us that she is only acting like a living creature, and yet there is some reason for that which we cannot fathom.[72]

But Nathaniel contradicts him, claiming to see a pure soul in her. This discrepancy between his perception and that of other people points to the uncertainty of his mental stability and presages his ultimate breakdown.

He finally decides to propose to Olympia and pays an unannounced visit to Spalanzani's home, only to be faced with a horrifying sight. Spalanzani and Coppola/Coppelius are fighting over the figure of Olympia while shouting at each other: "Let go! . . . Is this what I staked my life on? . . . I made the eyes! . . . I made the clockwork!—Poor fool with your clockwork!—Damned dog of a clock-maker! . . . Away with you! . . . Puppet showman!"[73] Nathaniel finally sees that Siegmund's presentiment was correct and Olympia is nothing but an automaton. As Coppola takes the object and leaves the room, the sight of Olympia's face with black holes where her eyes should be drives Nathaniel insane: "Ha, ha ha! Circle of fire, circle of fire! Spin, spin, circle of fire! Merrily, merrily! Puppet, ha, lovely puppet, spin, spin!"[74] His cries attract some people who barely prevent him from killing Spalanzani. Later on he seems to recover from the breakdown, but he eventually relapses into it, first by trying to throw Clara off a tower and then by jumping off himself when he sees Coppola in the crowd below.

Spalanzani is obliged to leave the city to avoid criminal persecution for his deception, but in a brilliant comic ending to the story, Hoffmann describes the effect the whole affair has on the town's citizens.

> To be quite convinced that they were not in love with a wooden doll, many enamoured young men demanded that their young ladies sing and dance in a less than perfect manner, that while being read to they should knit, sew, play with their puppy and so on, but above all that they should not merely listen but sometimes speak too, and in such a way that what they said gave evidence of some real thinking and feeling behind it. Many love-bonds grew more firmly tied under this regime; others on the contrary gently dissolved.[75]

A natural explanation is provided for everything that has occurred: Spalanzani and Coppola have perpetrated a deception with their collaborative work; Olympia is nothing but an automaton; and Nathaniel's discovery of her true nature pushes his already disturbed mind over the edge. Yet the aura of the uncanny and the supernatural linger on at the end, with some unresolved questions on the exact identity of Coppola/Coppelius and with the fulfillment of Nathaniel's original presentiment of being somehow controlled by forces beyond his understanding.

> Everything, the whole of life, had become for him a dream and a feeling of foreboding; he spoke continuously of how each of us, thinking himself free, was in reality the tortured plaything of mysterious powers: resistance was vain; we had humbly to submit to the decrees of fate. He went so far as to assert that it was folly to think the creations of art and science the product of our own free will: the inspiration which alone made creation possible did not proceed from us but was effectuated by some higher force from outside.[76]

The fear Nathaniel carries with him from childhood is of becoming an automaton, an utterly helpless thing that is moved and used by others—a fear the entire town comes to share in light of the Olympia incident. This is also the sense of the uncanny Freud experienced when he realized himself to be a bungling automaton who nearly put his elderly patient at risk at the behest of some unknown force that controlled him. To the late Enlightenment view of the automaton-man as an image of humanity without freedom these Romantic stories added the horror of some mysterious supernatural power that might be at work in the blurring of the boundary between the animate and the inanimate, to the extent of inducing madness.

The sinister nature of people associated with automata, like Professor X——— and Coppola/Coppelius, is a prominent feature of these stories, as the characters are depicted as uncanny as their creations. This harkens back to the premodern image of the magician who dwells on the margins of society and possesses the power to bring life to the lifeless. In one case, such a character was based on a real-life person who drew the attention of the central literary figure of the era and provided the inspiration for the imagination of another writer. One of the last accounts of the fate of Vaucanson's flute-player and duck can be found in Goethe's journal from the year 1805, when he visited the house of a man named Gottfried Christoph Beireis in Helmstädt. Beireis was a prominent and popular professor of natural science at the university, a respected medical doctor, and something of a local celebrity, but Goethe describes him as "an eccentric, problematic man, already for many years notorious in many respects."[77] The notoriety came from his reputation as a modern-day sorcerer, originating from the story that as a young and impoverished medical student at the university in Jena, with a rather eccentric interest in alchemy, he had disappeared for three years and then returned to the city as a fabulously wealthy man.[78] Even after he settled into his career as an academic and doctor at Helmstädt, he neither confirmed nor denied rumors that he had traveled all the way to Egypt during that time and had discovered the ancient art of transmuting base metals into gold. And he deliberately enhanced his magical reputation by playing such tricks as showing up at a dinner party dressed in a red coat that turned black and then fell to pieces before disappearing altogether, and turning the wine in a dignitary's glass into vinegar.[79] What drew Goethe to him was his great collection of wonders in the Renaissance mode, which included works of fine art (by Titian, Raphael, Corregio, Rubens, and Dürer), rare minerals, stuffed animals, and mechanical devices, like the Vaucanson automata, which were in a deplorable condition by then. When Goethe describes his personality as that of "an earlier transitory epoch," he was referring not just to his age (Beireis was nearly twenty years older than Goethe) but his self-fashioned image as a premodern magus.[80]

This is not to say that Goethe considered Beireis to be an actual alchemist and sorcerer, as he regarded him more as a trickster in the manner of the modern magician-entertainer. In the journal entry itself, Goethe utilizes the narrative strategy of the fantastic genre, initially presenting Beireis as a figure with a magical aura, but then offering a natural explanation for the reputation, pointing out that he had probably accumulated his wealth in such a short time not through alchemical transmutation but by means of a mundane version of such an operation—the discovery of a new chemical method of creating dye material, which he sold to manufacturers, evidenced by the fact that he was in communication with industrialists in the dye business.[81] On this basis Goethe delivers a harsh judgment of those who are gullible enough to believe in Beireis's mysterious powers.

> Any one living in a remote place like Swedenborg, or in a small university like Beireis, had always the best opportunity of wrapping himself up in a mysterious obscurity, invoking spirits and labouring at the philosopher's stone. Have we not in modern times seen how Cagliostro, sweeping swiftly across large spaces, could now in the south, now in the north, now in the west, carry on his juggleries and everywhere find adherents? Is it then too much to say that a certain superstitious belief in demoniac men never dies out, so that at all times

a place is to be found where the problematic true, for which in theory alone we have respect, may most conveniently in practice associate with a lie?[82]

The case of Beireis embodies many of the themes that are important to the understanding of the automaton symbolism of this period, including the return of the magical aura to the mechanical object and the view of the mechanically minded sorcerer as an illusionist who wields possibly supernatural power. So despite Goethe's condemnation of the backward mentality that allowed such fraudulent claims of magic to flourish even in the modern age, Beireis was exactly the kind of figure that the fantastic writers among the Romantics liked to write about. In fact, Ludwig Archim von Arnim turned Beireis into a fictional character in his long novel *Armut, Reichtum, Schuld und Buße der Gräfin Dolores* (The poverty, wealth, guilt and atonement of Countess Dolores [1810]).

Arnim's greatest contribution to Romantic literature is the three-volume collection of German folksongs *Des Knaben Wunderhorn* (The youth's magic horn [1804–1807]), which he edited with his brother-in-law Clemens Brentano, another major writer of the period, later inspiring the Grimm brothers to produce such a collection of folktales.[83] In *Gräfin Dolores*, the central character, Count Karl, receives a vision through a mesmeric hallucination of his wife Dolores's infidelity, which torments him and drives him to leave her. In one of twelve side-episodes that symbolically mirror the count's condition and dilemma, he finds himself in the town of H———, where he is led to the house of the place's famous Wondrous Doctor. He is greeted graciously by the strange figure but becomes uneasy at the sight of all the odd things he finds in the dwelling, "full of wondrous but quite wretched appliances."[84] Among them is Vaucanson's flute-player and duck, the sight of which fills him with the uncanny feeling "of being entirely strange and alone under the power of unfeeling machines, which, created by man, could easily get the upper hand over him."[85]

As the Doctor must excuse himself to see a patient, Count Karl is left by himself in this intimidating place, where he is frightened by a voice from a device known as the "Invisible Girl." This was a real-life trick object that was put on show in London in 1803 and elsewhere, mentioned by Wordsworth in the lines quoted earlier from *The Prelude*, featuring four small trumpets in a suspended framework that could answer questions from people (a girl in another room spoke through a hidden rubber tube).[86] In responding to the count's astonished inquiries about her identity, she tells him about her love for the automaton flute-player, which, in this story, can sing as well as play the instrument. After a beautiful performance by the figure, the Wondrous Doctor returns to claim that the whole thing is done by machinery and goes so far as to mimic Vaucanson in opening up the automaton to expose its gears and cylinders.

Later on, at a public house, the count learns that the "Invisible Girl" is none other than Arnika Montana, a famous clairvoyant of the time, who has spoken to him from another room. He also hears the story of how a flute-player named Florio fell in love with her voice during a performance but when he came to see her was disillusioned by her plain appearance. He was nevertheless captivated enough by her voice and the beauty of her vacuous sister Divina to follow their traveling show. At this point, Arnim makes a significant departure from the real-life automaton of Vaucanson, as he combines its external appearance with the trickery of Kempelen's chess-player.

In the story, the Wondrous Doctor's flute-player turns out to be fraudulent, as it is revealed that Florio was eventually hired to play and sing in a separate room and the automaton made to move as if it were performing the music. When the count returns to the mysterious house and speaks to Arnika again, she tells him how she and Florio are resigned to their places in the Doctor's magical-mechanical games and that the "old man curses us, when he becomes addled in the wondrous circles of his machines and experiments, and they become more powerful than he."[87]

Once again, the uncertainty about the exact nature of the uncanny devices is prolonged until there is the revelation of a mundane explanation. As with other stories discussed above, however, the aura of the magical lingers in the unresolved questions of Arnika's actual psychic abilities, the Doctor's strange dominance over her and Florio, and the rather ominous intimations at the end of strange forces at work that have pushed the Doctor's mind to the edge of madness and endowed the machines around him with supernatural powers.

Mary Shelley's Frankenstein and Heinrich Heine's Runaway Automaton

Romantic literature of the fantastic teems with other kinds of artificial being than the automaton, featuring the same themes of the uncanny appearances of boundary-crossing entities, the darker implications of *Naturphilosophie*, and the expression of anxiety and uncertainty in a time of political and social crisis. These beings include the animate broomstick in Goethe's 1797 poem "The Sorcerer's Apprentice," made famous by the Mickey Mouse cartoon in the Walt Disney film *Fantasia,* and the homunculus episode in part 2 of Goethe's *Faust* (1832);[88] the golem of Jewish legend and an animate mandrake in Arnim's fairytale novella "Isabella of Egypt" (1812);[89] the toys that come alive at night in "The Nutcracker and the King of Mice" (1816), by Hoffman, best known from Tchaikovsky's ballet;[90] and a reanimated mummy in Jane Webb's futuristic novel *The Mummy! A Tale of the Twenty-second Century* (1827), also featuring automata.[91]

The most famous artificial being from this period is, of course, the monster in Shelley's *Frankenstein, Or the Modern Prometheus* (1818), whose characters, both the creator and the created, have become iconic figures in the modern imagination. Although the creature Frankenstein creates is not a mechanical automaton but a reanimated being consisting of patched-up pieces of corpses, it is of interest here for several reasons. The novel is commonly considered the first work of science fiction because in the preface to his wife's work, the Romantic poet Percy Bysshe Shelley, distinguishes its story from those that merely weave "a series of supernatural terrors." The events described in the story are "not of impossible occurrence" considering the ideas of such scientists of the era as Dr. Darwin and "the physiological writers of Germany."[92] Mary Shelley was familiar with much of the advanced scientific ideas of her time, including the botanical and zoological ideas of Eramus Darwin (a vitalist and early theorist of evolution, and the grandfather of Charles Darwin) and the chemistry and physics of Humphry Davy (the British counterpart to the German *Naturphilosophen* and the model for Frankenstein's teacher Waldman).[93] It is never revealed in the novel what exactly Frankenstein has discovered about the essential nature of life or how exactly he uses that discovery to reanimate the corpse, but there is a strong implication of the use of electricity. In an early episode in Frankenstein's story, the destruction of an oak tree by lightning is explained by his father as an electric phenomenon, the properties of which he demonstrates through the use of a "small electric machine" and a kite with wire and string.[94] In the early nineteenth century one could attend spectacular demonstrations of the electrical ideas of Luigi Galvani in which parts of dead animals and the corpses of recently executed criminals were made to move through the application of electricity, which Galvani considered the essential fluid of vital force.[95] It is also significant that before Frankenstein receives his training in modern science, he studies the works of Albertus Magnus, Cornelius Agrippa, and Paracelsus premodern figures associated with artificial beings.[96]

This story of an artificial creature that turns into a murderous monster due to the hubris and neglect of its creator is also the best example of a narrative that embodies the political, social, and intellectual uncertainties and anxieties of the era. I have already noted the idea of the monster as a representation of the discontented masses who are liable to rise up in rebellion. It also points to the dark side of Romantic science by imagining an uncanny event that occurs as a result of crossing the boundary between the animate and the inanimate. Chris Baldick has made the revealing observation that Frankenstein takes care to select the most beautiful parts of corpses for his creation, but when the creature comes alive everyone, including its creator, is horrified by its ugliness.[97] The significance of this is that the image of the segmented body of the monster, also representing the segmented society of a nation in crisis, is based on the mechanist vision of the corporeal entity consisting of so many discrete parts. Frankenstein's attempt to bring forth a new being by putting together parts of its body like so many pieces of a machine leads to the creation of an unnatural monster. So even though he has been trained in the ideas of *Naturphilosophie*, he ultimately fails to understand that the beauty of a being lies in its harmonious whole, not in the qualities of its individual parts.

In addition, in the passage on the exact moment the creature comes alive, there is a vivid description of the feeling of horror experienced before the sight of a boundary-crossing event—of a dead thing coming alive. While Frankenstein engages in the sickening task of stealing parts of dead bodies from cemeteries and charnel houses and patching them together in his room, he never expresses disgust at the work, but that changes when he witnesses the success of his work.

> I started from my sleep with horror; a cold dew covered my forehead, my teeth chattered, and every limb became convulsed; when, by the dim and yellow light of the moon, as it forced its way through the

window-shutters, I beheld the wretch—the miserable monster whom I had created.

Oh! no mortal could support the horror of that countenance. A mummy again endued with animation could not be so hideous as that wretch. I had gazed on him while unfinished; he was ugly then; but when those muscles and joints were rendered capable of motion, it became a thing such as even Dante could not have conceived.[98]

As awful as the thing was before it came alive, it was safely in the territory of dead matter. But the acute feeling of the uncanny that is aroused when the distinction between the animate and the inanimate collapses leads to the sense of absolute horror. And it is Frankenstein's failure to control that emotion and face up to his responsibilities toward what he has created that leads to the transformation of the creature into a homicidal monster.

As I have shown in this chapter, these stories of uncanny automata and other artificial beings were informed by the political and social anxieties of the era, which had little to do with concerns over the Industrial Revolution, which did not commence in continental Europe until the 1830s. In fact, none of the narratives discussed here takes place in an industrial setting with steam-driven technology or engages with the problems of industrialization. Almost all the references to automata are to the clockwork devices of the eighteenth century, and their creators are portrayed as isolated geniuses wielding uncanny powers rather than as industrialists presiding over factories. This situation changed with the continued progress of industry in England and concomitant developments in France and Germany, significantly altering the nature of automaton symbolism in the midcentury and beyond.

A major German literary figure who struggled directly with issues raised by industrial modernity was Heinrich Heine. He began his writing career as a Romantic poet of exquisite lyrical verses, but after his move to France in 1831, in the wake of the second revolution there, he gained a new literary and political perspective, inspired by the socialist ideas of the Saint-Simonians. So his career serves as a convenient bridge between the world of uncanny automata and that of the industrial automata that I explore in the next chapter. In the 1830s he produced two major works of criticism, *Towards a History of Religion and Philosophy in Germany* and the companion piece *The Romantic School*, in which he lashed out against the movement as a whole and the Schlegel brothers in particular, who had taken a conservative political turn in their later years. A major distinguishing aspect of Heine's works is his willingness to fully engage with modern life, including the consequences of rapid industrialization, urbanization, commercialization, and the resulting materialistic outlook of modernity. Many of his comments on such topics were made in reference to England, which he visited in 1827 on an unpleasant trip.[99]

In *Towards a History of Religion and Philosophy in Germany* he draws a stark distinction between the English and the German characters based on the essential materialism of the former and the idealism of the latter. In his description of the English character, he points to the empirical ideas of Locke and claims that he "turned the human mind into a kind of calculating machine; the whole human being became an English machine."[100] Later in the essay, he elaborates on this point by telling a fantastic story.

> The tale goes that an English inventor, who has already invented the most ingenious machines, finally hit on the idea of constructing a human being. In the end he succeeded; the work of his hands could behave and act just like a man; it even bore within its leathern breast a sort of human feeling differing not too greatly from the usual feelings of the English; it could communicate its emotions by articulate sounds, and it was precisely the noise of the wheels inside, of springs and screws, which was then audible, that lent these sounds a genuinely English pronunciation. In short, this automaton was a perfect gentleman, and nothing was wanting to make it a real human being except a soul.[101]

The mechanical creature, feeling the lack of the soul, demands that its creator provide him with one, with such persistence that its creator is compelled to flee. The automaton follows him, and the two appear all over the world, the machine crying out to him "Give me a soul."

The basic story appears to be a reference to *Frankenstein*, with crucial differences in the creature being a mechanical automaton instead of an animated corpse, its success at being a perfect simulacrum of an English gentleman instead of appearing as a hideous monster, and the creator running away from its impossible demand rather than fleeing from its uncanny nature. S. S. Prawer has pointed to a letter in Prince Hermann von

Pückler-Muskau's writings on his visit to England in 1826–1828 as the probable source of the story.[102] In the entry for October 13, 1826, the prince reports attending performances of "hideous melodrames," including "Frankenstein, where a human being is made by magic."[103] Since the brief description indicates none of the major changes made by Heine, it can be considered a more or less original story. He apparently had a penchant for reworking narratives in such a manner, as in one instance when he appropriated Hoffmann's "The Sandman" to amuse a visitor with a tale of another Englishman who fell in love with a woman who turned out to be an automaton, and had to travel to Italy to get over his grief.[104]

The significance of his reworked version of the Frankenstein story is in the moral he draws from it, which has to do with the impact of living in the modern world of mechanized work. As Heine explains,

> we see [in the story] how one part of the English people is weary of its mechanical existence, whereas the other part, out of anxiety at such a desire, is driven about in all directions, and neither can endure things at home any longer.
>
> This is a dreadful story. It is terrifying when the bodies we have created demand from us a soul. It is far more dreadful, terrifying, uncanny, however, when we have created a soul and it demands from us its body and pursues us with this demand. The thought we have conceived is such a soul, and leaves us no peace until we have helped it to become a material phenomenon. Thought strives to become action, the word to become flesh.[105]

This conceptual use of the automaton on the part of Heine, a Romantic poet who eventually rejected many of the tenants of Romanticism, which he came to associate with political conservatism, marks the next transition in the cultural use of the object. The satirical and uncanny automaton of the earlier period was informed by ideas of *Naturphilosophie* and embodied the anxieties of the historical era. In the age that followed, the themes Heine explored of the mechanization of humanity through industry, the fear of machinery as the fear of the working class, and the terrifying notion of the great, powerful machines of the industrial age taking on lives of their own emerged as the central concerns in the use of the automaton in the culture's meditation on the human condition in the modern age.

6
The Living Machines of the Industrial Age, 1833–1914

The Age of the Living Machine

The Victorian writer Samuel Butler came from an illustrious line of English clergymen, including a grandfather and namesake who was the bishop of Lichfield and a major classics scholar. The younger Samuel experienced a crisis of faith following his graduation from Cambridge that made it impossible for him to follow in the footsteps of his line. Instead he emigrated to New Zealand, where he became a successful sheep farmer at the Canterbury Settlement. In 1863, four years after the publication of Charles Darwin's *On the Origin of Species*, Butler published an article entitled "Darwin among the Machines" in the Christchurch newspaper the *Press*. In this work he applied evolutionary ideas to the development of machinery, pointing to advancements in technology made in the last centuries from simple tools like the lever and the pulley to the locomotive.

> We have used the words "mechanical life," "the mechanical kingdom," "the mechanical world," and so forth, and we have done so ad-

visedly, for as the vegetable kingdom was slowly developed from the mineral, and as in like manner the animal supervened upon the vegetable, so now in these last few ages an entirely new kingdom has sprung up, of which we as yet have only seen what will one day be considered the antediluvian prototypes of the race.[1]

Given the speed of mechanical evolution, machines will soon become humanity's successors as the masters of the world, as "we are daily giving them greater power and supplying by all sorts of ingenious contrivances that self-regulating, self-acting power which will be to them what intellect has been to the human race. In the course of ages, we will find ourselves the inferior race."[2] The article ends with a call to arms as well as a dire warning about the coming age of autonomous machines:

Our opinion is that war to the death should be instantly proclaimed against them. Every machine of every sort should be destroyed by the well-wisher of his species. Let there be no exceptions made, no quarter shown; let us at once go back to the primeval condition of the race. If it be urged that this is impossible under the present condition of human affairs, this at once proves that the mischief is already done, that our servitude has commenced in good earnest, that we have raised a race of being whom it is beyond our power to destroy, and that we are not only enslaved but are absolutely acquiescent in our bondage.[3]

Butler returned in 1864 to England, where he eventually became an antagonist of Darwin over scientific issues, but he continued to elaborate his idea on the rise of living machines, most comprehensively in his 1872 novel *Erewhon*. This unusual work combines the utopian genre (Erewhon is an approximate backward spelling of "nowhere"—Thomas More created the name Utopia from the Greek *ou-*, "not," and *topia*, "place") with a satire of contemporary society, as in the writings of Jonathan Swift. An Englishman named Higgs travels to an unnamed colony where he comes across a civilization that has rejected machinery so completely that they are displeased when they see his pocket watch.[4] After the traveler learns the local language, he finds out that centuries before a prophet predicted that machines were ultimately destined to supplant the race of man, and to become instinct with a vitality as different from, and superior to, that of animals, as animals to vegetable life. So convincing was his reasoning, or unreasoning, to this effect, that he carried the country with him; and they made a clean sweep of all machinery that had not been in use for more than two hundred and seventy-one years... and strictly forbade all further improvements and inventions.[5]

The prophet's ideas are elaborated in chapters entitled "The Book of the Machines," much of the contents of which was taken from "Darwin among the Machines" and another article by Butler entitled "The Mechanical Creation" (1865).[6]

Butler's ironic imagining of a disastrous future for mankind in the age of autonomous machinery is a representative case of the automaton motif in the second half of the nineteenth century, a period Thomas Carlyle in his influential essay "Signs of the Times" (1829) anxiously declared the mechanical age: "It is the Age of Machinery, in every outward and inward sense of that word.... Nothing is now done directly, or by hand; all is rule and calculated contrivance.... On every hand, the living artisan is driven from his workshop, to make room for a speedier, inanimate one."[7] As the Industrial Revolution transformed the physical, economic, and social environment of western Europe in a rapid and radical fashion, there was an outpouring of literary portrayals of steam- or electricity-driven machines taking on characteristics of living creatures. While some of them featured the sense of the uncanny, the central ideas behind these visions of living machines had a different significance from those of the Romantics. In the fantastic literature of the earlier period, the possibility of artificial constructs coming alive was presented in the language of *Naturphilosophie*—the possibility of latent, supernatural Spirit realizing itself through Matter. In contrast, later works were informed by scientific ideas of the midcentury, especially thermodynamic theory and Darwinian evolution. While many of the more disturbing stories by the Romantics reflected the political and social upheavals of the late eighteenth to the early nineteenth centuries, the central issue of concern in the following decades was the ongoing advancement of industrial technology and its impact on human society and identity.

To highlight the most important themes in the automaton writings of this period, another look at the writings of Butler is warranted. In his meditations on modern technology, several features stand out as essential characteristics of the era's view of self-moving machines. Unlike the clockwork devices in Romantic writings, what Butler describes are steam-driven machines. And the language of evolution figures prominently in his meditation on the rapid advancement of technology in his time, reflecting the importance of Darwinian ideas in the era's intellectual and cultural discourse. Furthermore, the continuum drawn from the vegetable and the animal kingdoms to those of humanity and machinery, seeing in all of them the same process of work and transformation, points to ideas from the other major scientific achievement of the midcentury: the formulation of the laws of thermodynamics. These themes need to be elaborated on for a full understanding of the view of living machines in the industrial era.

From Enlightenment Automata to Industrial Technology

When Captain Mirvan in Burney's novel *Evelina* denigrates automata on the basis of their uselessness, the French guide defends the marvelous devices by pointing to their aesthetic value.[8] But he might have done better by pointing out, as Samuel Johnson did, that

> to collect the productions of art, and examples of mechanical science or manual ability, is unquestionably useful, even when the things themselves are of small importance, because it is always advantageous to know how far the human powers have proceeded, and how much experience has found to be within the reach of diligence.[9]

Johnson may have been referring to the automata on display in England at the time he wrote the passage in 1751, as he further notes:

> It may sometimes happen, that the greatest effort of ingenuity have been exerted in trifles; yet the same principles and expedients may be applied to more valuable purposes, and the movements, which put into action machines of no use but to raise the wonder of ignorance, may be employed to drain fens, or manufacture metals, to assist the architect, or preserve the sailor.[10]

An example of such a trifling "wonder of ignorance" leading to utilitarian technology occurred some thirty years later when a clergyman named Edmund Cartwright attended a demonstration of Kempelen's chess-player. Later on, he found himself in a conversation with some men from Manchester discussing Richard Arkwright's cotton-spinning machine. When they expressed skepticism about whether such a complex device as a weaving mill could be automatized, Cartwright responded that

> there had lately been exhibited in London an automaton figure which played at chess. Now you will not assert, gentlemen, said I, that it is more difficult to construct a machine that shall weave, than one which shall make all the variety of moves which are required in that complicated game.[11]

Inspired by this exchange, Cartwright went on to invent the power-loom, which played an important role in industrial development in England.[12]

It was not until the mid-nineteenth century, however, with industrialization fully under way both in Britain and on the Continent, that major scientists gave explicit credit to eighteenth-century automata and their makers for providing inventions that contributed to useful technology. The Scottish physicist and inventor David Brewster in his 1831 work *Letters on Natural Magic* discusses the mechanical devices of the previous era before pointing to instances of their practical contributions in terms similar to Johnson's.

> The same combination of the mechanical powers which made the spider crawl, or which waved the tiny rod of the magician, contributed in future years to purposes of higher import. Those wheels and pinions, which almost eluded our senses by their minuteness, reappeared in the stupendous mechanism of our spinning-machines, and our steam-engines. The elements of the tumbling puppet were revived in the chronometer, which now conducts our navy through the ocean; and the shapeless wheel which directed the hand of the drawing automaton has served in the present age to guide the movements of the tambourine engine. Those mechanical wonders which one century enriched only the conjuror who used them, contributed in another to augment the wealth of the nation; and those automatic toys which

once amused the vulgar, are now employed in extending the power and promoting the civilization of our species.[13]

Similarly, Hermann von Helmoltz, a central figure of thermodynamic theory, in his essay "On the Interaction of Natural Forces" (1854) points to the achievements of the automaton-makers of the past, including Vaucanson, whose duck was the "marvel of the last century." Their drive to understand the problem of animate motion and to replicate it mechanically led to the later breakthroughs in industrial technology.

> The object, therefore, which the inventive genius of the past century placed before it with the fullest earnestness, and not as a piece of amusement merely, was boldly chosen, and contributed not a little to enrich the mechanical experience which a later time knew how to take advantage of.[14]

A significant case of the impact of automaton technology on nineteenth-century machines can be found in the works of Charles Babbage, the designer of calculating machines as well as the writer of the major work of industrial planning *On the Economy of Machinery and Manufactures* (1835). I have already shown how his encounter with Kempelen's chess-player led him to consider whether it was possible to build an actual game-playing machine. In his memoir *Passages from the Life of a Philosopher* (1864) he reveals that two other eighteenth-century automata provided inspiration for him—their distinct paths in his life converging at his famous salon, attended by many intellectual luminaries of his day.

Vaucanson's automata were powered by spring-driven cylinders that turned to produce motion in the artificial figures. When he went on to build the mechanical loom that gained him membership in the Académie Royale des Sciences in 1746, he used the same technology, with spokes on the cylinder determining the pattern weaved by the machine, like such a device in a player-piano (patterns could be altered by using cylinders with different configurations of spokes). During the Napoleonic era, a talented silk-weaver named Joseph-Marie Jacquard improved this device by replacing the cylinder with loops of cardboard paper with holes punched through them, making it easier to change the patterns. These punch-cards, in turn, provided the crucial innovation in Babbage's design of the Difference Engine, the first of his calculating machines, which was only partially built.[15] Babbage's fascination with machines began early, from an episode in his childhood when he saw several exhibitions of automata, including those of John Joseph Merlin. As he was quite taken with the mechanical objects, his mother took him to Merlin's workshop, where the Belgian showed him around, demonstrating two twelve-inch female figures, one of them a silver dancer with a moving bird on her right hand.[16] Many years later, he encountered the automaton again at the auction of Thomas Weeks's museum, where it went up for sale in a dilapidated condition. Babbage bought it, repaired it himself, and took care to dress it properly before displaying it in a glass case for visitors to his salon.[17]

In one of the most interesting episodes in the history of automata, Babbage relates that at a dinner party he hosted he displayed both the silver lady and the partially constructed Difference Engine in two separate rooms. The result of the showing was that the crowd surrounded the automaton while the calculating machine was ignored by all but two foreign visitors.[18] Despite the world-changing possibilities latent in the ancestor of the modern computer, it was the life-imitating power of the "unproductive" automaton that drew the attention of the guests, the calculating machine's utilitarian promise proving no match for the silver lady's mechanical allure.[19] The inspiration that the works of both Vaucanson and Merlin provided Babbage neatly illustrates the impact of automaton technology on the scientists and engineers of the industrial age.

The appreciation by the scientists of the industrial era of eighteenth-century clockwork devices, in terms of their technological contributions, takes on a new dimension when considered in the light of thermodynamic theory. In the new physics the midcentury, through the new scientific language of work and energy, there occured a return of the man-machine analogy that established a nexus among machinery, industry, and biology.

The Steam Engine and the Language of Thermodynamics

During the 1840s and 1850s, there was a major scientific reaction against Romantic *Naturphilosophie*, initially in Germany on the part of younger figures like Karl Ludwig, Karl Vogt, Emil du Bois-Reymond, Hermann von Helmoltz, Jacob Moleschott, Rudolf Clausius, and Ludwig Büchner (brother of Georg Büchner). In direct response to the metaphysical excesses of idealist philosophy as well the conservative turn of some of the prominent Romantics, the new generation of intellectuals, many of liberal or radical bent, adopted a materialist stance with a renewed effort toward the quantification of natural phenomena.[20] The greatest scientific achievement of this generation, which became a pan-European phenomenon of central importance to the entire cultural and intellectual milieu of the period, was the formulation at midcentury of the laws of thermodynamics, beginning with Helmholtz's theory of the conservation of energy and followed by a series of simultaneous discoveries by Clausius and, in Britain, by William Rankine and, most important, William Thomson (Lord Kelvin).[21] In a general sense, their endeavor represented a return to the mechanistic and mathematical spirit of the classical Enlightenment but with several essential differences in ideas and methodology.

Advances in physics from the time of Newton through that of Lagrange and Laplace made it impossible to envision the world in a simple mechanistic manner with the clockwork machine as the model of nature. With the Romantic vital force proving to be an unsatisfactory concept in the further elucidation of the workings of the dynamic cosmos, the scientists of the new era concentrated on the rigorous quantification of work in terms of heat expenditure through the burning of fuel-material. Their success in establishing calculable formulas for the purpose with increasing sophistication allowed them to build on their findings to construct a universal theory of energy. With the clockwork automaton no longer suitable as the representation of their ideas, the role of the central emblem of the thermodynamic worldview was filled by the steam engine. As Helmholtz explains:

> To the builders of automata of the last century, men and animals appeared as clockwork which was never wound up, and created the force which they exerted out of nothing. They did not know how to establish a connexion between the nutriment consumed and the work generated. Since, however, we have learned to discern in the steam-engine this origin of mechanical force, we must inquire whether something similar does not hold good with regard to men.[22]

This passage is significant not only for its replacement of the classical automaton with the steam-driven device as the primary conceptual object but also for Helmholtz's explicit linkage of machine and body under uniform principles of energy.[23] The importance of thermodynamic theory in the larger intellectual culture lay in the fact that it provided a set of universally applicable ideas that went beyond the field of physics to those of physiology, industry, and even economy. Crosbie Smith and M. Norton Wise in their comprehensive biography of William Thomson have analyzed the impact of thermodynamics not only on British technology but also on industrial and economic planning, which led to the rise of the steam engine as a metaphor for "work, wealth, and progress."[24]

In the area of physiology, a prominent proponent of the application of thermodynamic principles to the study of natural bodies was the French physiologist and pioneering practitioner of chronophotography (the use of photography for the study of motion) Étienne-Jules Marey.[25] In his *Animal Mechanism* (1873) he points to the limits of Enlightenment mechanistic physiology and goes on to assert that "modern engineers have created machines which are much more legitimately to be compared to animated motors; which, in fact, by means of little combustible matter which they consume, supply the force requisite to animate a series of organs, and to make them execute various operations."[26] In their ideas, Helmoltz and Marey did not try to reduce natural bodies to machinery but saw both biological and mechanical entities as working under the same principles of work and heat expenditure. As Anson Rabinbach puts it, the energeticist theory showed that "work performed by any mechanism, from the fingers of a hand, to the gears of an engine, or the motions of the planets, were essentially the same."[27] This led to the proliferation in the larger culture of the description of people as thermodynamic machines and machines of the industrial age as living beings.

The linkage of the operation of the human body with that of the steam machine can be found in the industrial discourse of the eighteenth century. Works by Adam Ferguson and Adam Smith and the industry-related articles in the *Encyclopédie* are prominent examples. Simon Schaffer has shown the continuation of the theme in nineteenth-century British works like Andrew Ure's *Philosophy of Manufactures* (1835), Babbage's *On the Economy of Machinery and Manufactures* (1835), and William Cooke Taylor's *Tour in the Manufacturing Districts of Lancashire* (1842), which also feature the depiction of machines as living creatures—in Ure's characterization of a spinning mule as an "*Iron Man* sprung out of the hands of our modern Prometheus" and in Taylor noting that "the machines can do everything but speak."[28]

In such writings, the notion of human beings as parts and extensions of machinery and of machines as animate creatures, forming a symbiotic whole, was often portrayed in a celebratory manner as the description of the smooth and productive functioning of the industrial system. But such imagery was also deployed for negative purposes. Those critical of unfettered industrialization used it to express their concerns, some of them imagining the machines as powerful monsters representing the inexorable and destructive force of modernity itself and others seeing the mechanized worker as the representation of dehumanization in the industrial environment.

In Benjamin Disraeli's 1844 novel *Coningsby*, the title character has a sublime vision of the living machine after a visit to a Manchester factory.

> A machine is a slave that neither brings nor bears degradation; it is a being endowed with the greatest degree of energy, and acting under the greatest degree of excitement, yet free at the same time from all passion and emotion. It is, therefore, not only a slave, but a supernatural slave. And why should one say that the machine does not live? It breathes, for its breath forms the atmosphere of some towns. It moves with more regularity than man. And has it not a voice? Does not the spindle sing like a merry girl at her work, and the steam-engine roar in jolly chorus, like a strong artisan handling his lusty tools, and gaining a fair day's wage for a fair day's toil?[29]

This depiction of the machine as a happy worker ("merry," "jolly") working under a rational economic system leads to a generally positive view of the working conditions of female laborers at the weaving factory—"a thousand or fifteen hundred girls may be observed in their coral necklaces, working like Penelope in the daytime; some pretty, some pert, some graceful and jocund, some absorbed in their occupation; a little serious some, few sad."[30]

Eleven years after the publication of *Coningsby*, in the United States Herman Melville published his story "The Paradise of Bachelors and the

Tartarus of Maids," which exposes the deplorable environment of a paper factory in New England. Melville's description of the machines and workers found there overturns Disraeli's views on them: the happy and productive machine is transformed into a monster and the laborers into slaves. As the narrator describes the central machinery:

> Something of awe now stole over me, as I gazed upon this inflexible iron animal. Always, more or less, machinery of this ponderous, elaborate sort strikes, in some moods, strange dread into the human heart, as some living, panting Behemoth might. But what made the thing I saw so terrible to me was the metallic necessity, the unbudging fatality which governed it.[31]

While Coningsby saw the machine as a supernatural slave that happily served man, it becomes the master in the American factory with the female workers as its subordinants.

> Nothing was heard but the low, steady overruling hum of the iron animals. The human voice was banished on the spot. Machinery—the vaunted slaves of humanity—here stood menially served by human beings, who served mutely and cringingly as the slave serves the Sultan. The girls did not so much seem accessory wheels to the general machinery as mere cogs to the wheels.[32]

Less than a decade after the appearance of this story, Samuel Butler sent out his warning from New Zealand of the coming war with machines.

In the famous section on the division of labor at the beginning of Adam Smith's *Wealth of Nations* (1776), he often describes the work performed by a laborer as that of a disembodied hand, as if that most outwardly utilitarian part of the human body was sufficient to accomplish the repetitive task in a modern workshop.[33] This reflected the common practice of referring to industrial workers as "hands," which became a source of protest for those concerned with the human cost of industrialization in Victorian England. Charles Dickens's *Hard Times* (1854) takes place in the fictional industrial town of Coketown, where the animate steam engine that powers each factory moves its piston "up and down, like the head of an elephant in a state of melancholy madness."[34] The workers there are "a race who would have found more favour with some people, if Providence had seen fit to make them only hands, or, like the lower creatures of the seashore, only hands and stomachs."[35] Likewise, in Elizabeth Gaskell's *North and South* (1854), the sentimental protagonist Margaret Hale, who "does not like to hear men called 'hands'," offends the industrialist Thornton by reporting her encounter with a disgruntled worker who "spoke as if the masters would like their hands to be merely tall, larger children—living in the present moment—with a blind unreasoning kind of obedience."[36]

Marx picks up on this point in *Capital* (1867), where he analyzes the subordination of the workers to the machinery of the factory.

> The special skill of each individual machine-operator, who has now been deprived of all significance, vanishes as an infinitesimal quantity in the face of the science, the gigantic natural forces, and the mass of social labour embodied in the system of machinery, which, together with those three forces, constitutes the power of the "master." This "master," therefore, in whose mind the machinery and his monopoly of it are inseparably united, contemptuously tells his "hands," whenever he comes into conflict with them: "The factory operatives should keep in wholesome remembrance the fact that theirs is really a low species of skilled labour. . . . The master's machinery really plays a far more important part in the business of production than the labour and the skill of the operative."[37]

In other passages Marx characterizes the organized system of machines in factories as "a mechanical monster whose body fills whole factories, and whose demonic power, at first hidden by the slow and measured motions of its gigantic members, finally bursts forth in the fast and feverish whirl of its countless working organs"[38]—and notes that the extreme specialization of the worker in the modern factory turns him "into a crippled monstrosity. . . . Through the suppression of a whole world of productive drives and inclinations . . . the individual is divided up, and transformed into the automatic motor of a detailed operation."[39]

The living machines and the worker automata of the industrial age were viewed in such disparate ways as parts of a harmonious and productive system of wealth-producing industry and as elements of an oppressive and debasing system of mechanical mastery over enslaved humanity. They were, however, not the only visions of the fate of humanity in the age of high technology. In literary and artistic works of the late nineteenth and early twentieth centuries that were the products of the newly emerging aesthetics of modernism, transcendent images of the hybridization of humanity with machinery appeared. The works of writers like Auguste Villiers de l'Isle-Adam, Alfred Jarry, F. T. Marinetti, and Raymond Roussel, describe the creation of part-biological, part-mechanical beings. The impact of the very idea of the living machine in the industrial context on the modernist imagination needs to be explored at this point.

The Living Machine and Modernism

Among the many aspects of modernism that are the sources of endless scholarly attempts to define the culture that emerged in the last decades of the nineteenth century, the most relevant here is the technological aspect.[40] In this perspective, modernism can be broadly defined as the cultural and intellectual attempt to grapple with the implications of the industrial age in ways that go beyond the two traditional discourses of the materialist-utilitarian drive toward a high-tech utopia and of the Romantic obsession with the idealist, the spiritual, and the individual through which much of modernity was rejected, including industrialization, urbanization, and the quantifying methodology of the new science. As Torbjörn Wandel has pointed out, it is futile to search for a unified modernist position on the enormous changes brought on by industrialization in terms of either celebration or rejection. Modernity, rather, "is the simultaneous sense of exhilaration and anxiety about the now and what it might hold in store. Each modern expression is characterized by this inherently ambivalent attitude toward the present."[41] And once the steam-driven machine emerged as the central cultural emblem of the period, modernist writers and artists drew from it an imaginative inspiration that went beyond its utilitarian function.[42] As a result, one finds in the art and literature of the period varying images of machines as beautiful gods, terrifying monsters, sublime works of art, and hybrid beings of evolutionary transcendence.

A particularly vivid and early example of this can be found in the British caricaturist Robert Seymour's etching "The March of Intellect" (1828),

Robert Seymour, *The March of Intellect* (1828). Etching. Progress and reform is envisioned as a gigantic steam-driven automaton. Courtesy of Guildhall Library, City of London.

which portrays modern progress as a gigantic steam-driven automaton with gas lights for eyes, hot-air balloons coming out of its pipe, and a stack of books and the building of University College London on its head. Wielding a broom with the head of the Whig reformer Henry Brougham fixed to its upper end, the automaton sweeps away outdated laws, quack doctors, and a black devil of superstition, while proclaiming "I come I come!!" The illustration is a celebratory one of the rapid and exciting changes occurring in the political and social landscape of Britain in the period, but it also features the image of a superhuman entity of frightening dimensions and awesome power, reflecting the ambivalence felt in the era toward both industrialization and modernity in general.[43]

That ambivalence became much more pronounced later in the century, as can be seen in H. G. Wells's story "The Lord of the Dynamos" (1894), which presents an electrician, Holroyd, who is in charge of generators that power the railways of England, and his assistant, a dark-skinned, primitive man from "the mysterious East."[44] The latter regards the largest of the dynamos as a god and prays to it for deliverance from his alcoholic master, who abuses him at every turn. Holroyd, who "doubted the existence of the Deity but accepted Carnot's cycle," ridicules his subaltern for his superstition, but even he sees the religious potential of the machines: "'Look at that,' said Holroyd; 'where's your 'eathen idol to match 'im? . . . Kill a hundred men. Twelve per cent on the ordinary shares . . . and that's something like a Gord.'"[45] The assistant thinks that he must make a sacrifice to the deity to attain his freedom and murders his boss by throwing him into the machine. When he is foiled in an attempt to do the same to Holroyd's replacement, he commits suicide by the same means. Wells concludes: "So ended prematurely the worship of the Dynamo Deity, perhaps the most short-lived of all religions. Yet withal it could at least boast a Martyrdom and a Human Sacrifice."[46]

Wells displaces onto the primitive savage the idea of the machine as a supernatural being that, as already shown, can be found in many writings of the period. A famous example of this is Henry Adams's description of the great hall of dynamos at the 1900 Great Exposition in Paris. In the chapter "The Dynamo and the Virgin" in *The Education of Henry Adams*, he too falls into religious language describing "the forty-foot dynamos as a moral force, much as the early Christians felt the Cross. . . . Before the end, one began to pray to it;

inherited instinct taught the natural expression of man before a silent and infinite force."[47] Modernist imaginings on the living machine of the industrial age, however, were not limited to that of sublime awe, as the visions of autonomous machinery alternated from the most optimistic hope for utopia to the darkest fear before the intimidating creatures of metal and steam. An excellent illustration of this fluctuation can be found in imaginative descriptions of the locomotive, one of the key transformative technologies of the era.

Dolf Sternberger relates how in January 1875 a celebratory banquet commemorating the centennial of James Watt's perfection of the steam engine was held as a part of the founding festival of the Prussian Verein für Gewerbefleiss (Union for Business Diligence), attended by the most important industrialists of Berlin.[48] During the event, Ernst Engel, the director of the Prussian Statistics Office, gave a toast in which he extolled a truly happy and productive marriage. The husband was "a true child of Nature, of ancient stock," whose hot-headed ancestors went so far back as to have witnessed the creation of the world, while the wife's family was much younger, as ages ago "they were very simple and roughhewn people." Engel went on to describe their union.

> The marriage performed in 1775 is, despite the vast differences between the spouses, one of the happiest on the face of the earth and is still in its heyday. And it is also the most fruitful. Its offspring number in the hundreds of thousands. With very few exceptions, they are the best bred, hardest working, and most docile of creatures. They never rest by day or night and are veritable models of obedience and temperance.... Wherever we build huts for them and treat them properly, their entrance is followed by success and abundance close at their heels.[49]

The marriage was that of steam as the husband and engine as the wife, and their children were the myriad machines of the Industrial Revolution. Engel's fanciful story represents the essence of what can be called classical technophilia, featuring the optimistic vision of industrial progress and the conception of modern machines as obedient slaves that will free people from work and allow them to enjoy lives of bourgeois leisure. What many modernists appreciated about such machines, however, was not their utilitarian value but their aesthetic possibilities.

Almost a decade after Engel told his parable in Berlin, the French writer J. K. Huysmans described the fair daughters of steam and engine through the character of Jean des Esseintes, the protagonist of the Decadent novel *À Rebours* (1884). This solipsistic aristocrat, sitting alone in his apartment so as to indulge in his strange possessions and even stranger thoughts, denigrates Nature and all of her supposed splendors in an anti-Romantic manner.

> Nature has had her day; she has definitely and finally tired out by the sickening monotony of her landscapes and skyscapes the patience of refined temperaments. When all is said and done, what a narrow, vulgar affair it all is . . . what a tiresome store of green fields and leafy trees, what a wearisome commonplace collection of mountains and seas![50]

What he prefers are the artificial creations of man, which have surpassed, he believes, those of Nature in beauty. The example he thinks of is a "living, yet artificial organism"—the locomotive—like the two that recently went into service on the Northern Railroad of France. As he feverishly describes them:

> the Crampton, an adorable blonde, shrill voiced, slender-waisted, with her glittering corset of polished brass . . . the perfection of whose charms is almost terrifying when, stiffening her muscles of steel, pouring the sweat of steam down her flanks, she sets revolving a puissant circle of her elegant wheels. . . .

> the Engerth, a massively built, dark browed brunette, of harsh-toned utterance, with thick-set loins, panoplied in armour-plating of sheet iron, a giantess with disheveled mane of black eddying smoke, with her six pairs of low, coupled wheels, what overwhelming power when, shaking the very earth, she takes in tow, slowly, deliberately, the ponderous train of good wagons.[51]

Des Esseintes compares them favorably to actual women and thinks that man has done as well as God in creating these wondrous creatures of steel. Their value, in other words, lies not in their utilitarian function but in their sublime beauty.

The living machine also appeared in darker forms as a frightening monster, like the mad elephant in Dickens's Coketown that threatens to destroy not just the physical environment but the sanity of individuals. From Des Esseintes's appreciation of locomotives as sublime goddesses, the engineer Jean Lantier in Émile Zola's novel *La Bête humaine* (1890) takes the next step by falling in love with a beautiful daughter of steam and engine. A disturbed individual who feels an urge toward violence whenever he is aroused by an actual woman, Lantier harbors an intense passion for the train under his charge, which he has named Lison.

> She was gentle, obedient, moved off easily, kept up a regular, continuous pace thanks to her good steaming. Some made out that her getting under way so easily was due to the excellent tyres on the wheels and above all the perfect adjustment of the slide valves. . . . But [Lantier] knew that there was something more, for other engines, identically built, assembled with the same care, showed none of these qualities. There was the soul, the mystery in creation, the something that the chance hammering bestow on metal, that the knack of the fitter gives to the parts—the personality of the machine, its life.
> So he loved Lison with masculine gratitude.[52]

When the train is destroyed in an act of sabotage, Lantier sees its end as the death of a living creature.

> For a little while it had been possible to see her organs still working in her torn-open body, her pistons still beating like twin hearts, the steam circulating through her slide-valves like the blood in her veins; but as though they were arms in convulsion her driving rods were now only quivering in the last struggles for life, and her soul was departing along with the strength which had kept her alive, that powerful breath she still could not completely expel. The disemboweled giant quietened down still more, then gradually dozed off into a gentle sleep and final silence. She was dead.[53]

The loss of his beloved is so traumatic to Lantier that he loses his mind and murders a woman. He himself meets his death by being torn to pieces in the wheels of another train, falling into them after he gets into a violent struggle with another engineer, who is also killed. Consequently, the vehicle, carrying French soldiers into the disaster of the Franco-Prussian War, goes on its way with no one controlling its direction or restraining its power.

> Now out of control, the engine tore on and on. At last the restive, temperamental creature could give full rein to her youthful high spirits, like a still-untamed steed that has escaped from its trainer's hands and was galloping off across the country. The boiler was full of water, the newly stoked furnace was white-hot, and for the first half-hour pressure went up wildly and the speed became terrifying.⁵⁴

In the final paragraph of the work, a new image of the locomotive as a living creature emerges, this time not as an obedient servant or a beautiful goddess but a fearful monster.

> What did the victims matter that the machine destroyed on its way? Wasn't it bound for the future, heedless of spilt blood? With no human hand to guide it through the night, it roared on and on, a blind and deaf beast let loose amid death and destruction.⁵⁵

This image of the violent and terrifying child of steam and engine, the dark brother of the beautiful daughters, also appears in Gerhart Hauptmann's naturalist story "Lineman Thiel" (1887). The title character, Thiel, works for the railways, stationed at a post alongside the tracks, and regards the train as a fearful creature.

> From the distance a snorting and thundering came in waves through the air. Then suddenly the stillness was rent. A furious raging and roaring filled the whole place, the lines bent, the ground trembled—a mighty pressure of air—a cloud of dust, steam and smoke, and the black, snorting monster was past.⁵⁶

As in the case of Lantier, the dark vision of the locomotive is linked to the unstable mental state of the beholder. Thiel becomes increasingly disturbed by the fact that his second wife, Lena, is abusing his mentally handicapped son, Tobias, the child of his first marriage, but he feels powerless to intervene. The stress from the unbearable situation causes him to have frenzied dreams of the train passing by his station. Just before the tragic accident that kills Tobias, he has another prophetic vision of the machine as monster.

> Two round red lights pierced the darkness like staring eyes of some enormous monster. A blood-red glow preceded them, transforming the raindrops in its proximity into drops of blood. It was as though the heavens were raining blood.
>
> Thiel experienced a shudder of horror and, as the train came nearer, an ever-growing anxiety; dream and reality fused into one for him.[57]

After the death of his son, Thiel, like Lantier, becomes what he has beheld and turns into a crazed murderer.

What is significant about these negative portrayals of the living machine as a monster is that the image is related not only to the unstoppable progress of industry and modernity but also to an irrational and destructive will at the heart of the enterprise that is reflected in the madness of individual characters. The sociologist Jeffrey Herf has pointed to a twentieth-century political ideology he calls "reactionary modernism," with roots in Weimar Germany and espoused by such figures as Ernst Jünger, Oswald Spengler, Werner Sombart, and Nazi ideologues of the Third Reich. What he finds paradoxical in their view is that while theirs was essentially a conservative position that rejected the ideals of the Enlightenment, including its rationalist outlook and liberal democratic politics, which were associated with the bourgeois state, it embraced modern technology, the very product of Enlightenment science.[58] This aspect of the movement, which constitutes its modernism, distinguished it from traditional forms of conservatism that tended to repudiate all aspects of modernity, including technology and industrialization. The reactionary modernists made this move away from tradition by seeing in machines not as the tools of rational control of nature but as embodiments of irrational will. Herf describes this idea in the context of early and mid-twentieth-century political movements, but such depictions of modern technology were already widespread in western Europe in the second half of the nineteenth century, and not necessarily associated with reactionary politics.

The image of the industrial machine as an irrational, terrifying, destructive, and superhuman entity was envisioned by those who felt that the progress of industrialization had taken on a life of its own beyond human interest and control. While those like Ernst Engel, who celebrated the happy marriage of steam and engine, looked forward to a utopian future of mankind freed from work, others pointed out that the essential impulse behind the ceaseless mechanization of society and work seemed to have little to do with the pursuit of human happiness. For Samuel Butler this situation gave rise to frightening visions of machines rising up in a slave revolt. Many modernists, however, were more interested in the aesthetic and transcendent possibilities suggested by this new race of mechanical beings.

Works that explicitly feature the new automata and the man-machine of the industrial age can be divided between those of the 1886–1914 period, which deal primarily with ideas of the living machine, and those of the interwar period, which are more concerned with ideas of work, industry, and revolution. The automaton stories by the Symbolist, Surrealist, and Futurist writers Villiers, Jarry, Marinetti, and Roussel before World War I do not take place in industrial settings, nor do they feature the worker-automata of the factory. Their main interest lies rather in exploring the fantastic natures of the new beings, including biological-mechanical hybrids that represent the next step in evolution and autonomous objects of pure artistic value beyond the concerns of utility and industry. Inspired by thermodynamic ideas on the equivalence of natural and artificial systems, the rise of the steam engine as the central emblem of the new worldview, and the ubiquity of the machines of the industrial era, the more imaginative modernists dreamt up fantasies of living machines and man-machines that were celebratory and comic as well as prophetic and cautionary.

Woman as Machine: Villiers de l'Isle-Adam's *Tomorrow's Eve*

In 1878 the American inventor Thomas Edison became a celebrity in France after the demonstration at the Paris Exposition of his phonograph, for which he won the grand prize at the event. Along with the "talking machine," he presented the designs for a doll that would be equipped with a miniature version of the device, allowing it to utter words from the top of its head.[59] In the same year, the Symbolist writer Auguste Villiers de l'Isle Adam began writing a novel in which the inventor plays a central role.

Published in 1886 under the title *L'Eve future*, it concerns an English aristocrat, Lord Celian Ewald, who visits his friend Edison at his domain in Menlo Park, New Jersey.

Ewald is in great despair over an actress, Alicia Clary, whose appearance is of perfect and noble beauty but whose personality is shallow, vulgar, and materialistic. Having fallen in love with her external form, Ewald has become convinced that there must be a spirit as beautiful and noble as her outward appearance inside her, so he has taken her on a journey across Europe, to great cities, natural wonders, museums, and theaters, hoping to awaken that spirit. But she remained unmoved by all those things and made only trite and ignorant comments at every turn, confounding Ewald with the discrepancy between her body and her soul:

> You would think she was some dreadful mistake of the Creator's! . . . What right does anyone so beautiful have to be so devoid of spirit? By what right is this unparalleled form able to appeal at the deepest level of my spirit, to some sublime emotion—only to destroy my faith in it? My eyes are constantly imploring her, "Betray me if you will, but exist! Live up to the spirit of your own beauty!"—and she never understands me.[60]

The nobleman once helped Edison when he was destitute, so the inventor proposes to return the favor with a technological solution to Ewald's problem. Edison introduces his friend to Hadaly, an android whose form and personality can be molded to the specification of his desire. The bulk of the novel consists of discussions between Edison and Ewald on whether such an artificial creature, even if it could successfully simulate a real woman, could fulfill the role of a satisfying companion. In response to Ewald's skepticism, the inventor insists that all that a typical woman is can be reproduced in a machine driven by "electromagnetic power and Radiant Matter."[61] When Ewald doubts that such a woman-machine could be a real being since it would not possess a soul, and therefore consciousness, Edison answers by questioning whether that would be such an important lack in a woman:

> the "consciousness" of a woman! I mean, a *woman of the world!* . . . Oh! Oh! What a notion! It's an idea that once baffled a council of the church. A woman only sees things according to her personal inclina-

tions, and twists all her "judgments" to conform with the opinions of the man she's attracted to. A woman may be married ten times over, be sincere every time, and yet be ten different persons.⁶²

After Alicia Clary is brought to the inventor's realm under false pretenses to model her appearance and voice for Hadaly, the android is completed in her image. Ewald meets his mistress and is enchanted by what he takes to be a noble aspect of her he has never seen before and falls in love all over again, only to be told by her that she is in fact Hadaly, not Alicia. Ewald is startled at first, but the artificial woman explains that she is more than a mere machine. She is an otherworldly spirit who sought to be born in this world by occupying Edison's electromechanical construction. "I called myself into existence in the thought of him who created me, so that while [Edison] thought he was acting of his own accord, he was also deeply, darkly obedient to me. Thus, making use of his craft to introduce myself into this world of sense, I made every last object that seemed to be capable in any way of drawing you out of it."⁶³

When the two of them go to Edison's laboratory, he presents his final explication of the construction of the android without knowing what the others have just spoken of. Up to this point in the narrative, Edison has presented himself as a technoscientific genius capable of rivaling Nature, promising grandiosely "to raise from the clay of Human Science as it now exists, a Being *made in our image*, and who, accordingly, will be to us WHAT WE ARE TO GOD."⁶⁴ Yet he confesses now to Ewald that he has been in contact with a feminine supernatural spirit named Sowana. When Edison told this spirit of his android project, she was eager to help him in its construction and has proved invaluable in the work. In fact, he has discovered that there is an affinity among the psychic energy of the spirit, the electromagnetic energy that powers the machine, and the nervous energy of human beings (i.e., a uniform and universal energy), through which Sowana has been "able, *occasionally*, TO INCORPORATE HERSELF WITHIN [THE ANDROID], AND ANIMATE IT WITH HER 'SUPERNATURAL' BEING."⁶⁵

It occurs then to Ewald that Hadaly is in that state of incorporation by the spirit now, as she has told him about her origin in the supernatural realm and her strategy for entering into the material world. Edison, however, is under the impression that Sowana merely helped in the construction of the android, which is now working according to its electromechanical design. Ewald thinks of pointing out the presence of the spirit in the machine but remembers that Hadaly has asked him to keep their conversation a secret from her creator. At the end of the narrative, Ewald is convinced of Hadaly's nature as a true woman and takes leave of Edison to begin his new life with this wondrous being.

In this way, over two centuries after Descartes asserted that man was a machine occupied by a soul, Villiers turned the philosopher's machine-man analogy upside down by presenting an electromagnetic android that is controlled by a spirit, asking whether such a creature could be considered a living being. Ewald becomes convinced of Hadaly's humanity after a lengthy conversation with her and is happy to take her to his home as his new mistress, though the story concludes with a rather contrived ending of their ship capsizing in the sea. But the significance of his acceptance of the android as a true woman lies in the eradication of the line that separates the natural from the artificial, the biological from the mechanical, and finally the living from the dead. And so Hadaly becomes a beguiling representative of the age of living machines.

There are interesting gender implications in this story that envisions the creation of a whole new woman through the use of technology and spiritualism, but the misogyny in much of the narrative is subverted by the revelation of the crucial role played by a female being who is of supernatural origin.[66] As with the description of workers in the industrial discourse of the time, however, the mechanistic vision of humanity was employed for both men and women in the modernist imagination. Both Alfred Jarry and F. T. Marinetti created fantasies of the mechanization of the male in their works, describing the process in terms of masculine empowerment and transcendence over the natural and the feminine.

Man as Machine: Jarry's *Supermale*

Alfred Jarry, one of the most notorious enfants terribles of French literature, published his novel *The Supermale* in 1902. Despite its hilariously bizarre plot, it is one of his most accessible works, as it features a straightforward narrative, unlike his more experimental writings.[67] The story is set in a futuristic 1920 and begins with an evening gathering at the mansion of an aristocrat, André Marcueil, at which two outlandish claims are made. Marcueil asserts to his guests, which include a distinguished doctor, a scientist, an industrialist, and various ladies, that man has the natural capacity to engage in the act of love indefinitely, as a historical anecdote tells of a man who could have intercourse up to seventy times in a single day. His claim is understandably met with skepticism, but an American chemist, William Elson, considers the possibility and offers an equally strange boast that he has invented from an alcohol and strychnine base what he calls "Perpetual Motion Food," the regular consumption of which will allow a person to physically exert himself without needing to rest—"a fuel for the human machine that could indefinitely delay muscular and nervous fatigue, repairing it as it is spent."[68]

As both claims are demonstrated in the unfolding story, it becomes an extended satire of what Anson Rabinbach has identified as one of the central concerns of the physicomechanical thought of the period, namely the problem of fatigue.[69] If the natural body is indeed like a machine that converts energy to work, just like a steam engine, why can it not exert itself without rest? What is the nature of fatigue—is it the result of a natural and insurmountable limitation of the organic body or is it a problem that can be solved? If one understood its true nature, can a method be found to turn the body into an endlessly working machine? The character of Elson and his invention are also a remarkable presaging of the real-life German physiologist Wilhelm Weichardt and his claim in 1904 to have created a vaccine for fatigue, a chemical that could counteract it when sprayed on subjects.[70] The discovery was met initially with enthusiasm, but subsequent testing proved it to be ineffectual.

In a quintessential scene of modernist imaginings on physicomechanics, Elson organizes a ten-thousand-mile race between a locomotive and a tandem bicycle ridden by five men, who are instructed to ingest his Perpetual Motion Food at regular intervals.[71] The competition between the steam-machine and the man-machine takes a bizarre turn when one of the cyclists dies during the race, but the power of the chemical drives his body on, allowing the corpse to continue moving its legs on the peddles. In fact, once it regains its pace, the body works in a more efficient manner than before, having thrown off the superfluous burden of the soul.[72] In the end the cyclists defeat the train, but not before a mysterious stranger in a concealing garb overtakes them and disappears on the road ahead, proving that there is a force even more powerful than the perpetual motion human-machine.

The stranger is the aristocrat Marcueil himself, who is revealed to be an evolutionary "supermale" with a permanent erection with which he can indeed have intercourse without rest. An utterly amoral monster who probably raped a child to death at his estate, Marcueil seeks to demonstrate his sexual energy by informing his friends that he has found an Indian man capable of engaging in the sex act "threescore times and ten" in a day with the help of some herbs. He plans to dress up as that Indian and have intercourse with seven prostitutes in a chamber at his mansion while his friends watch through a peephole in the next room. On his way to the women, however, he is intercepted by Ellen Elson, the daughter of the American chemist, who has taken an interest in him. She unmasks him and seduces him into demonstrating his sexual energy on her. In the course of the following day, they exceed the Indian's record, the American woman turning out to be a match for the supermale. At the end, however, she collapses, and Marcueil thinks that he has killed her, which puts him in a melancholy trance as he realizes that he loved her.

> "The act of love is of no importance, since it can be performed indefinitely."
> Indefinitely . . .
> Yet there was an end.
> An end to the Woman.
> An end to Love.[73]

It turns out, however, that Ellen did not die but has only fainted from pleasure. After she regains consciousness, she goes home to her father in a happily intoxicated state. After learning of what occurred at the mansion, Elson's primary concern is the bourgeois one of whether Marcueil will make an honest woman of his daughter. He discusses the matter with his cohorts, the doctor Bathybius and the industrialist Arthur Gough, and decides that, given Marcueil's exploits, "He is not a man, he's a machine."[74] And since he is a machine, they construct a countermachine, an electromagnetic device designed to infuse him with a sense of love, so that he may indeed take Ellen as his bride.

If André Marcueil were a machine, or endowed with an iron constitution that enabled him to overcome machines, why then, the combined efforts of the engineer, the chemist and the physician would pit one machine against another for the greater good of bourgeois science, medicine, and morality. Since this man had become a mechanism, the equilibrium of the world requires that another mechanism should manufacture—a soul.[75]

They construct a "love machine" for the purpose, find Marcueil at his mansion, still in a stupor of despair, and hook him up to the device, which subjects him to a tremendous amount of electromagnetic energy. The supermale, however, turns out to be a creature of electromagnetic energy himself, which he pumps back into the device, submitting it to his will and making it fall in love with him. The machine, in the throes of passion for him, overheats and explodes, fatally burning Marcueil as he runs howling out of the mansion, to meet his death among the twisted steel bars of the gate.

The theme of the interconnection between artificial machinery and the organic body in this bizarre tale of the supermale as an indefatigable sex automaton is established in an early scene in which Marcueil comes across a dynamometer at a zoo. It is described as "an iron thing ... squatting, with things that looked like elbows on its knees, and armored shoulders without a head," and Marcueil calls it a beast and one that is "a female ... but a very strong one."[76] When he touches it, the power of his electromagnetic energy is so great that the machine is utterly demolished. The destruction is depicted as the death of a living creature, like that of the locomotive in *La Bête Humaine*, and presages Marcueil's own end. "His phrase ended in a terrible crashing of twisted steel, the broken springs writhed on the ground as if they were the beast's entrails; the dial grimaced and its needle raced madly around two or three times like a hunted creature looking for a way of escape."[77]

The theme of the man-machine is established in the race between the locomotive and the cyclists, which is not a competition between the artificial and the natural in the Romantic mode. Elson's Perpetual Motion Food effectively turns each cyclist into a mechanical being, his body capable of working endlessly as long it is given fuel and even if its soul should depart. The cyclists emerge victorious over the train, but they are inferior to the supermale, who has achieved a mechanical transcendence through evolution, as demonstrated by his limitless sexual energy. In essence, the principles under which the locomotive, the cyclists' bodies, and the supermale's sexuality operate are uniform, all three depicted as dynamic motors that have overcome the limitations of fatigue. As in the physiological ideas of Helmholtz and Marey, the natural body is not simply reduced to a machine but shown to be equivalent to it in dynamic function. In the scenes featuring the race between the machine and the man-and-corpse-machine, the demonstration of the man-machine as a sex-machine, and the battle between the electromagnetic machine and the electromagnetic man-machine, Jarry effectively and gleefully destroys all distinction between the organic and the mechanical. It is fitting, then, that the supermale, as the perfect organic automaton, should meet his death following a competition of love with a machine, destroying them both in a conflagration of electromagnetic energy.

In the decades following the publication of Jarry's novel, the theme of the masculine superman as a dynamic and suprarational machine was carried out most fully and obsessively by F. T. Marinetti, the leader of the Futurist movement in Italy.

Superman as Machine: Marinetti's *Electric Dolls* and *Mafarka the Futurist*

At the end of the nineteenth century there was a revival of a modernist version of vitalism that featured the philosophical concepts of the Will, the Unconscious, and the Reality Flux, in the works of Arthur Schopenhauer, Edouard von Hartmann, Friedrich Nietzsche, and others. While their ideas appeared in the context of fin-de-siècle skepticism toward rationalism and new materialism, the advanced thinkers of the era also avoided retreating to the Romantic alternative with its adulation of nature and teleological cosmology of Spirit and Matter. The modernists, in their search for a new language to describe the postrational and post-Romantic world of living machines and thermodynamic humanity, drew their ideas from such disparate sources as Darwinian evolution, the physics of Ernst Mach (especially his epistemological notion of the sensory flux), and Sigmund Freud's pioneering psychological works on the irrational and the unconscious. In the first decade of the new century, Henri Bergson's elaboration of the élan vital in *Creative Evolution* (1905) had a significant impact on the larger European culture.[78]

The Futurist movement was inspired by Bergsonian neovitalism as well as Nietzschean concepts of the *Übermensch*, the Dionysian impulse, and the transvaluation of all traditional values. It was inaugurated in 1909 with the publication of its manifesto, by the Italian writer Filippo Tommaso Marinetti, which features a conspicuous obsession with the power of modern technology, the dynamic products of which were often described as wondrous living creatures. In Marinetti's 1908 poem "To My Pegasus," which originally appeared under the title "To the Automobile," he praises his vehicle in animistic terms:[79]

> Vehement god of a race of steel,
> space-intoxicated Automobile,
> stamping with anguish, champing at the bit!
> O formidable Japanese monster with eyes like a forge,
> fed on fire and mineral oils,
> hungry for horizons and sidereal spoils,
> I unleash your heart of diabolic puff-huffs,
> and your giant pneumatics, for the dance
> that you lead on the white roads of the world.
> At last I release your metallic reins . . . You leap,
> with ecstasy, into liberating Infinity! . . .[80]

In his 1912 work "The Pope's Monoplane," the narrator envisions his liberation from an earthbound existence with the help of a living aircraft:[81]

> Horror of my room like a coffin with six walls!
> Horror of the earth! Earth, dark lime
> trapping my bird feet! . . . Need to break free!
>
> Ecstasy of climbing! . . . My monoplane! My monoplane!
> In the breach of the blown-out walls
> my great-winged monoplane sniffs the sky.
> Before me the crash of steel
> splits the light, and my propeller's cerebral fever
> spreads its roar.[82]

On one occasion, he described even the machine gun in such a manner after he witnessed as a reporter its use in the 1911 battle of Tripoli fought by Italy and Turkey.

> Ah yes! you, little machine gun, are a fascinating woman, and sinister, and divine, at the driving wheel of an invisible hundred horsepower, roaring and exploding with impatience. Oh! soon you will leap into the circuit of death, to a shattering somersault or to victory![83]

Considering the theme of machine-human connection established in the previous century, it should be no surprise that Marinetti also envisioned humanity in energetic-mechanical terms. In fact, his manifestoes feature some of the most overt expressions of the convergence of the notions of the living machine and the man-machine. In "Multiplied Man and the Reign of the Machine," from his larger work "War, the World's Only Hygiene" (1911–1915), Marinetti describes what he calls "mechanical beauty"—how those who work with machines naturally regard them as living beings:

> Have you never seen a mechanic lovingly work on the great powerful body of his locomotive? His is the minute, knowing tenderness of a lover caressing his adored woman. . . .
>   You surely must have heard the remarks that the owners of automobiles and factory directors commonly make: motors, they say, are truly mysterious. . . . They seem to have personalities, souls, or wills. They have whims, freakish impulses. You must caress them, treat them respectfully, never mishandle or overtire them.[84]

From this point, Marinetti leaps to a vision of the merging of man and machine, the creation of a mechanical superman, reminiscent of Jarry's supermale, to whom the future belongs:

> we must prepare for the imminent, inevitable identification of man with motor, facilitating and perfecting a constant interchange of intuition, rhythm, instinct, and metallic discipline. . . . It is certain that if we grant the truth of Lamarck's transformational hypothesis we must admit that we look for the creation of a nonhuman type in whom moral suffering, goodness of heart, affection, and love, those sole corrosive poisons of inexhaustible vital energy, sole interrupters of our powerful bodily electricity, will be abolished. . . .

This nonhuman and mechanical being, constructed for an omnipresent velocity, will be naturally cruel, omniscient, and combative. He will be endowed with surprising organs: organs adapted to the needs of a world of ceaseless shocks. From now on we can foresee a bodily development in the form of a prow from the outward swell of the breastbone, which will be the more marked the better an aviator of the future becomes.[85]

Such imagery came out of the ideas and language of energy theory and neomechanistic physiology of the second half of the nineteenth century, but despite his celebration of animate machines it would be a mistake to regard Marinetti as a classical technophile like Ernst Engel. The central program of Futurism was the radical destruction of all things bourgeois—in art, literature, and social mores as well as politics. Consequently, the Futurists were anarchist in political alignment and overtly anti-Romantic in their aesthetics—their art filled with dynamic images of motion, speed, transformation, and cataclysmic destruction, followed by an equally awesome rebirth. What appealed to Marinetti about modern technology was its ability to bring about rapid change in the artistic, social, and political landscape by forcing people to reconsider the very notions of space and time, a process that was under way in the general culture of western Europe, through the expanding use of machines such as trains, automobiles, airplanes, and the telegraph.[86] Marinetti, therefore, was as far as was possible from a utopian who hoped that the machinery age would set people free from work to lead genteel bourgeois lives. What he saw in modern technology was a tireless dynamism culminating in the fusion of humanity with machinery, which would become the source of radical transformation at all levels of human life and culture.

For a classical technophile, the rational human self was an unproblematic being, as its role was that of a master-controller of the obedient machine-servants. In an age when these very devices seemed to be taking on lives of their own and threatening to overwhelm their creators, modernists like Marinetti thought that humanity must also change in order to survive and thrive in the age of awesome and irrational machinery. And he envisioned this transformation in the birth of the man-machine hybrid as a new transcendent being. Jeffrey T. Schnapp points out in his essay on Marinetti's

"Technical Manifesto of Futurist Literature" (1912) that his vision of the living machine was a volatile, ceaselessly changing, and supremely individualist one that was the very opposite of the quantified and obedient worker of the Taylorist factory.[87] Interestingly, in the two works by Marinetti in which actual human-machines are featured, the play *Poupées électriques* (Electric Dolls) and the novel *Mafarka the Futurist* (both written originally in French and published in 1909), they appear in contrasting images as the conformist bourgeois machine and the dynamic mechanical superman.

In contrast to the experimental nature of Futurist drama, *Poupées électriques*, written in 1905–1907, before the official formation of the movement, is a fairly traditional play with a clear story line in three acts, notwithstanding the enormous controversy its performance aroused.[88] Dedicated to the aviation pioneer Wilbur Wright, "who knew how to elevate our wandering hearts beyond woman's captivating lips,"[89] the work reveals its modernist sensibility in the introduction of the four principal characters, who make up two couples of contrasting cultural types. Paul de Rozières and Juliette Duverny are familiar figures from traditional Romantic drama: the dashing rake who seduces an impressionable young woman only to abandon her, after which she pines for him until she destroys herself. Their friends John Wilson and his wife, Mary, are a thoroughly modern and unconventional pair. John is a successful American engineer who is regarded as a creative genius on account of the automata, or "electric dolls," his firm produces, and Mary is an exotic and highly sensual woman from Egypt. As the initial interaction among them demonstrates, the couples exemplify the contrast between the traditional and the new, the European and the foreign, the bourgeois and the modernist.

The romantic drama between Paul and Juliette, which John regards with disdain and Mary with sensual delight, unfolds in the first act and ends with the suicide of the abandoned Juliette. The second act, the most interesting and provocative part of the play, takes place a year later, when John finds himself restless, perhaps because he has been affected by the events of the previous year. To recover from this state, the result of an "infection" of bourgeois sentiments, he persuades his wife to participate in a bizarre act of sexual catharsis. In the attic, there are two electric dolls, named Mother Prunelle and Mister Prudent, that resemble Mary's aunt and father, representing, in John's words, "the ghastly reality of duty, money, virtue, old age, monotony, emotional boredom, physical exhaustion, engrained stupidity, social laws, and who knows what else."[90] To rekindle the passion they felt for each other during their courtship, when they had to meet in secret because Mary's guardians disapproved of their love, John has sex with her behind the backs of the automata. At first Mary enjoys the act, even commenting that her husband perhaps regards her as one of his dolls, which he does not deny: "The most beautiful of them all. In fact, your mechanics are identical. . . . And it is electricity that makes our nerves vibrate like conducting wires of voluptuousness."[91] But Mary becomes uneasy about the game and her husband's strange passion. She tells him that she wants to be loved for herself alone, not as a part of the elaborate mechanical scenario he needs to arouse himself. At the end of the act, he is worn down by her demands and eventually succumbs to her wishes, throwing the automata out the window. In the third act, the demoralized John has turned into a typical bourgeois husband who becomes jealous when the rake Paul shows up again to confess his love to Mary. In his rage, John bullies his wife until he makes her life unbearable and drives her to suicide.

The significance of the narrative is in the depiction of a creative genius whose supremely unconventional attitude toward life, love, and society is assailed first by the insipid romantic drama that unfolds before him and then by his wife's refusal to participate further in the sex act that proves inspirational for him.[92] As for the electric dolls, they turn out to be symbols of a complex nature. As products of modern technology as well as the embodiment of the machine-human connection, revealed in John's equating of his wife's bodily mechanism with those of the machines, their creation is what makes their inventor a high-tech superman. In the second act, however, when the automata make their appearance, it is clear that they represent everything that John loathes, both as stand-ins for Mary's relatives and as symbols of bourgeois respectability. The situation is a paradoxical one, in that John is portrayed as an antibourgeois genius by the virtue of his creation of the bourgeois machines. This is complicated even further when he finds that it is through the use of the automata, in an act of sexual transgression, that he is able to regain his passion. Both as products of supreme artistic genius and as representations of stifling conventionality, the automata function as provocative and volatile symbols at the heart of the drama.

A grandiose and less ambivalent creature of artificial construction makes its appearance toward the end of Marinetti's novel *Mafarka the Futurist*, a work that got him indicted for obscenity in Italy.[93] The story takes place in a fantastic North Africa and concerns a warlord by the name of Mafarka-el-Bar, who engages in an epic war to dethrone his uncle, King Boubassa. He achieves victory but then loses his beloved brother Magamal, who dies after being bitten by a rabid dog. Mafarka falls into despair, which threatens to destroy him utterly, until he has a vision of bringing to life a new being, a son he will call Gazourmah. The latter is a gigantic winged automaton, animated by both high technology and magic. As Mafarka describes the glory of the artificial being, he emphasizes that his "offspring" will be a manmade creature of mechanical parts and masculine energy and not a being born of a woman, who is associated with weakness, sentimentality, and corruption.[94]

> For I tell you that I have given birth to my son without the help of the vulva! . . . One night I suddenly asked myself: "Does it take gnomes to run like sailors on the deck of my chest to raise my arms? Does it takes a captain on the poop of my forehead to open my eyes like two compasses?" So I concluded that without the support and stinking collusion of the woman's womb, it is possible to produce from one's flesh an immortal giant with unfailing wings! Our will must come out of us so as to take hold of matter and change it to our fancy. So we can shape everything around us and endlessly renew the face of the earth. Soon, if you appeal to your will, you will give birth without resorting to the woman's vulva. That's how I killed Love, by replacing it with the sublime voluptuousness of Heroism![95]

In the course of bringing the mechanical superman to life, Mafarka faces the temptation of the feminine, first from two slave girls who try to seduce him and then from a lover from his youth who tries to claim the place of mother of the artificial son. Mafarka thwarts their designs and protects the construction of Gazourmah. When the giant is completed, he climbs up to its face and kisses it, transferring his life force into it. The son comes alive, as the first sign of which his "leathery, copper-colored penis stiffened like a sword."[96] He then casts aside the now superfluous body of his father, and sets himself free.

> Then the huge, orange-hued wings thundered, like temple gates, in that huge arc of cliffs. Gazourmah launched himself forward between the broken jaws of the cage. His impetuous feet trampled on the seaweed lining of the huge womb of rock; then suddenly his breast cut into the wavy, changing silk of the sea. A huge guffaw of foam splashed his face, and with a leap he soared into space.[97]

Nature herself tries to bring him down with her elements out of fear and jealousy of his power and beauty, but just as his father has predicted, Gazourmah proves invincible as he dominates the sky in the fullness of his glory.

The final apotheosis of the artificial son in Marinetti's novel can be regarded as the culmination of fantastic imaginings on living machines. Villiers dreamt of an electromagnetic android whose combination of mechanics and guiding spirit could make it indistinguishable from a real woman, while Jarry portrayed the evolutionary supermale as an indefatigable sex machine. Marinetti took a step further in creating a whole new creature that was the embodiment of high technology as the source of the life force. While Villiers's spirit-operated Hadaly mimics a natural woman and Jarry's Marcueil is a product of natural evolution, Marinetti's Gazourmah has transcended nature altogether, as it has been created and animated by masculine will alone, avoiding birth through the "inefficient vulva" that subjects one to decay and death.[98] In the world of the living machines, what occurs in *Mafarka the Futurist* is the final divorce of biology and technology and the emergence of irrational machinery as the source of a new reality superior to that of nature.

Artistic Machines: Roussel's *Impressions of Africa* and *Locus Solus*

The writings of Raymond Roussel were greatly appreciated by both Surrealist artists and later French *nouveau romanciers* like Alain Robbe-Grillet and Michel Butor but have only recently received widespread recognition as some of most original literary works of the period.[99] Like many of the more experimentally inclined figures of modernist literature, Roussel, in his two major writings, *Impressions of Africa* (1910) and *Locus Solus* (1914), wrote against the traditional novel by abandoning conventional narrative structures and filling his plots with fantastic elements that defy realism. In both works, he provides a basic background story from which he unfolds a series of episodic scenes that sometimes read like catalogues of discrete incidents, characters, and objects. As examples of the latter, Roussel describes some of the strangest machines ever imagined, many of nonutilitarian and utterly mysterious nature.

Impressions of Africa begins with a detailed description of a grand gala performance in the African city of Ejur for the pleasure of an emperor, Talu VII. We learn only in the middle of the book that the performers, a group of talented Europeans, including inventors, entertainers, artists, and scientists, were on a ship that departed from Marseilles and headed for Buenos Aires when it ran into a hurricane that sent them to the shores of Africa.[100] The natives they encountered there conducted them to Ejur, where they are to stay as guests of the emperor until their ransom money, sent for through an envoy, arrives from Europe. To while away the time, they form what they call the Incomparables Club to entertain their host during the celebration of his recent conquests. Among the numerous performances they offer as exemplars of the best of European culture, several fantastic machines are presented.

The tool-maker La Billaudière-Maisonnial displays a foil-wielding machine, which he sets against the fencing master, Balbet.[101] The object is directly under the control of its inventor, but it becomes clear that it is much more than a simple mechanical device since it is able to devise its own strategies in fighting its human counterpart: "From time to time, La Billaudière-Maissonal, by pushing a long, serrated rod repeatedly backwards and forwards, completely changed the disposition of the various mechanisms, thus producing a new series of feints unknown even to himself."[102] With its capacity for infinite improvisation, an ability not usually associated with a machine, it is able to defeat Balbet. The chemist, Bex, then presents a massive machine in a glass cage that he explains is a mechanical orchestra run by an electric motor and a special temperature-sensitive chemical he calls bexium. Like the fencing machine, it is more than a mere imitation device, as it is capable of performing all manners of music by means of the quick adjustment of its heat: "The repertory of orchestral pieces was infinitely rich, and Bex presented all sorts of dances, medleys, overtures and variations."[103] And the inventor, Bedu, constructs an automatic loom powered entirely by the flow of a river. Far from a utilitarian device made for a factory, it is an artistic machine that creates a panoramic tableau of scenes from the biblical Flood for the entertainment of the spectators.

> There was no limit to the number of variations which could be obtained by the raising of certain groups of threads, coinciding with the lowering of others. Together with the host of multi-coloured shuttles, this multiplicity of successive patterns depending on the way the warp was divided, made practicable the production of fairy-like fabrics, resembling the pictures of the old masters.[104]

In the final performance, the woman explorer, Louise Montalescot, presents a metal plate on a tripod that is in fact a painting and drawing machine. Provided with the right paints and equipment, it can work independently to reproduce on a canvas such a close likeness of the subject that it might as well be a photograph.[105]

Such fantastic machines also appear in *Locus Solus*, which narrates a tour given by a genius inventor, Martial Canterel, to a group of visitors who have come to his estate to view its marvels. The first mechanical device they witness is perhaps the single most bizarre automaton imagined in literature—a paving-beetle (a construction device) with working arms, attached to a balloon that allows it to float around, the entire thing being powered by the sun and the wind.[106] Its purpose is even stranger than its appearance. Among Canterel's countless achievements is his discovery of a method of pulling teeth without causing any pain to the patient. After receiving a stream of toothache sufferers, he ended up with an enormous number of extracted teeth that he did not know what to do with. After considering the matter, he created the flying automaton, which was set to the work of first sorting the teeth by their colors and then using them to create a mosaic of a scene from a Scandinavian legend.

In a gigantic glass tank in the shape of a diamond filled with *aqua micans*, Canterel's special water in which people and animals can breathe, the visitors are shown a famous dancer, a Siamese cat, the animated head of the French revolutionary leader Danton, and many little diving automata, or bottle-imps, that enact scenes from mythology and history.[107] The most unsettling of his works appears when Canterel takes his guests to a path around a large glass cage that is divided into sections that look like so many stage sets representing different scenes. In each of the sets is an individual who enacts a suggestive scene over and over again. Canterel explains that the person is actually a corpse that has been animated by two elements of his invention called resurrectine and vitalium. Reminiscent of the Perpetual Motion Food in *The Supermale*, the two substances, when infused into the dead body, "released a powerful current of electricity at that moment, which penetrated the brain and overcame its cadaveric rigidity, endowing the subject with an impressive artificial life."[108] It also has the effect of awakening certain significant memories in the corpse and making it reenact them repeatedly, thus turning it into an organic automaton. The purpose of the displays is for the bereaved to visit their deceased loved ones as they go through the motions of replaying the most important moments of their lives.

Roussel was an admirer of the works of Jules Verne and modeled Canterel on such famous technological masters as Nemo and Robur, but the nature of the fantastic machines described in both *Impressions of Africa* and *Locus Solus* are of a completely different order from that of the submarine, the airplane, and the rocket imagined by Verne.[109] While Roussel's machines are clearly imagined as the products of modern industrial technology, from their use of complex mechanics, chemicals, and electricity, all of them are objects of artistic rather than utilitarian value. In other words, the devices, rather than fulfilling practical functions, were designed to either create art or to be objects of art themselves. Roussel delights mainly in giving detailed descriptions of the machines themselves, as their central purpose is simply to be the things that they are and nothing more, in the manner of "art for art's sake." For instance, as he minutely describes the painting and drawing machine in *Impressions of Africa*:

> The lid of the rectangular plate, with a ring in the centre by which it could be grasped, was directly exposed to the light of the dawn; from its back, completely unconcealed, sprang a myriad of exceedingly fine metal wires, giving it the appearance of a head of hair growing too evenly, which seemed to connect every imperceptible area of the surface with a kind of machine charged with a supply of electric energy.[110]

And so on. On the mosaic-creating automaton in *Locus Solus*:

> Canterel drew our attention to the various components of the apparatus, sparing no pains to provide us with a wealth of fascinating commentary. Right at the top of the aerostat, left bare by the netting which formed a kind of flat collar at this point, there was an automatic aluminum valve, consisting of a circular opening with a plug, and next to it a little chronometer whose dial was visible to us. Under the balloon, the slender vertical cords forming the lower part of the net, made entirely of light, fine, red silk, were attached to holes perforated in the very low, straight rim of a round aluminum tray which served as a gondola.[111]

And he goes on for over ten pages.

These descriptive passages reveal that the uses of the machines and the pseudoscientific explanations of their functions are really secondary to the true point of the devices, which is to be unique and independent objects in themselves, apart from any consideration of utility and symbolism. The *nouveau romancier* Robbe-Grillet theorized that it was the purpose of all of Roussel's fiction to point to the inherent mystery of the external appearance of objects, without explaining what they stand for, the art of his prose lying in the way he describes things.

> Since there is never anything beyond the thing described, that is, since no supernature is hidden in it, no symbolism (or else a symbolism immediately proclaimed, explained, destroyed), the eye is forced to rest on the very surface of things: a machine of ingenious and useless functioning, a post card from a seaside resort, a celebration whose progress is quite mechanical, a demonstration of childish witchcraft etc. A total transparency, which allows neither shadow nor reflection to subsist, this amounts, as a matter of fact, to a *trompe-l'oeil* painting.[112]

From Villiers's woman-machine, Jarry's man-machine, and Marinetti's superman-machine, Roussel imagined devices whose use and meaning are independent from the concerns of humanity and whose primary function is to be autonomous in all their strange uniqueness. It should be no surprise, then, that the first admirers of Roussel's works were the Surrealists, who sought to create a new reality in their art, often by playing with the boundary between what is animal and human, inanimate and animate, dead and living, and artificial and organic.

Roussel, frustrated by the failure of *Impressions of Africa* to reach a large audience, took the advice of his friend Edmond Rostand and wrote a theatrical version of the novel, which was performed in Paris in 1911. Although it was both a financial and critical disaster, Guillaume Apollinaire, Marcel Duchamp, and Francis Picabia, three key figures of the Surrealist movement, saw it together in 1912 and were greatly inspired by it. Duchamp in particular was fascinated by all the machines that appeared on stage and acknowledged the direct influence of the play on the creation of his most celebrated work, the massive construction *Large Glass or The Bride Stripped Bare by Her Bachelors, Even* (1915–1923).[113] Duchamp's explanation of the various elements of the work reveals the impact not only of Roussel's artistic machines but of the prevalent ideas on the machine-human connection. He refers to the predominant mechanical figure on the top panel of the work as the "bride," which

> basically is a motor. But before being a motor which transmits her timid-power.—she is this very timid-power—This timid-power is a sort of automobiline, love gasoline, that distributed to the quite feeble cylinders, within reach of the *sparks of her constant life*, is used for the blossoming of this virgin who has reached the goal of her desire.[114]

The figures on the bottom panel, based on a chocolate-grinding machine, are called the "bachelor machines."[115] Other major modernist artists, including Picabia, Fernand Léger, George Grosz, Kurt Seligmann, Benjamin Péret, and Hans Bellmer, used images of living machines and artificial humanity in their works to explore the artistic possibilities of the merging of the human with the machine.[116]

What is interesting about the late nineteenth- and early twentieth-century works by Villiers, Jarry, Marinetti, and Roussel is that while they were written in the context of industrialized Europe, the automata and mechanized people in their writings do not appear in dehumanized forms in their works. They are portrayed rather as unique, dynamic, and captivating spirit-

and-machine woman, indefatigable sex machine, vital superman, and artistic machines. In their attempts to assimilate the implications of modernity through their considerable imaginative powers, these early modernist writers of automaton literature consciously went beyond the classic nineteenth-century positions of technophile celebration or Romantic rejection. They not only considered the merging of the organic and the artificial as a positive development in human history that could usher in a wondrous new age in which the distinction between the living and the artificial would be finally erased but also saw the driving force of the great change, namely high technology, in aesthetic terms, with the automaton as the most compelling example of the beauty of the machine and of the mechanical in the organic. These modernists were able to explore the issue in more radical and speculative ways than the Romantics since their imagination was not constrained by the notion of the supremacy of Spirit and Nature. In fact, some of them wrote in direct opposition to traditional reverence toward the natural world, seeing in the artificial a new reality that could be shaped by human imagination at its most free and chaotic play. Neither classical technophiles who hoped that machines would one day set us free from work and create a leisurely heaven on earth nor nature lovers who thought that mankind's salvation lay in rejecting modernity and returning to a pre-mechanical state, these modernists sought to create in their works a novel world of counterreality. And what better creatures to fill such an artificial world than living machines and mechanical organisms that can no longer be distinguished from one another.

As the experiences of technological warfare, revolution, and continued industrial dehumanization forced the later modernists to consider machines in a much darker light, one might regard the early modernist writings on automata as products of a special historical and cultural moment when technology was regarded as a potential source of a marvelous aesthetic transformation of humanity, when the merging of human and machine was seen as something transcendent and beautiful. The inevitable war between man and machine feared by Samuel Butler in 1863 was not the only vision of our fate in the age of living machines.

7
The Revolt of the Robots, 1914–1935

The Mechanical Subaltern

In G. K. Chesterton's story "The Invisible Man" (1911), one of his idiosyncratic murder mysteries involving the priest-detective Father Brown, automata make an unusual appearance. A diminutive but successful industrialist, Isidore Smythe, goes missing, possibly the victim of foul play. The basis of his fortune is Smythe's Silent Service, or domestic automata that do household chores, advertised as "a butler who never drinks," "a housemaid who never flirts," and "a cook who is never cross."[1] When the machines are first described in Smythe's house, they appear to be utilitarian devices with nothing intimidating about them.

> Like tailor's dummies they were headless; and like tailor's dummies they had a handsome unnecessary humpiness in the shoulders, and a pigeon-breasted protuberance of chest; but barring this, they were not much more like a human figure than any automatic machine at a station that is about the human height. They had two great hooks like arms, for car-

rying trays; and they were painted pea-green, or vermilion, or black for convenience of distinction; in every other way they were only automatic machines and nobody would have looked twice at them.[2]

That changes when Smythe disappears, leaving behind a bloodstain on the floor of his house. As the mystery deepens, an acquaintance of Smythe's, John Angus, returns to the scene, where the automata take on a sinister aspect.

> Angus looked round at the dim room full of dummies, and in some Celtic corner of his Scotch soul a shudder started. One of the life-size dolls stood immediately overshadowing the blood stain, summoned, perhaps, by the slain man an instant before he fell. One of the high-shouldered hooks that served the thing for arms, was a little lifted, and Angus had suddenly the horrid fancy that poor Smythe's own iron child had struck him down. Matter had rebelled, and these machines had killed their master.[3]

When Father Brown arrives at the solution, that Smythe's archenemy entered the place posing as a postman and carried out his corpse in a mailbag, the reader is surprised that the automata had nothing to do with the murder and that their function in the narrative is merely to heighten the uncanny atmosphere of the mystery. The modern reader is predisposed to thinking of such automatic machines as capable of killing due to the proliferation of out-of-control robot narratives in popular culture.

But Martin Gardener points out that the objects play a significant symbolic role in the story. Smythe's disappearance is confounding because his house was under watch at the time he vanished, those who were guarding it claiming that no one came in or out of the place. In actuality, a postman (i.e., the murderer in disguise) appeared, but he was "an invisible man" to the watchers because they ignored him as a person of no importance. As Father Brown explains,

> Have you ever noticed this—that people never answer what you say? They answer what you mean—or what they think you mean. Suppose one lady says to another in a country house, "Is anybody staying with you?" the lady doesn't answer "Yes; the butler, the three footmen, the parlourmaid, and so on," though the parlourmaid may be in the room,

or the butler behind her chair. She says "There is *nobody* staying with us," meaning nobody of the sort you mean.[4]

The story then functions as a critique of the English class system in which servants are hardly regarded as humans by their masters. And this takes us back to the automata, the headless mechanical dummies, which represent the members of the working class as viewed by their employers, as useful but unnoticed things existing only to serve.[5]

I have pointed to instances in which the idea of self-moving machines suggested that of mechanized subalterns as the most useful application of the technology. The stories of Daedalus's moving statues, for instance, gave Aristotle visions of the self-weaving shuttle and the self-playing lyre in a world in which "chief workmen would not want servants, nor masters slaves."[6] And during the Industrial Revolution that there were extended considerations of machines as workers and workers as machines. In the second half of the nineteenth century, when the steam engine and thermodynamics became central to intellectual discourse, the living machine came to represent three related phenomena: industrialization itself; modernity in general (not just industry but also the capitalist economy, the modern political system, and the expanding urban world with its commerce, crowds, and inevitable tensions); and the working class (as part of the seamless mechanical organism of the factory system or as the dehumanized victims of an oppressive order that treats them as nothing but utilitarian "hands").

A prominent theme in such writings is the sheer intimidating power of the machines. During the classical Enlightenment, when the automaton was deployed as a metaphor for the world, the state, and the body, the mechanical traits that were emphasized were those of harmonious function, intricacy, and even beauty, as in the well-made clock. In the industrial period, however, when the massive steam engine took the place of the clockwork device as the central emblem of the era, the machine was often regarded as a superhuman entity of enormous and sometimes mysterious force. This association of the machine with immense power, as it was articulated in the industrial era, is very much present today in the colloquial description of a person who demonstrates relentless strength and unerring performance at a task as a machine. This is reflected in the much-used phrase "mean, lean, fighting machine" from the 1981 comedy movie *Stripes,* and popular science fiction films like *The Terminator* (1984), *Robocop* (1987), and *Iron Man* (2008), in which mechanization is depicted as empowerment.

But darker visions of the fate of humanity in the age of living machinery characterized the machine's power in exactly the opposite way, seeing in the mechanization of the world both the debasement of people into mere devices of production and the helplessness of individual human beings before the unstoppable rush of modernity. As I showed in the previous chapter, such concerns can be found in nineteenth-century works that grappled with the full consequences of the great changes that were occurring in the period. Yet when writers like Villiers, Jarry, Marinetti, and Roussel wrote imaginative stories of the living machine, they chose to explore the creative and transcendent possibilities suggested by the thermodynamic and the electromagnetic automaton, apart from the industrial context from which it emerged. This was in direct contrast to anxious comments by people like Thomas Carlyle, Samuel Butler, and others on the fearful implications of the emergence of the machine age.

The experience of World War I greatly amplified these anxieties, as works produced in the interwar period that prominently feature automata emphasized the destructive, dehumanizing, and maddening aspects of modern technology. Several elements of these narratives stand out in contrast to those of the previous period. While few of the works by earlier modernists take place in the industrial context, most of the ones published in the 1920s unfold in the factory, overtly featuring the themes of exploitation, dehumanization, and class conflict. In addition, the interwar stories are much darker narratives that end in great catastrophes of massive human casualty and the utter destruction of civilization. These disasters are brought on by the revolt of the machines, the revolt of human workers against machines, or their masters' effort to control these uprisings through callously destructive means. Furthermore, the machines are portrayed as irrational entities, and in a much more terrifying manner than in the prewar works. The irrationality or madness of the machines is also presented as a reflection of the madness of their creators and their relentless drive toward industrial perfection no matter what the human cost, as well as of the enraged madness of the working class, driven to becoming a rampaging mob of chaotic power after it is pushed too far by inhumane exploitation.

The automaton stories of Villiers, Jarry, Marinetti, and Roussel were informed by many of the cultural anxieties and scientific ideas of their era, but they take place in environments that are removed from the historical events of the time, in the private realms of eccentric scientists, some times in faraway exotic places. This allowed the writers to explore the nature of their strange mechanical creations in terms of their aesthetic and transcendent qualities, apart from the concerns of the everyday world. The postwar narratives, even when they take place in the future or some other fantastic world, explicitly point to the devastating events of the new century. This development represents an "intrusion of history" in automaton literature—or of the mechanical nightmare of history from which the later modernists could not awake.

Before looking at representative works of this kind, including Karel Čapek's play *R.U.R.: Rossum's Universal Robots* (1921), Romain Rolland's film script *The Revolt of the Machines* (1921), Ruggero Vasari's play *The Anguish of the Machines* (1925), and Fritz Lang's film *Metropolis* (1927), we must take into account the historical developments that brought about the interwar culture of technophobia.

Fearing the Machine

For the writers and artists pondering the living machine in the interwar period, three historical events informed their imagination in significant ways: the experience of technological warfare in World War I, the adoption of Taylorist regime in European factories in the course of the war, and the triumph of the Bolshevik revolution in Russia. The devastation of the Great War, with its unheard-of casualties, was due not only to the adoption of cutting-edge technology in its battles but also the use of outdated strategy and tactics completely inappropriate to the new warfare. Lateral lines of charging soldiers were mowed down by machine guns, which put all the advantage on defensive positions.[7] In addition to significant improvements to heavy artillery and firearms, the appearance of tanks and airplanes gave the soldiers who faced them the impression of fighting against monstrous autonomous machines. Prior to the war, concerns about modern technology had to do mainly with industrial dehumanization, the shock of the transition from the agrarian to the industrial economy, and the negative impact of the change on the natural and social environment. The 1914–1918 conflict added the spectacle of terrifying machines built specifically for the purpose of mass killing.

Like the oppressed workers in the modern factory, the soldiers in the trenches were often described as human beings reduced to mindlessly obeying and killing machines, liable to break down at any time due to the stress on their mechanical systems from shell shock. In a powerful passage in Erich Maria Remarque's classic novel *All Quiet on the Western Front* (1928), the narrator describes a battlefield occupied by devastated human machines.

> The brown earth, the torn, blasted earth, with a greasy shine under the sun's rays; the earth is the background of this restless, gloomy works of automatons, our grasping is the scratching of a quill, our lips are dry, our heads are debauched with stupor—thus we stagger forward, and into our pierced and shattered souls bores the torturing image of the brown earth with the greasy sun and the convulsed and dead soldiers, who lie there—it can't be helped—who cry and clutch at our legs as we spring away over them.
> We have lost all feeling for one another. We can hardly control ourselves when our glance lights on the form of some other man. We are insensible, dead men, who through some trick, some dreadful magic, are still able to run and to kill.[8]

On the other side of the conflict, the French veteran and pacifist Henri Barbusse expressed the psychological impact of the war on soldiers in similar terms of mechanization in his novel *Under Fire* (1916).

> You are always waiting, in wartime. We have become machines for waiting.
> For the moment what we are waiting for is grub. Then it will be letters. But everything in its own time; we'll think about the letters when we've finish the meal. Then we'll start to wait for something else.[9]

Toward the end of this novel, veterans discuss their experiences and become dismayed at the sense that even the horrors they went through will eventually be lost in memory.

"When I was on leave I saw how I'd forgotten lots of things from my life before. There were letters from me that I read like opening a book. And yet, in spite of that, I also forgot what I'd suffered in the war. We are machines for forgetting. Men are things that think a little but, most of all, forget. That's what we are."

"Not them, then, or us! So much misery lost!"

This prospect, coming on top of the desolation of these beings like the news of an even greater disaster, depressed them even further on the shore where they were stranded from the flood.[10]

In the years prior to the war, there were agonizing debates in western Europe on the adoption of the industrial organization advocated by the American engineer and theorist Frederick Winslow Taylor, as outlined in his *Principles of Scientific Management* (1911).[11] In addition to the streamlining of the factory system through a stricter administrative coordination of production, the use of standardized tools and machinery, and the determination of wage by output, the work performed by a laborer was to be divided into ever more specific parts, into mind-numbingly simple tasks that required regular motion with minimal steps. To maximize productivity, a "time-and-motion expert" was to be present on the floor, measuring workers' efficiency with a stopwatch for the purpose of analysis and improvement. For many, Taylorism represented the ultimate step in exploitation and dehumanization, with the presence of the time-and-motion expert as a particularly odious feature, while others saw in it the most scientific method of increasing efficiency and boosting productivity. The debate lasted until the outbreak of the war, when the needs of the war economy necessitated the adoption of the method.[12] So the war not only turned soldiers into machines, fighting actual machines on the battlefield, it also took a further step in turning workers into automatic devices whose monotonous movements were to be constantly timed and measured. The implementation of Taylorist organization and the spectacular rise of the United States as a major industrial power in this period also accounts for the frequent appearance of "American" images in revolt-of-the-machine narratives.

The 1917 Bolshevik triumph in Russia was greeted alternatively as the great hope and as the nightmarish consequence of the class conflict that had been brewing across Europe throughout the industrial era. The explicit purpose of the newly founded Soviet state's leaders was to turn it into an industrialized utopia, with the efficient but humanely run factory as the model for the entire society.[13] Prior to the revolution, Marxists were ambivalent about Taylorism, concerned that the system could further exploit workers and weaken their effort at collective action, but the Bolshevik leaders adopted it wholeheartedly, with Lenin asserting that the method was oppressive in a capitalist economy but positive in a socialist setting since it increased productivity and shortened the duration of necessary labor.[14] In this vein, the poet Alexei Gastev celebrated the modernized industrial system and the mechanization of the worker in many of his works.[15] In his 1914 poem "We Grow out of Metal," he envisions the transformation of the Soviet mechanical superman in a manner reminiscent of Marinetti.

> Into my veins runs a new, metallic blood.
> I have grown some more.
> I myself am growing steel shoulders and infinitely strong hands.
> I have merged with the metal of the edifice.
> I have raised myself up.
> With my shoulders, I push out the rafters, the beams, the roof.
> My feet are still on the ground, but my head is above the building.
> I am still gasping from these inhuman exertions, but I shout [already]:
> —I ask for the floor, comrades, the floor!
> A metal echo drowns out my words, the entire construction trembles with impatience.
> But I am raised still higher, already even with the smoke stacks.
> And not a story, not a speech, but only my iron word:
> I'll yell:
> "We will be victorious!"[16]

In direct contrast, Yevgeny Zamyatin's dystopian novel *We* (written in 1920–1921) portrays a society in which the Taylorist regime has been imposed to the extent that its citizens have been reduced to obedient working automata with names like D-503, I-330 and O-90. The story was originally aimed at the advanced industrial nations of the West, but it also expressed concern over the socioeconomic developments in the Soviet Union.[17] With the Russian Revolution playing a significant role in the continuing meditations in the West on the development of the modern industrial system, the motif of the workers' revolution figured prominently in the automaton literature of the interwar period.

As a consequence of these historical developments, there were many expressions of fear and revulsion toward modern industry and machinery in the larger culture of the interwar decades, many of them featuring the theme of the mechanization of humanity and society. To examine the technophobia of the time as a background to its automaton stories, it will be useful to examine a few representative cases of such expressions from the disparate fields of psychiatry, art, and cinema.

The Influencing Machine, Republican Automata, and the Steel Animal

In the nineteenth century great advancements were made in the scientific understanding and technological application of electricity, from the pioneering works by the *Naturphilosoph* Hans Christian Oersted to the full elaboration and formulation of electromagnetic force by Michael Faraday and James Clerk Maxwell, and ultimately to the world-changing inventions of Thomas Alva Edison. A controversial application of electricity was in the field of medicine, with many people of varying degrees of intellectual respectability claiming to have found ways of curing both physical and mental ailments through the use of electric devices. Their ideas, which harken back to the practice of mesmerism, were often met with skepticism by the medical community, not only because no verifiable science of electromedicine emerged at the time but also because many of the demonstrations of such applications took place outside established experimental practices and institutions, giving their practitioners the air of the impresario and the charlatan.[18] In the last decades of the nineteenth century, however, there were renewed attempts by legitimate doctors to use electrotherapy and other methods in the treatment of the mentally ill.[19] One of the devices that appeared in this context was the so-called influence machine, a spectacular instrument that gave off showers of sparks from static electricity, supposedly creating a force field with beneficial effects on those with nerve-related illnesses. The famed French neurologist Jean-Martin Charcot doubted the claims of its efficacy but considered it to have positive value in the treatment of hysterics through the power of suggestion.[20]

During the interwar period, the psychologist Victor Tausk, a disciple of Freud, wrote a pioneering work on schizophrenia, "On the Origin of the 'Influencing Machine' in Schizophrenia" (1919).[21] He points to a prevalent fantasy among psychotics of a machine that is used to manipulate their bodily and mental functions from a distance. As Tausk describes such a device that is identified as the "influencing machine":

> It consists of boxes, cranks, levers, wheel, buttons, wires, batteries, and the like. Patients endeavor to discover the construction of the apparatus by means of their technical knowledge, and it appears that with the progressive popularization of the sciences, all the forces known to technology are utilized to explain the functioning of the apparatus. All the discoveries of mankind, however, are regarded as inadequate to explain the marvelous powers of this machine, by which the patients feel themselves persecuted.[22]

A woman named Natalija A. suffers from the delusion that a suitor she has rejected possesses an electrical machine in the shape of her body with batteries inside that "represent the internal organs."[23] Whatever he does to this automaton double happens to her, including the loss of feeling in her genitals after her tormentor removes the corresponding part in the machine. Tausk explains the hallucinations in terms of schizophrenics' lack of "ego-boundary," as they have difficulty distinguishing thoughts and sensations occurring in their minds from events taking place in the outside world. What is interesting about this case is that what was originally demonstrated in the prewar era as a device that could cure people suffering from nervous illnesses was imagined as an uncanny machine that caused torment through remote manipulation and sexual abuse.[24]

This image has a wider cultural significance, especially in the context of postwar Germany, in that an outpouring of artistic works dealt not only with fearful and maddening machines but also with a mesmerized subject who is manipulated into committing horrific acts. Among the great expressionist films of the period, there were *The Cabinet of Dr. Caligari* (1919), about a hypnotized man who is forced to murder people by his master; *Nosferatu: A*

Symphony of Horror (1921), based loosely on Bram Stroker's *Dracula*, about a vampire with hypnotic powers; *The Golem* (two versions, 1914, 1920), featuring an artificial being that goes out of control; *The Student of Prague* (two versions, 1913, 1926), involving an uncanny double; and *The Hands of Orlac* (1924), about disembodied body parts that turn against the protagonist. Various scholars have examined such works in the contexts of the trauma of World War I, the dissolution of the German and Austro-Hungarian empires, and the ongoing crises of the unstable Weimar state. There was also a sense among many, especially war veterans, of having been manipulated by incompetent and callous political and military leaders who were directly responsible for the disasters of the era.[25] They felt themselves, in other words, to be the mesmerized victims of the influencing machine of the state controlled by the diabolical Caligaris of authority. Just as in the Romantic period, anxieties

George Grosz, *Republican Automatons*. 1920. Watercolor and pencil on paper, 23 ⅝ x 18 ⅝" (60 x 47.3 cm). Advisory Committee Fund. (120.1946). The Museum of Modern Art, New York, NY, USA © Estate of George Grosz/licensed by VAGA, New York, NY. Digitial Image © The Museum of Modern Art/Licensed by SCALA / Art Resource, NY.

George Grosz, *Daum Marries Her Pedantic Automaton "George" in May 1920, John Heartfield Is Very Glad of It*. 1920. Watercolor, pencil, pen collage on cardboard. Galerie Nierendorf, Berlin. Photo Credit: Ericj Lessing / Art Resource, NY. © Estate of George Grosz/licensed by VAGA, New York, NY.

aroused by sociopolitical instability were often expressed through such stories of uncanny and supernatural horrors.

As mentioned previously, machines of all kinds were represented in significant modernist paintings and sculptures. Mechanical images that appeared after the war tended to be darker ones of either the irrational machine or the dehumanized human being as a machine, whether as a worker, a soldier, or a mindless conformist. The paintings of George Grosz, a veteran of the war, provide some of the best examples. Grosz is most famous for his drawings of Berlin life in the interwar period, including scathing caricatures of the city's citizens, from the rich and the powerful to the petty bourgeoisie and prostitutes. In 1920 he produced a series of provocative works with mechanized images of people, one scholar calling it Grosz's year of robots.[26] Among them is "Republican Automata," a painting that features two figures: a flag-waving veteran with missing limbs and the number twelve on his head in lieu of a face and an upper-class figure in a dinner jacket also with hands missing, with a medal and a series of gears at its side, uttering "1, 2, 3, Hurrah" out of its empty head. The war and mindless nationalism have reduced them both to machines, with numbers and slogans in place of real identities. In another work, "Daum Marries Her Pedantic Automaton 'George' in May 1920. John Heartfield Is Very Glad of It," a half-naked woman turns away from a collage image of a robotic figure consisting of machine parts. Since Daum was the nickname of Eva Peter, whom Grosz married that year (a small photograph of her appears in the upper left corner of the work), the automaton George represents the artist himself, with numbers on his head manipulated by a pair of hands, representing the alienation he feels from his environment and the sense of his own self.[27] Grosz explains in an essay on his 1920 works: "In my so-called artistic works, I set out to build an entirely real platform. Man is no longer shown as individual, with psychological subtleties, but as collectivistic, almost mechanistic concept. The fate of the individual no longer counts."[28] The satirical depiction of people as mindless automata carries with it the inherent danger that it may reveal the artist himself to be a machine as well.

A later but highly significant example of the fearful and irrational machine can be found in an episode described by Leni Riefenstahl, the director of the propaganda film *Triumph of the Will* (1935). Sixty years after Ernst Engel toasted the marriage of steam and engine, Germany, now under the Nazi regime, reached another centennial that occasioned a celebration—the hundredth anniversary of the building of the first railways across the nation, which had initiated the so-called second industrial revolution. To mark the occasion, the directors of the Reich Office of Railway Transportation commissioned a film on the history of the German railways. The man chosen for the task, the avant-garde director Willy Zielke, produced a work entitled *Das Stahltier* (The steel animal). When it was screened for the officials in 1935, however, they were horrified by what they saw. As Riefenstahl describes the movie:

> [Zielke's] locomotive looked like a living monster. The headlights were its eyes, the instruments its brain, the pistons its joints, and the oil dripping from the moving pistons looked like blood. This overall impression was intensified by the revolutionary sound montage.... The railway cars smashed together so violently when they were being switched that the viewers jumped out of their seats. It was a shock for the railway directors who wanted train travel to be smooth and gentle.[29]

Riefenstahl herself thought it was a brilliant film, "a grand visual symphony, such as I hadn't experienced since Eisenstein's *Potemkin*," that was decades ahead of its time.[30] But the directors were so offended by it that they tried to destroy all prints of the film, which compelled Zilke to appeal to Riefenstahl, who was well connected with the Nazi leadership. She attempted to save the work by arranging a screening for the propaganda minister, Joseph Goebbels. After the viewing, Goebbels thought the director talented but said his work was "too modern and too abstract; it could be a Bolshevist film," and refused to intervene on his behalf.[31]

Given Jeffrey Herf's characterization of "reactionary modernism" as an ideology that rejects the rationalist and liberal values of the Enlightenment but embraces modern technology as the embodiment of irrational will, Zilke's portrayal of the locomotive as a mad, bloody monster-machine can be regarded as a representation of the Third Reich itself.[32] In the ongoing debates about the nature of the Nazi Holocaust, some scholars have asserted that while there have been large-scale ethnic genocides in the past, one aspect of the Nazi extermination effort that made it unique was its technological nature—not just in the use of poison gas and other products of modern science and technology but in the way the death camps were run like factories with the prisoners as so many devices that were made use of and then systematically destroyed.[33] In the view of the victims of that most monstrous of machines, the Holocaust survivors Primo Levi and Elie Wiesel both evoked the image of the automaton in their searing descriptions of both the victims and the oppressors caught up in this ultimate form of dehumanization. Levi writes in *Survival in Auschwitz* (1947):

> When this music plays we know that our comrades, out in the fog, are marching like automatons; their souls are dead and the music drives them, like the wind drives dead leaves, and takes the place of their wills. There is no longer any will; every beat of the drum becomes a step, a reflected contradiction of exhausted muscles. The Germans have succeeded in this. They are ten thousand and they are a single grey machine; they are exactly determined; they do not think and they do not desire, they walk.[34]

Wiesel writes in *Night* (1960):

> The SS made us increase our pace. "Faster, you swine, you filthy sons of bitches!" Why not? The movement warmed us up a little. The blood flowed more easily in our veins. One felt oneself reviving....
> "Faster, you filthy sons of bitches!" We were no longer marching; we were running. Like automatons.[35]

For the late Enlightenment writers who used the automaton-man image for satirical purposes, the figure of the person lacking freedom deserved contempt or pity. In the modern age, however, in the wake of the spread of industrial machinery and the horrors of technological warfare, the man-machine became a much darker entity that appeared in the context of the total loss of humanity, systematic destruction on a massive scale, and modernity itself as a bloodthirsty monster of necromantic hunger.

What is striking about the major works from the era that feature the automaton is that they are so similar to one another in theme, plot, and even characters. They take place in a world that is bifurcated into two realms—a leisurely paradise of the privileged few and an industrial hell of laborers. The people of the former world are described as cerebral masters of great machines, keeping their control over both realms through their intellects and technological ingenuity, while those of the latter are dehumanized workers who have been conditioned into obeying and toiling. A crisis develops, born out of the tensions within the binary system, consisting variously of the workers' revolt, the machines' malfunction, or some unforeseen consequence of overdevelopment. The masters' effort to correct the situation through extreme and callous means brings about a catastrophe of massive destruction. These stories also feature a female character, a woman of nature and sentiment who represents humanity's natural self. She also functions as a boundary-crosser who traverses the realms of the workers and masters to mediate between the two. The gender implication here is that the drive behind technological modernity is seen as essentially a male one that needs to be moderated by the feminine if a great disaster for all of humanity is to be avoided.

These industrial automaton narratives are filled with overt references to contemporary developments of modern warfare, revolution, and dehumanization, revealing their authors' struggles with the great and urgent issues of the day. They are the darkest visions of the fate of humanity in the age of autonomous machinery.

The Revolt of the Machines: Čapek and Rolland

The word "robot" (which has outstripped "automaton" as the most commonly used word denoting a self-moving machine that mimics a living creature) was first coined by the Czech artist and writer Josef Čapek from the Slavic word *robota*, "drudgery," derived from its medieval sense of the unpaid labor a vassal was obliged to perform for his feudal lord.[36] It was adopted by his brother Karel for his celebrated 1921 play *R. U. R.: Rossum's Universal Robots*.[37] The story takes place in an industrial setting, a factory complex on a remote island that produces artificial beings designed to be workers and servants of humanity. Harry Domins, the central director of the company, explains to Helena Glory, the visiting daughter of the nation's president, that a materialist philosopher named Rossum (from *rozum*, Czech "reason") has discovered a living matter on the island that he has shaped into animal-like beings.[38] To prove that God was unnecessary in the process of the rise of life on Earth, he made humanoid creatures out of them. Rossum's more practical son transformed them into obedient and efficient workers, which the company now mass-produces for the benefit of the entire world.

It turns out that Helen (the female character of nature and sentiment in this work) is a secret activist for the League of Humanity, an organization that wants to set the robots free. When her purpose is revealed, Domin assures her that the League's efforts to emancipate the artificial beings are in vain since the creatures have no will of their own. But Dr. Hallemeier, a psychologist in charge of educating the creatures, admits that every once in awhile a robot malfunctions and "breaks whatever it has in hand, stops working, and gnashes its teeth," so that it has to be sent to the "stamping-mill."[39] Domin dismisses Helen's insistence that it must be the sign of a soul in rebellion against servitude and explains the great benefits the robots will bring to the world. It is true that they are depriving human workers of their occupations, but

> within the next ten years, Rossum's Universal Robots will produce so much wheat, so much cloth, so much everything that things will no longer have any value. Everyone will be able to take as much as he needs. There'll be no more poverty. Yes, people will be out of work, but by then there'll be no work left to be done. Everything will be done by living machines. People will do only what they enjoy. They will live only to perfect themselves.[40]

The next act takes place ten years later, with Helen living on the island as Domin's wife. The situation with the robots begins to unravel. A revolt by human workers against the artificial creatures who have replaced them in their jobs has forced the robots' owners to give them weapons to defend themselves. When they proved themselves to be able fighters, they were used as soldiers in the wars that followed. But the number of robot malfunctions has increased, sometimes resulting in the murder of people. Eventually the rebellious creatures have come together to form a union to collectively stand up to their masters. Even more disturbing is the news of widespread sterility among humans.

When Helen asks various people for their opinions on its cause, her uneducated and religious nurse, Nana, says she believes it to be punishment from God for the blasphemy of creating new life; Alquist, the chief builder of the robots, thinks that procreation is impossible in a cursed paradise where no one does any work; and Dr. Gall, the head of physiological research, speculates that Nature herself has been offended by the production of robots and is now wreaking her vengeance on humanity. Ultimately, the robots rise up in general revolt against humanity and begin to exterminate everyone. Their leader, Radius, shouts out their revolutionary slogan, which is reminiscent of Marx's *Communist Manifesto:* "Robots of the world! Many people have fallen. By seizing the factory we have become the masters of everything. The age of mankind is over. A new world has begun! The rule of Robots!"[41]

In the final act, all humans have been wiped out except for Alquist, who has been spared because the robots consider him one of them, as he is the only one on the island who has worked with his hands—in building his house. But then the triumphant creatures reach a crisis and look to him for help. The robots were designed to last for a limited period of time with no capacity to reproduce themselves, and the formula for creating new ones was destroyed by Helen during the revolution. They beg Alquist to piece together what knowledge remains of their construction so their kind can live on.

FOURTH ROBOT: Teach us to make Robots.

DAMON: We will give birth by machine. We will build a thousand steam-powered mothers. From them will pour forth a river of life. Nothing but life! Nothing but Robots!

ALQUIST: Robots are not life. Robots are machines.

SECOND ROBOT: We were machines, sir, but from horror and suffering, we've become—

ALQUIST: What?

SECOND ROBOT: Something struggles within us. There are moments when something gets into us. Thoughts come to us which are not our own.

THIRD ROBOT: Hear us, oh hear us! People are our fathers! The voice that cries out that you want to live; the voice that complains; the voice that speaks of eternity—that is their voice![42]

Alquist falls into despair, not knowing whether he should help these creatures who killed off humanity, and whether he could help them even if he wanted to. He becomes convinced that there is no hope, that all sentient life will disappear from Earth. He then awakes from a nap to find two robots, Primus and Helena, acting very much like they are in love. He tests them by claiming that he needs to dissect one of them for his research. When each of them offers to die in place of the other, Alquist realizes that their feelings are genuine and that the robots are becoming human. The play ends with Alquist's joyous conviction that as long as love survives there is always hope that life will be reborn and continue somehow, even if in a different form from that of humanity.

It is significant that the robots increasingly take on human characteristics in the course of the play, no longer acting like machines after their victory over mankind. They want to feel, procreate, and love, insisting that they now possess souls. It turns out, then, that Helen was right to see in the early breakdowns of robots the stirring of a budding humanity, their rebellion itself a sign of their transcendence of their artificial origin. The themes of class conflict, revolution, and the dangers of the unbridled industrialization in the play are explored in an uncompromising fashion as the human race dies out. But for Čapek, a staunch supporter of the liberal democracy of Czechoslovakia in the interwar period and an outspoken antifascist, there was always hope that even through a devastating high-tech war, a bloody revolution, and the wholesale destruction of humankind, the most precious aspects of humanity would find a way to survive.[43]

Despite the fact that *R. U. R.* was Čapek's most successful work, he never liked it much and considered it the least of his dramatic writings.[44] During a 1924 performance in Prague, he sat in the president's box with the French writer and Nobel laureate Romain Rolland, providing a translation of the dialogue for him even while pointing to the weaknesses of his own work.[45] Both during and after World War I, Rolland was one of the most outspoken pacifists in the European literary world. While he believed that radical change needed to take place in the Western political and economic landscape to avoid another catastrophe, he was ambivalent about the Bolshevik revolution in Russia. He defended the new regime's right to experiment with a new kind of state, but he became increasingly concerned with the overly dogmatic way it was developing, as he expressed in his public debate of 1921–1922 with another major French pacifist, Henri Barbusse, who had become a committed Marxist.[46]

In the same year Čapek's play was first performed, Rolland published a work with a similar theme, *The Revolt of the Machines*—a "motion picture fantasy," or screenplay for a film that was never produced. While its plot also features machines that rise up against their masters, it emphasizes the irrationality of technology in the behavior of the mechanical creatures. The story begins on the opening day of the Palace of Machinery, an immense escalator-ringed hall for the display of the inventions of Martin Pilon, the Master of Machines, an aloof but passionate engineering genius. In the grand ceremony that takes place, attended by illustrious politicians, generals, foreign sovereigns, and ambassadors, the president delivers a speech on the progress of civilization. He relates how humanity was able to rise out of its pastoral barbarity through the inventions of tools and machinery, allowing man to grow from a "prehistoric primate to the modern demigod."[47] He then points to the image of the latter:

> See the ultra-modern type of the American demigod, who from his office chair rules the sun and moon and the elements. An army of machines obeys the negligent pressure of his fingers on an electric keyboard.
>
> Let us do homage, gentlemen, to this magnificent sight: man, King of machinery! We celebrate today his victory, the apogee of human progress and human genius.[48]

Pilon is accompanied by his wife, Félicité, who is a simple peasant woman, but he lusts after a famous actress, Hortense. The machines are turned on, and their creator gives the people a tour while paying special attention to Hortense. The spectators are impressed by gigantic machines that can lift enormous weights and stretch out their arms like great spiders and a psychological machine that can display images of people's thoughts. When some young people play a prank on Pilon, his vexation causes the machines to act on his subconscious desires, playing tricks on those he secretly despises. The escalator malfunctions, causing the dignitaries on it to bounce up and down; a machine pinches Hortense in the rear; another winds its rubber pipe around a diplomat who is flirting with her; and others spray cement and blow on other guests. When one of the devices snatches up the president and dangles him by the seat of his pants, "The Master storms at the machine and pleads with it, at the same time he is subconsciously so amused that he cannot help laughing at the grotesque poses of the President."[49] The crowd then turns against Pilon, and he is subsequently put under arrest. It is only when he is led off to prison that the machines become still.

That night the machines come alive again and murder the night watchmen. At first they act like wild beasts: "We hear the whistling, howling, strident laughter and braying of monsters. We see a hundred steel arms stretch and bend, leather belts begin to move, wheels to turn, boilers to steam, exhaust pipes to hiss."[50] But soon they organize themselves to systematically wreak havoc on a "city of colossal American skyscrapers ... with factory chimneys and towers and one solitary church steeple."[51] As other machines join the fight, buildings are pulled down, people are slaughtered, and the few survivors, including the people who were at the Palace of Machines, gather on a hill outside the city. In the new situation, the old order is turned upside down, as Pilon's peasant wife Félicité (the female character of nature and sentiment) takes charge and proves to be a capable leader, while the president and his ministers turn out to be useless. As the survivors flee up a mountain, they see that the people who have surrendered are being enslaved by the machines, made to labor as the machines once did for them.

In the last act, Pilon comes to a crucial understanding about the rebellious machines. While the devices may be the products of science and technology, they themselves are hardly rational creatures, acting on recognizably human emotions. Their creator goes down among them and turns them against one another.

> He sows discord among the Machines. Proud and stupid as they are, they admire themselves and long to be admired by others. He therefore shows admiration for some, to excite the jealousy of the others. He persuades them that they are the most beautiful, the strongest, that the supreme authority belongs to them. ... Soon they are at each other's throats. ... Neighing, lowing, bounding, kicking, firing, they rush upon one another.
>
> Airplanes against hydroplanes, tanks and engines of war against machines of peace, giant cranes, mechanical saws, perforators, etc.— they seize each other and roll on the rocky slopes, or come crashing

through the air, disemboweled, smashed, blown up, or they are sunk to the bottom of the sea.[52]

The machines destroy themselves in general warfare, and the surviving humans descend from the mountain to form an agrarian society devoid of autonomous devices, with Félicité as their leader. After a harvest scene, the people gather for a celebration during which the comical president delivers another speech, the content of which is the exact reverse of the one he gave at the Palace of Machines. Humanity has reached the highest level by freeing itself of the laws of scientific barbarism and machine civilization. The ideal human is no longer the American industrialist surrounded by machinery but a tranquil sage, watching his flocks and piping on his reed.[53] The people play music and dance, and lovers walk off together, but Martin Pilon, the ex–Master of the Machines, stands by himself. The story ends in an ominous note:

> He cannot reconcile himself to this life of nature, this life without machines. He talks to himself and he gesticulates. He is tense. . . . He draws feverishly, covering the surrounding stones with geometric figures and calculations.
>
> Suddenly we see, projected against the gold of the sunset sky, the formidable shadow of Machines far more monstrous than those of yore, the dreams of the inventor.
>
> The finished cycle begins anew.[54]

Rolland portrays the disaster brought on by the revolt of the machines as the manifestation not of a new mechanical consciousness but of the primordial human urge to fight, destroy, and dominate. Modern technology is problematic in its capacity to multiply our power, but the root of its destructive force lies in ourselves, and its danger will continue as long as we are unable to control our own violent impulses. In many stories of machines revolting that have been told from the interwar period to the present time, the mechanical beings are often presented simply as the Other, acting on some mysterious and diabolic impulse to destroy their creators. But in Čapek's and Rolland's narratives, their behavior is linked directly to human nature, their rebellion and will to power reflecting their creators' inner desires. The revolt of the machines becomes, then, a representation of the revolt of humanity's dark side (or the Freudian notion of the "death instinct") against its own civilization, an issue of central concern in the postwar period.

The Breaking of the Machine World: Vasari, Lang, and Harbou

During World War I, the Futurist movement lost many of its most significant members, including its greatest artist Umberto Boccioni and the architect Antonio Sant'Elia. In the aftermath, the movement's leader, F. T. Marinetti, who was also wounded during the conflict, gathered a new group of artists and writers before fatally throwing his support to Mussolini. Among the postwar Futurists was the Sicilian dramatist Ruggero Vasari, who conceived his play *The Anguish of the Machines* in 1921. It was put on in 1923 after he saw a performance in Berlin of *R. U. R.*, about which he wrote a mixed review.[55] The work features some of the most explicit technophobic themes in the literature of the period, so it is surprising that Marinetti, the great advocate of the irrational machine and the merging of man with machine, not only approved of the play but considered it one of the most important works Futurism produced. This may point to a dampening of his enthusiasm for modern technology after the experience of the war.

The story takes place in a bizarre world of three superman-figures: the dictator Bacal, his second-in-command, Singar, and the technological genius Tronchir, who rule over a kingdom of machines, inhabited by both worker-automata and human workers who have been turned into machines. This society is kept in order by the Brain-Machine, which embodies the three rulers' collective minds and keeps absolute control over the realm. There are no females in this world to deplete the masculine energy of the rulers since the women have all been exiled to the "old continent." The first act begins with the return of the women in airplanes and the appearance of their representative, Lipa (the female character of nature and sentiment), who comes to Bacal and Singar. She tries to persuade them to return to the natural state by reuniting with women so they may procreate and continue the human race, but Bacal rejects her: "What did you do for us? You have ensnared us in the softness of your arms—sucked every drop of virility—shrouded us with the noxious smoke of your opium."[56] He then shows her the destruction of the women's airplanes by the superior technology of the machine world. So far, this is in line with Futurist themes of the prewar era, technology as the product of sheer male will, woman as a debilitating entity to the life force, and the dream of absolute power and control through the rejection of bourgeois sentiment.

In the second act, things take a drastically different turn as Lipa meets Tonchir, the engineering genius whose inventions have made the kingdom of machines possible. When the technological master first appears, he is in the throes of despair, as he realizes that the world he has created is a horrific one of tyranny and slavery that is bound to destroy itself sooner or later. As he expresses his anguish in a semicoherent rant—

> I can't go on . . . It is the machine that prolongs agony. I can't go on. To dehumanize mankind—to turn men into machines—machines that are not machines—men that are no longer men. And yet . . . maybe the last convulsion of this aberrant world will be born here? After us—nobody else! The last inhabitants—the last tyrants! Then—why not speed it up? More fierceness—mutilations![57]

As Tonchir plans to accelerate the process of destruction, Lipa tries to persuade him to choose the way of nature by loving her and finding his humanity again. He is tempted, but he is then visited by three female spirits from his past that drive him further into the conviction that he has committed an unforgivable offense against nature.

The Brain-Machine is connected to the minds of the three rulers, so everything in the mechanical world functions smoothly as long as they are in perfect harmony of purpose. As Tonchir's mind disintegrates, however, the machine also becomes deranged, and the entire order starts to fall apart. Bacal and Singar try to prevent the dissolution of their kingdom by urging Lipa to kill Tonchir, but she refuses to do so, making instead one last attempt to persuade him to choose the way of life and spirit. It is all to no avail, however, as he makes his farewell to his beloved Brain-Machine and walks off to his death. The machine, now reflecting its creator's madness, loses control over itself, and all is destroyed in the end.

While this work retains many of the Futurists' themes and aesthetics, it also reads as a damning critique of the utopian vision of the prewar Futurists with their dream of the reign of irrational and superhuman man-machines that would transcend the world of nature and femininity. Vasari makes a reference to the gigantic winged automaton that appears at the end of Marinetti's *Mafarka the Futurist* in a scene in which Lipa gazes on Tonchir's machines and feels an erotic attraction toward them.

> Divine Machine—embrace me! Embrace me with your rounded, shining rods—your arms bend me in a red spasm. Let me kiss your belly, polished as that of an adolescent! My hair is caressed by your convulsed flywheels! Divine one—hug me—hold me tightly—take me entirely! I'll give you a mighty son—the winged son—the son-god![58]

As one can see in the overt references in the play to technological warfare, the dehumanization of industrial laborers, and the threat of a breakdown of the entire social and economic system, Vasari speculates that if the ultimate fantasy of the Futurists is realized, the world will not be a dynamic utopia of mechanical supermen but a sterile and hellish factory of debased humanity ruled by irrational and destructive despots. For the despairing Tonchir, the only feeble hope that remains is in the knowledge that such a world, no matter how powerful its instruments of control, can never last, as it will inevitably destroy itself in the fire of its own lunacy.

The most famous work of the era that features such a crisis in a machine-world is Fritz Lang's expressionist film *Metropolis*, which was first screened in 1927, though the novel version—by his wife, Thea von Harbou, and based on the screenplay, on which they had collaborated—was published the year before. The initial inspiration came from Lang's view of the Manhattan skyline during his 1924 visit to the United States.[59] The narrative takes place in a great city of skyscrapers and factories, but divided once again into a luxurious upper world of the privileged and a hellish underground of workers. Among the fortunate living in the higher stratum is Freder, the son of Joh (i.e., Jehovah) Fredersen, the absolute Master of Metropolis, who leads a leisurely, idyllic existence until he encounters Maria (the female character of nature and sentiment) leading a group of children from below. Freder is so smitten by her that he goes down to the machine-world, where he is appalled by the suffering of the people. One of the most striking scenes in the film is the march of workers during shift change, endless lines of them walking in step, with their heads bowed down, into what looks like the gaping mouths of machines. As Harbou describes it in the novel:

> Men, men, men—all in the same uniform, from throat to ankle in dark blue linen, bare feet in the same hard shoes, hair tightly pressed down by the same black caps . . . they all had the same faces. And they all appeared to be of the same age. They held themselves straightened up, but not straight. They did not raise their heads, they pushed them forward. They planted their feet forward, but they did not walk. The open gates of the New Tower of Babel, the machine center of Metropolis, gulped the masses down.[60]

In the film, Freder has a vision of the so-called Pater Noster machine as an open-mouthed face of a pagan god into whose fiery mouth the workers are thrown as sacrifices, which some have seen as a presaging of the Nazi death camps.[61] While the laborers have been reduced to obedient machines, the industrial machines are described in animistic terms, as superhuman creatures. Freder goes to his father and explains what he has seen of them in a manner reminiscent of Henry Adams at the hall of dynamos:

> I went through the machine-rooms—they were like temples. All the great gods were living in white temples. I saw Baal and Moloch, Huitzilopochtil and Durghas . . . all machines, machines, machines, which, confined to their pedestals, like deities to their temple thrones, from the resting places which bore them, lived their god-like lives . . . And near the god-machine, the slaves of the god-machines: the men who were as though crushed between machine companionability and machine solitude. . . . They have nothing else to do but eternally one and the same thing, always the same clutch at the same second, at the same second.[62]

For the Master of Metropolis, however, the problem is that the workers are not machine-like enough, as they are subject to fatigue. As he coldly explains to his son:

> That men are used so rapidly at the machines, Freder, is no proof of the greed of the machines, but the deficiency of the human material. Man is the product of change, Freder. A once-and-for-all being. If he is miscast he cannot be sent back to the melting-furnace. One is obliged to use him as he is. Whereby it has been statistically proved that the powers of performance or the non-intellectual worker lessen from month to month.[63]

The solution to this problem is to replace them with machine-men.

Unable to persuade his father to take a different course, Freder returns to the underground in search of Maria, while the Master of Metropolis visits Rotwang, a sorcerer-scientist with an artificial hand who lives in a gothic house decorated with a pentagram. In their conversation, it becomes clear that Joh has long been planning to replace human workers with machines, as he demands that Rotwang deliver his order.[64] Rotwang instead shows him a robot in female shape, which, like Villier's Hadaly, can be molded into the shape of one's desire. Two men then go down a secret passage to the underground, where they listen in on the workers, who are planning a revolt. They are restrained by Maria, who urges patience, as she hopes for a peaceful rapprochement between the two worlds. It is at this

Fritz Lang, *Metropolis* (1927). The robot as a woman. Illustration courtesy of Photofest NYC.

Fritz Lang, *Metropolis* (1927). The workers as robots. Illustration courtesy of Photofest NYC.

point that Joh makes an enigmatic move. He tells Rotwang to transform the robot into another Maria and send it down to wreak havoc among the workers.

Freder, meanwhile, tries to help an exhausted worker he encounters by taking his place at his machine. In this scene, Lang comments directly on the Taylorist regime and its time-and-motion study by portraying the machine as a gigantic, illuminated clock on which Freder has to move the arms faster and faster until he nearly collapses from fatigue.[65] When the shift is finally over, he trudges further into the underground with other workers to attend a speech given by Maria, in which she portrays those of the upper world as the brain and those of the lower as the hands: "That the Brain and Hands no longer understand each other will one day destroy the New Tower of Babel.... Brain and the Hands need a mediator. The Mediator between Brain and Hands must be the Heart."[66] Freder and Maria finally meet, and after he expresses his earnest desire to become that mediator, they fall in love.

In the rather complex plot that follows, Rotwang kidnaps Maria and sends out the robot posing as her, which visits a high-class entertainment club in the upper world to beguile the men there and then goes down to the underground to incite the workers to revolt. Although it seems as if it is acting under the orders of Rotwang, who in turn is following Joh's

Fritz Lang, *Metropolis* (1927). Freder tries to help a worker at a clock-like machine—a commentary on the adoption of Taylorism in European factories. Illustration courtesy of Photofest NYC.

command, the reasons for much of its actions are unclear, unless the artificial being is following its own mysterious impulses. Critics have noted that the version of the film initially released in the United States does not explain why Joh would want the robot to raise up the masses below. The novel does provide an answer of a sort but not an altogether satisfactory one. As Freder confronts his father again with the catastrophe that would result in a revolution, Joh replies:

> "Death has come upon the city by my will."
> "By your will—?"
> "Yes."
> "The city is to perish?"
> "Don't you know why, Freder? . . . The city is to go to ruin so that you may build it up again. . . . For your sake, Freder; so that you could redeem them."[67]

This could be interpreted as Joh wishing his son to carry out his final plan, to have the city destroyed and its workers exterminated so that Freder can rebuild it with machine-men instead of human workers. The workers' rebellion he has secretly provoked through the false Maria robot begins in earnest, as the masses rise out of the underground, shouting their slogan:

> We've passed sentence upon the machines!
> We have condemned the machines to death!
> The machines must die—to hell with them!
> Death!—Death!—Death to the machines![68]

The uprising, however, causes the machines of the city to be neglected, resulting in their coming alive with fury, in an orgy of destruction.

In the ensuing confusion, Maria manages to escape from Rotwang and goes underground to meet up with Freder again to save a group of children from drowning in a flood. The masses above, thinking that their children have perished, turn against the robot Maria and burn it on a great pyre as a witch, where it laughs insanely as it is destroyed. The children and their saviors appear, to everyone's relief, but Rotwang kidnaps Maria again, resulting in a fight on top of a church, from where Freder sends the evil scientist flying to his death. The novel concludes on an uncertain note, with Freder vowing to serve as the mediator between the two worlds. The ending of the film version, regarded by many as the weakest part of the work, features a scene of class reconciliation, as Joh and the chief engineer of the workers are urged by Freder and Maria to shake hands in front of all.

Peter S. Fisher has demonstrated the great popularity of narratives of the future *(Zukunftsroman)* in Weimar Germany, featuring ideas that ran the gamut from right-wing fascist to liberal pacifist.[69] Some scholars have interpreted *Metropolis* as a protofascist work (Thea von Harbou was an ardent supporter of the Nazi regime after it took power, but Fritz Lang divorced her and emigrated to the United States in 1934).[70] It is true that the story shares some prominent themes with the more reactionary narratives of the time, including that of the Mediator, a mythic figure who through the sheer power of his personality is able to bring about a reconciliation of the classes and put an end to their conflict, which is tearing the society apart.[71] This reading of the work, however, is an overdetermined one that misses some of the central themes that emerged from the general technophobia of the period, in writings of both the left and the right.[72] After the trauma of technological warfare and the humiliation of defeat in World

War I, ambivalence toward machinery and industry was particularly pronounced in German culture.[73] In some of the most nationalistic narratives of the future, technology was portrayed as an instrument of greedy and corrupt capitalists, often Jews, who were responsible for the disastrous war. Jeffrey Herf has shown that in Germany, "die Streit um die Technik" (the debate on technology) was carried out intensely throughout the Weimar period, as "hundreds of books, lectures, and essays emerged from both the technical universities and nontechnical intellectuals from all points along the political spectrum dealing with the relation between Germany's soul and modern technology."[74] Considerations of its ideological alignment aside, *Metropolis* is one of the most overt expressions and certainly the most visually powerful one of that debate.

In the two-tiered city, the workers have been reduced to automata even as the machines they serve have grown to the dimensions of gods and monsters. It is clear, however, that the Master of Metropolis has failed to completely deprive them of their spirit since the contradictions inherent in the system cause them to rebel. The ultimate revolution is incited by a robot, with the workers unaware that it was all planned by the Master, who is plotting their destruction, possibly so that they may be replaced by machines. In this narrative the most beguiling image is that of the robot itself. It is a rather shadowy entity in the novel, whose actions are often told in secondhand accounts, but Brigitte Helm gives a stunning performance as both the humane Maria and her destructive double. Scenes of the metallic construction's transformation into a woman, its performance at the entertainment club, and its lunatic laughter at its fiery end are some of the most memorable moments of the film. As Andreas Huyssen has pointed out, the robot becomes the concrete symbol of both the fascinating and fearful faces of technology itself—an artificial entity that seduces men into obedience, thus making automata out of them, but that also causes wholesale destruction out of an impulse of its mysterious, irrational essence. The fear of modern machinery is then connected to the fear of female sexuality in the figure of the female robot, so it must be destroyed for the story to come to reassuring end:

> the destructive potential of modern technology, which the expressionists rightfully feared, had to be metaphorically purged. After the dan-

gers of a mystified technology have been translated into the dangers an equally mystified female sexuality poses to men, the witch could be burnt at the stake and, by implication, technology could be purged of its threatening aspects.[75]

In another essay, Huyssen has described a tendency in modernist culture of describing the masses and mass culture in feminine terms.[76]

What is in operation here is a binary opposition with a series of linked entities on either side. On one side are industrial technology, female sexuality, mass culture, and the working class (represented by the hand), all of which are described in terms of the irrational, sensual, dark, and feminine; on the other side are male power, the ruling elite, and the leisure class (represented by the brain), described as rational, cerebral, luminous, and masculine. The story begins with the latter's absolute domination over the former—the harnessing of the chaotic energy of nature and the lower order of humanity for the efficient production of work—but the narrative is ultimately about the futility of maintaining that state due to the unbearable tensions within the system. Through the Master's own less than rational actions, all that unruly energy below is ultimately released, erupting in a revolution and a flood. Machines, women, and the working class were all entities that were both necessary and beneficial to the smooth functioning of the world as long as they were kept under the power of the male authorities (i.e., within controlled parameters). But they are also the source of profound anxiety since they are perceived as having natures that are essentially irrational and chaotic, with the possibility of unleashing themselves to wreak havoc on the status quo.

The happy ending of the story, with the reconciliation of the master and the workers, the brain and the hand, reflects a hope, however feeble, for a solution to the essential problem at the heart of modernity. Lang did envision a more apocalyptic conclusion in which Freder and Maria escapes the utterly destroyed city in a spaceship.[77] The ending, however, is of less significance than the image of modern technology as a female robot that even as it obeys the dictates of its master joyfully sets about beguiling and then devastating the world around it. In this figure Lang created a captivating representative of the uncanny and mysterious essence of autonomous machines.

These stories produced in the interwar era by Čapek, Rolland, Vasari, and Lang in Czechoslovakia, France, Italy, and Germany not only referred to one another but functioned as the recurrent nightmares of an entire culture suffering from the trauma and crisis of modernity. This point is evidenced by the similarity of plot, ideas, and character types in the narratives. As I have shown in this chapter, the themes of technological warfare, industrial oppression, violent revolution, and catastrophe on an apocalyptic scale figure prominently in these dark narratives of the fate of humanity in the machine age. The mechanical figure sometimes represented the dehumanized worker and at other times the threatening Other of the mechanical being that replaces and renders useless the fragile biological being born of Nature. But these meditations on the industrial age led to a number of conclusions, from the bleakest apocalypse of complete destruction to a measure of hope through lessons learned. The modern robot also emerged as a complex, multifaceted entity that displayed such disparate qualities as the debased impotence of blind obedience and pitiful fragility and the monstrous power of immense productivity and sublime beauty. In terms of its behavior as well, to quote Jean Baudrillard again, if the image of the robot is that of the slave, we must not forget that "the theme of slavery is always bound up—even in the legend of the sorcerer's apprentice—with the theme of *revolt*."[78] As I have endeavored to demonstrate in this chapter, the historical significance of these nightmares of living machines must be understood in the context of the interwar period's anxieties over the tumultuous events of the time. From the battlefields of World War I and the factory floor run under the Taylorist regime to the sites of violent revolutions, there was an agonized and fearful refrain of questioning what would be left of humanity in the age of mad and monstrous machinery, the superhuman children of technological modernity.

Conclusion

I came here with a simple dream. A dream of killing all humans.

—**The robot Bender of the Fox television show** *Futurama*

"How can they fear robots?"
"It's a disease of mankind, one of which it is not yet cured."

—**Issac Asimov, "The Bicentennial Man"**

"What are you?"
"I already answered that," snapped the machine, clearly annoyed.
"I mean, are you man or robot?" explained Klapaucius.
"And what, according to you, is the difference?" said the machine.

—**Stanislaw Lem,** *The Cyberiad*

To end this historical study at the interwar period might seem arbitrary, but there are a number of good reasons for concluding the narrative proper here. In the decades between one devastating war in Europe and another, the ongoing meditations on the nature of the automaton reached an extreme point in the visions of violent confrontation between humanity and machinery. The apocalyptic nature of the stories represents a culmination of a centuries-long ambivalence toward the living machine. This does not mean that there were no new developments in the conceptual use of it in the following eras. On the contrary, since World War II there has been a veritable explosion of works dealing with the automaton or, more commonly, robot, embodying many of the themes I have explored. They include its countless appearances in science fiction literature, television shows, and movies; its continued use as a symbol of human conformism and dehumanization as well as transcendent power and hybridity; and its representation of breakthrough technologies in the digital, cybernetic, and virtual media. The sheer bulk of the material that would have to be covered for a comprehensive study of the robot symbolism in the postwar era would require another full volume. Such a work would also need to shift its focus geographically, predominantly to the United States; intellectually, to discussions on artificial intelligence and cybernetic theory; and historically, to the political and cultural contexts of the Cold War and beyond, as well as the cultural impact of the development of major technology from the atom bomb to the digital computer.

Significant European writings on the robot have been produced in the second half of the twentieth century—in fiction, Ernst Jünger's fascinating and underrated 1957 novel *The Glass Bees* and numerous works by the Polish writer Stanislaw Lem and the British writer John Sladek, among many others.[1] But a cultural study of the robot motif would have to be centered on the United States, where there has been an extraordinary output of literary, artistic, and scientific works on the topic—in science fiction stories by representative figures like Isaac Asimov (who coined the word "robotics," the study of robots), Philip K. Dick, and William Gibson (who coined the word "cyberspace"); classic films like *The Day the Earth Stood Still* (1951), *Forbidden Planet* (1956), and *2001: A Space Odyssey* (1968), and more recently *Blade Runner* (1982), the *Terminator* series (1984, 1991, 2003, 2009), and the *Matrix* series (1999, 2003, 2003); and scientific works in the fields of cybernetics, artificial intelligence, and contemporary cyborg theory.

Advances in biology and medicine in the twentieth century made the vitalist-mechanist question in physiology a dead issue, so the central focus of the machine-human comparison shifted to the nature of the mind. Some of the most interesting intellectual debates on the subject are on the questions whether the human brain works in some sense like a digital computer and whether a computer of sufficient complexity and power could achieve consciousness. A study of modern writings on such ideas would have to begin with their historical origins, somewhat neglected in this study, starting from various attempts to create machines that fulfill certain functions of the mind—like the calculating machines of Blaise Pascal, Gottfried Leibniz, Charles Babbage, and others. For the postwar period, one would need to look at the pioneering works of Alan Turing on artificial intelligence and the ensuing controversy over his idea for a test to determine the presence of sentience in a machine.

On the issue of human and artificial intelligence, much of the contentious discussions on the topic have been informed by ideas from the fields of cybernetics and contemporary cyborg theory. Bruce Clarke, in the introductory essay to his coedited volume on the subject, charts out the shift from energy and thermodynamics to information and cybernetics as the central areas of interest in the scientific and technological culture from the late nineteenth through the twentieth centuries.[2] Thermodynamic theorists established a set of universally applicable formulas that quantified the functioning of all dynamic entities, the steam engine as well as the natural body, in terms of work and heat expenditure. In the 1950s, Norbert Wiener, Claude Shannon, John von Neumann, Warren McCulloch, and others, in their new field of cybernetics, created a language for analyzing both biological and artificial systems in terms of information control, communication, and proliferation.[3] Regarding both the body and the mind as essentially data carriers (genes in the body and cultural and intellectual information, or "memes," in the mind) raised the implication that they and the digital computer operated under analogous principles, opening up the possibility of mergence in the form of a cybernetic organism, or cyborg.[4] Such ideas were revived in postmodern discourse of the 1980s as cyborg theory, starting from the now classic essay by Donna Haraway, "Manifesto for Cyborgs" (1985).[5] There was a concomitant development in science fiction literature in the subgenre of cyberpunk, in works by such figures as William Gibson, Bruce Sterling, Pat Cardigan, and Neal Stephenson featuring human-computer interface, virtual reality, and cyborg characters. Among the many artificial beings imagined over the centuries, from the clockwork automaton, the spiritual golem, the magical homunculus, and the industrial robot to the organic android, the one that is most relevant to contemporary meditations on humanity's relationship to autonomous technology is the biodigital hybrid.

Given the richness and complexity of such topics, it really is beyond the scope of this study to elaborate on them at any length, placing them in such historical contexts as the Cold War and its aftermath at the end of the century that ushered in an era of international uncertainty; the possibility of a literally apocalyptic technological warfare in the age of nuclear technology; the spectacular development of computing technology, from the corporate application of devices invented during World War II for decryption purposes to the personal computer revolution in the 1980s and then to the spread of the internet in the 1990s; the rise of the military-industrial complex and multinational corporations involved in large-scale technological, medical, and genetic research; and the cultural hopes and anxieties over such developments. While the topic must wait for future works on the cultural and intellectual use of the robot idea in recent times, I would like to conclude this study by pointing to some general ways the automaton concept is being deployed in current debates and imaginings on the machine-human question.

In our time, there are three contending views on the fate of humanity in the age of digital technology in which the robot is often used for conceptual and imaginative purposes. I will call these notions the theories of inevitable confrontation, of equivalence through sentience, and of cybernetic mergence. The theory of inevitable confrontation is the one with the oldest history, dating back to 1863, when Samuel Butler first raised the alarm about the emerging race of living machines. Since the visions of the revolt of industrial robots in the interwar period, the story of humanity's conflict with machinery for mastery of the world has been told so many times in science fiction literature and cinema that it is a veritable cliché of our time. In non-fiction works, the consideration of the scenario centers on the question of artificial intelligence and how computers in the near future could achieve the human brain's level of complexity. Kevin Warwick, a British expert on cybernetics and robotics, in his book *March of the Machines: The Breakthrough in Artificial Intelligence* (1997), summarized his ideas in three points:

1. We humans are presently the dominant life form on Earth because of our overall intelligence.
2. It is possible for machines to become more intelligent than humans in the reasonably near future.
3. Machines will then become the dominant life form on Earth.[6]

Earlier, Hans Moravec, the director of the Mobile Robot Laboratory at Carnegie Mellon University, made a similar prediction in his *Mind Children* (1988) but in a tone of resignation that regards the future situation in terms of the older generation passing in order to make room for the younger. While that would mean the end of the human race, Moravec thinks that we will live on in the memories and in the conceptual pattern of our artificial descendants.

CONCLUSION

Unleashed from the plodding pace of biological evolution, the children of our minds will be free to grow to confront immense and fundamental challenges in the larger universe. We humans will benefit for a time from their labors, but sooner or later, like natural children, they will seek their own fortunes while we, their aged parents, silently fade away. Very little need be lost in this passing of the torch—it will be in our artificial offspring's power, and to their benefit, to remember almost everything about us, even, perhaps, the detailed workings of individual human minds.[7]

A discussion of such a development in evolutionary terms can be found in Philip K. Dick's story "Second Variety" (1953), which takes place during a future war to the death between humans and robots. As two soldiers talk during a lull in the fighting, in a manner reminiscent of Samuel Butler's writings—

"It makes me wonder if we're not seeing the beginning of a new species. *The* new species. Evolution. The race to come after man."

Rudi grunted. "There is no race after man."

"No? Why not? Maybe we're seeing it now, the end of human beings, the beginnings of the new society."

"They're not a race. They're mechanical killers. You made them to destroy. That's all they can do. They're machines with a job."

"So it seems now. But how about later on? After the war is over. Maybe when there aren't any humans to destroy, their real potentialities will begin to show."

"You talk as if they were alive!"

"Aren't they?"

There was silence. "They're machines," Rudi said. "They look like people, but they are machines."[8]

This difference between the characters on whether machines should indeed be considered at least potentially living beings leads us to the second major position on the topic.

The theory of equivalence through sentience asserts that once machines do reach the human level of consciousness, they should indeed be treated as living, thinking beings and that granting them status as such, including even political and legal rights, might be the best way to prevent a catastrophic confrontation with them. This idea affirms the difference between natural and artificial creatures, human and digital intelligence, but it also asserts that the question of active consciousness, not its nature or origin, should be the essential factor in judging who or what should be accorded the dignity due a sentient being. Isaac Asimov, one of the central figures of American science fiction from the 1940s through the 1970s, was an outspoken technophile, especially on the subject of robots. He famously imagined that in the future, all robots will be equipped with what he calls a "positronic" brain that is bound to three absolute laws:

1. A robot may not injure a human being or, through inaction, allow a human being to come to harm.
2. A robot must obey the orders given by it by human beings except when such orders would conflict with the First Law.
3. A robot must protect its own existence as long as such protection does not conflict with the First and Second laws.[9]

These laws were created to demonstrate that the violent confrontation between humanity and robots is not inevitable, that with proper safeguards humankind and intelligent machines can peacefully coexist. In Asimov's celebrated story "The Bicentennial Man" (1976), a robot that has transcended its original programming seeks to gain recognition as a person through a series of lawsuits in the course of the two centuries of its life and by systematically replacing its mechanical parts with organic ones, ultimately choosing mortality in its quest.[10] The recent Will Smith action movie *I, Robot* (2004) loosely adapted some of Asimov's ideas for a retrograde revolt-of-the-robot story line. Closer to the spirit of his views is the Disney/Pixar animated film *WALL-E* (2008), which features a confrontation with an automated spaceship's machines but concludes with the friendly coexistence of humans and humanized robots.

The third contemporary position on the machine-human question is the theory of cybernetic mergence, which posits that in the future there will occur a gradual fusion of the two—to the extent that the very distinction between what is natural and artificial, organic and mechanical, biological and digital, will cease to have any meaning. While the theory of equivalence through sentience still maintains a distinction between the natural and artificial, in the sense of "different but equal," this latest view rejects the idea that the boundary between the two will remain intact when methods are discovered of freely and directly exchanging parts and information across the line. Like the cochlear implants for deaf people that are routinely used today, other defective or failing organs in the human body will be replaced by artificial ones (à la the bionic man), including parts of the brain; we will be able to interface with computers to send and receive data seamlessly between machines and minds, and even travel through the Web as a disembodied consciousness; artificial minds will be equipped not with crude mechanical parts but with more flexible and powerful devices made through nanotechnology; and most radical of all, humans will one day be able to download their entire minds, including memories, into computers to achieve a kind of digital transcendence and immortality. The most famous proponent of such ideas is the artificial intelligence expert Ray Kurzweil, who in his most recent book points to the moment of "singularity" in the near future when we will be able to leave behind the biological prison of our bodies and transcend to the infinite and infinitely malleable world of virtuality.[11]

This first decade of the twenty-first century has seen the publication of numerous books on the latest developments in robotic and cybernetic technologies with discussions of the possible consequences of their advancement to both human identity and survival.[12] The tension and ambivalence was brought out earlier in Haraway's "Cyborg Manifesto," in which she explored the possibilities of liberation and empowerment for women and others seeking to counter the political, social, and cultural status quo through the use of and mergence with modern technology. She is far from a naïve technophile, as she is very much aware of the destructive and dehumanizing aspects of technology: "the final imposition of a grid of control on the planet . . . the final abstraction embodied in a Star Wars apocalypse waged in the name of defence . . . the final appropriation of women's bodies in a masculine orgy of war."[13] She asserts, however, that a proper understanding of the emerging cybernetic world must take into account both its negative consequences and positive potentialities. What she sees as a possible outcome to look forward to in the world of cyborgs is that once the distinction between the natural and artificial is blurred and even made meaningless, other traditional Western dualisms that have been "systemic to the logics and practices of domination of women, people of colour, nature, workers, animals—in short, domination of all constituted as others" might be challenged.[14] Those dualisms include self/other, mind/body, culture/nature, male/female, civilized/primitive, reality/appearance, whole/part, agent/resource, maker/made, active/passive, right/wrong, truth/illusion, total/partial, God/man. The use of the automaton idea, with which people have meditated for centuries on the fearful consequences of the collapse of such binary oppositions as well as the positive possibilities of transcendences of such strictures, has been one of the central themes of this study. The cybernetic organism is just the most recent figure in that ongoing meditation in the Western imagination.

I began this study by presenting a theory of why the automaton is an inherently captivating and disturbing object at the same time, of how it can arouse such widely ranging emotions as fascination and awe, contempt and horror, and of the conditions under which the viewer's reaction can change from one to the other. Since the explanation addressed only the question of why the automaton is such a powerful thing in itself, by virtue of its capacity to cross the boundaries between the antithetical categories of the natural and the artificial, the animate and the inanimate, the living and the dead, I asserted that the understanding of how people of any given period reacted to and made conceptual use of it must be pursued historically, through the analysis of writings on automata in specific cultural and intellectual contexts. I have shown how medieval and Renaissance writers expressed their ambivalence toward the automaton in considering questions of its magical nature, some seeing in it the danger of demonic agency and others the marvelous potential of natural magic. During the scientific revolution and the classical Enlightenment, mechanistic thinkers stripped away the magical aura from the automaton but made full use of its captivating power as the central emblem of the rational order of the world, the state, and the body. For many late Enlightenment writers who were disillusioned with the promises of the age of reason, the automaton appeared neither marvelous nor rational, becoming a shabby representation of people lacking freedom either through oppression or conformism. In the Romantic era, the fear arising from the uncertainty of the automaton's nature arose again in the context of the crisis period of the late eighteenth and early nineteenth centuries, the iconic figure of the uncanny automaton appearing in literature. During the industrial era, when the machines of the age of steam and electricity were commonly portrayed as living superhuman creatures, the automaton appeared in modernist visions of human transcendence through the hybridization of man with machine. These visions became significantly darker and cautionary in the period after World War I, the Bolshevik revolution, and the adoption of Taylorist regime in European factories, when the robot became the representation of humanity turned into machinery through the inexorable forces of modernity, often resulting in a catastrophic revolt.

The narrative I have unfolded in the course of this study demonstrates the powerful hold the automaton has on the Western imagination, the object's capacity to arouse a wide spectrum of emotional reactions in different eras, and the protean nature of its role as a concept that was used to represent an enormous variety of meanings throughout the centuries, many of them embodying complex and sometimes contradictory views of humanity itself. In many shorter works dealing with the automaton idea, its significance is usually explained in starkly positive or negative terms—as the representation either of human empowerment and transcendence through technological control of nature or of debasement, dehumanization, and enslavement in the industrial and postindustrial world.[15] One of the central purposes of this study has been to demonstrate that for a full understanding of the automaton motif in the Western imagination as a whole, one must take into account both aspects of the mechanical entity, to see how the object functioned in different historical contexts as the representation of *both* human empowerment *and* oppression, liberation *and* subjugation, transcendence *and* debasement. In fact, it is precisely this capacity of the automaton to hold such disparate meanings that makes it such a powerful and enduring conceptual object.

To revisit some of the themes I have explored in this study, the negative ideas on the automaton have envisioned it in the following forms:

1. A diabolical idol of demonic possession or necromantic knowledge
2. A piece of trickery designed to fool the gullible and the superstitious into believing in supernatural entities
3. The idiot human being incapable of original thought or independent action
4. The contemptible conformist or tyrant who has given away freedom in order to gain the trappings of respectability and power
5. The downtrodden person lacking freedom due to oppression

6. An uncanny object that may be animated by a supernatural force that could endanger the sanity of an individual and the stability of a society

7. A destructive and frightening monster of superhuman power and irrational will

8. The dehumanized worker in the modern factory system

9. The worker in revolt, bringing about a revolution

10. A symbol of human hubris that would bring forth some unnatural, inhuman thing that cannot be controlled and is liable to turn on its creator and all of humanity

The positive visions of the automaton have taken the following forms:

1. A product of neutral and potentially beneficial natural magic or the manipulation of occult forces inherent in the world

2. The vision of the wonderfully constructed mechanical universe made by a rational God

3. The image of the well-functioning government built on rational principles

4. The representation of the human body, built by God in all of its intricacy, efficiency, and beauty

5. The rational human mind

6. The image of industrial power and its ability to harness the forces of nature for the benefit of humanity

7. The efficient and productive worker in the modern factory

8. The image of human transcendence of limits imposed by nature and of the evolutionary birth of the biomechanical superman

9. The new image of human transcendence of limits imposed by nature and the evolutionary birth of the biodigital cyborg

10. A general symbol of human curiosity, the drive toward knowledge, and mastery over the self as well as the natural environment

Given the sheer abundance and richness of all these uses of the automaton idea, I can alter the famous phrase from Claude Lévi-Strauss's *Totemism* to assert that Western culture has found the automaton not just "good to make" but also "good to think."[16] And what makes it such a good object to think about is precisely the fact that it is such an unstable and malleable symbol, whose volatility and flexibility are the very source of its powerful allure.

Uncertainties, anxieties, and ambivalences toward the artificial being continue to be expressed today in meditations on the consequences of the emergence of machine intelligence, biodigital mergence, and related issues of cloning and other forms of genetic manipulation, which some view as attempts to create custom-made biological automata.[17] The arguments fall into one of the three positions I have laid out: inevitable conflict, equivalence through sentience, and cybernetic mergence. To give the very last example, one of the most interesting recent uses of the automaton concept I have come across is in *The Robot's Rebellion* (2004), by Keith E. Stanovich, a cognitive psychologist. The title of his book refers to the biologist Richard Dawkins's famous depiction (in his 1976 book *The Selfish Gene*) of all living organisms as basically carriers of genetic replicators, biologically programmed to proliferate them as its main goal of existence. Stanovich, in picking up on a call by Dawkins to rebel against the selfish genes, explains his use of the robot image.

> In this book I use the term "robot's rebellion" to refer to the package of evolutionary insights and cognitive reforms that are necessary if we want to transcend the limited interests of the replicators and define our own autonomous goals. We may well be robots—vehicles designed for replicator propagation—but we are the only robots who have discovered that we have interests that are separate from the interests of the replicators. We indeed are the runaway robot of science fiction stories—the robot who subordinates its creator's interest to its own interest.[18]

What is emphasized repeatedly in the course of this book on how we, as robots, might rebel against our replicator-masters, is our essential capacity for conscious autonomy, reminiscent of Kant's argument against Leibniz's idea of God's pre-established harmony of the world in which we are "spiritual automata" and William James's critique of Thomas Huxley's depiction of human beings as "conscious automata." There is a striking irony here in that Stanovich uses the idea of our own artificial creation as a conceptual representation of ourselves (the created as a representation of the creator) and then places it in the common narrative trope noted by Baudrillard of the theme of robotic slavery leading to the theme of rebellion to point to the contentious relationship between our biogenetic nature and our autonomous consciousness. With the struggle between the robot slave and the human master internalized into our very beings, our own minds are likened to the machines that have the potential to rise up against its genetic masters. This rather complex use of the rebellious robot theme

On the occasion of Japanese prime minister Junichiro Koizumi's visit to the Czech Republic in August 2003, the advanced robot Asimo, made by Honda, paid tribute to Karel Čapek, who popularized the word "robot" through his play *R.U.R.: Rossum's Universal Robots* (1921) by presenting flowers to the writer's bust. Photograph by Dan Materna, AFP/Getty Images.

points to the extent to which the mechanical object has become an integral part of the language with which we ponder our own nature. The French thinkers Gilles Deleuze and Félix Guattari have also made fruitful use of the machine-human idea in the concept of people as "desiring machines" in their poststructuralist classic *Anti-Oedipus* (1972).[19]

Even as the images of the robot, the android, and the cyborg are evoked in such debates for both hopeful and cautionary purposes, we continue to enjoy watching films and television shows and reading stories about the encounters between human beings and their creations as wondrous dreams as well as horrific nightmares; children amuse themselves by playing with self-moving toys, while adults, too, are captivated by more sophisticated examples of such objects—even human beings acting like them; media pundits commenting on issues of politics, economics, and education constantly evoke the automaton and the robot in discussions of human oppression and inequity, brainwashing and conformism; and scientists and engineers, despite dire warnings from even among themselves, work toward building the perfect human simulacrum, the ever more efficient and productive machine-worker, and the truly sentient artificial intelligence. At one time or another, everyone is bound to experience the myriad feelings of delight, amusement, contempt, disgust, horror, enlightenment, and transcendence aroused by the self-moving machine that imitates life.

All of this points to the undeniable fact that the automaton remains a rich and relevant idea in our time. For now and the foreseeable future in the cybernetic age, we still find ourselves dreaming the millennia-long dream of sublime and uncanny living machines.

> How could the *anthropos* be threatened by machines? It has made them, it has put itself into them, it has divided up its own members among their members, it has built its own body with them. How could it be threatened by objects?
>
> —Bruno Latour, *We have never been modern*

Notes

Introduction

1. Alexander Pushkin, "The Bronze Horseman," in *Collected Narrative and Lyrical Poetry*, trans. Walter Arndt (Dana Point, California: Ardis, 1984), 425–438, 438.

2. See Prosper Mérimée, "The Venus of Ille," in *Carmen and Other Stories*, trans. Nicholas Jotcham (Oxford: Oxford University Press, 1989), 132–161; E. T. A. Hoffmann, "The Sandman," in *Tales of Hoffmann*, trans. R. J. Hollingdale (London: Penguin Books, 1982), 85–125. For a variation of the Don Juan story, see Molière's "Don Juan," in *The Miser and Other Plays*, trans. John Wood and David Coward (London: Penguin Books, 2000), 126–128. For full-length works on the animated statue in literature and the arts, see Theodore Ziolkowski, *Disenchanted Images: A Literary Iconology* (Princeton: Princeton University Press, 1977), and Kenneth Gross, *The Dream of the Moving Statue* (Ithaca: Cornell University Press, 1992).

3. Bruce Craig, "History Defined in Florida Legislature," *Perspectives* 44, 6 (September 2006): 13.

4. Charlotte Bronte, *Jane Eyre* (Oxford: Oxford University Press, 2000), 253; Honoré de Balzac, *Old Goriot*, trans. Marion Ayton Crawford (London: Penguin, 1951), 37.

5. Arthur Conan Doyle, *The Adventures of Sherlock Holmes and the Memoirs of Sherlock Holmes* (London: Penguin, 2001), 3; and T. H. Huxley, "A Liberal Education; And Where to Find It," in *Lectures and Lay Sermons by Thomas Huxley* (London: Dent, 1926), 60.

6. Henry James, *The American* (New York: Norton, 1978), 179–180.

7. For discussions of the importance of the "self-mover" as a philosophical concept in Aristotle and others, see Mary Louise Gill and James G. Lennox, eds., *Self-Motion: From Aristotle to Newton* (Princeton: Princeton University Press, 1994).

Also Sylvia Berryman, "Ancient Automata and Mechanical Explanation," *Phronesis* 48, 4 (2003): 344–369.

8. Henry Cornelius Agrippa von Nettesheim, *Three Books of Occult Philosophy*, trans. James Freake (St. Paul, Minn.: Llewellyn, 2000), 233.

9. I make the argument that the success of Vaucanson's works was directly responsible for the emergence of this more specific definition of the word automaton in Chapter 3.

10. Mark Seltzer's *Bodies and Machines* (New York: Routledge, 1992), 109.

11. For the origin of the word "robot" see Chapter 6.

12. The most comprehensive is Alfred Chapuis and Edouard Gélis, *Le Monde des automates: Étude historique et technique* (Paris: E. Gélis, 1928), translated by Alec Reid as *Automata: A Historical and Technological Study* (New York: Central Book, 1958).

13. See Chapter 3. Perhaps the first historian to consider the automaton as a subject worthy of examination in its own right was Derek De Solla Price, "Automata and the Origins of Mechanism and Mechanistic Philosophy," along with Silvio Bedini, "The Role of Automata in the History of Technology," both in *Technology and Culture* 5, 1 (Winter 1964): 9–42. The most recent book, Gaby Wood, *Living Dolls* (2002; U.S. title: *Edison's Eve*), examines several episodes in the history of automata, including those of the Vaucanson works, the so-called chess-player of Wolfgang von Kempelen, and Thomas Edison's talking doll.

1. The Power of the Automaton

1. See for instance, Alfred Chapuis, *Les Automates dans les oeuvres d'imagination* (Neuchâtel: Editions du Griffon, 1947), 11–17; John Cohen, *Human Robots in Myth and Science* (London: Allen and Unwin, 1966), 15–26; Harry M. Geduld and Ronald Gottesman, eds., *Robots, Robots, Robots* (Boston: New York Graphic Society, 1978), 3–31; and Sidney Perkowitz, *Digital People: From Bionic Humans to Androids* (Washington D.C.: Joseph Henry, 2004), 17–21.

2. Derek de Solla Price, in his 1964 essay on automata in the history of technology, argues for the importance of studying the automaton not only by pointing to important technological breakthroughs that developed from it but also by discerning in its early history a "strong innate urge toward mechanistic explanation" that led to the scientific and industrial revolutions. This is an outdated, teleological view that this study seeks to correct by showing how the automaton's imaginative significance in the premodern period had little to do with mechanistic ideas, and how even in the modern period there is a major nonrational, uncanny component to the entity that is antithetical to the rationalist worldview of mechanistic philosophy. Derek De Solla Price, "Automata and the Origins of Mechanism and Mechanistic Philosophy," *Technology and Culture* 5, 1 (Winter 1964): 9–23.

3. Sylvia Berryman makes a similar caution against identifying these ancient automata as modern robots. Sylvia Berryman, "Ancient Automata and Mechanical Evolution," *Phronesis* 48, 4 (2003): 344–369.

4. Homer, *The Iliad*, trans. Richard Lattimore (Chicago: University of Chicago Press, 1951), 385 (bk. 18, ll. 369–377). Homer also mentions golden women at the forge who are the god's attendants. See 386 (bk. 18, ll. 417–421).

5. Apollonius of Rhodes, *The Voyage of the Argo*, trans. E. V. Rieu (London: Penguin, 1959), 191 (bk. 4, l. 1638).

6. For the most famous version of the story in Ovid, see *The Metamorphoses*, trans. Horace Gregory (New York: Mentor Books, 1960), 281–282. For more on the different versions of the myth see Essaka Joshua, *Pygmalion and Galatea: A History of a Narrative in English Literature* (Aldershot, England: Ashgate, 2001).

7. Pindar, "Diagoras of Rhodes," in *Olympic Odes, Pythian Odes*, trans. William H. Race (Cambridge, Mass.: Harvard University Press, 1997), 127.

8. Aulus Gellius, *The Attic Nights of Aulus Gellius*, trans. John C. Rolfe (London: Heinemann, 1927), 37. Kathleen Freeman points out that Archytas, whose love of children was proverbial, was also credited with the invention of the rattle, so the flying dove may have been another toy he made for them. Freeman, *The Pre-Socratic Philosophers* (Oxford: Blackwell, 1966), 234.

9. On the dating and other information on Hero, see A. G. Drachmann, *Ktesibios, Philon and Heron: A Study in Ancient Pneumatics* (Copenhagen: Munksgaard, 1948), 74–77.

10. For more on all three figures, see Chapuis and Gélis, *Automata*, 71–73. See also Hero of Alexandria, *The Pneumatica of Hero of Alexandria*, trans. Joseph Gouge Greenwood (London: MacDonald, 1971). The *Automatic Theaters* has not been translated into English, but a German translation, along with the original Greek, can be found in Herons von Alexandria, *Druckwerke und Automatentheater*, trans. Wilhelm Schmidt (Leipzig: Teubner, 1899), 339–453.

11. Aristotle, *Politics*, trans. Benjamin Jowett, in *The Complete Works of Aristotle*, vol. 2, ed. Jonathan Barnes (Princeton: Princeton University Press, 1984), 1989.

12. Sarah P. Morris provides an enlightening analysis of this myth in her *Daidalos and the Origins of Greek Art* (Princeton: Princeton University Press, 1992), 215–237.

13. Aristotle, *On the Soul*, trans. E. S. Forster, in *The Complete Works of Aristotle*, vol. 2, ed. Jonathan Barnes (Princeton: Princeton University Press, 1984), 648; Diodorus Siculus, *Diodorus of Sicily*, trans. C. H. Oldfather (Cambridge, Mass.: Harvard University Press, 1939), 57 (bk. 3, pt. 4). For discussions of these passages see Morris, *Daidalos*, 240–244.

14. Maurizio Bettini, *The Portrait of the Lover*, trans. Laura Gibbs (Berkeley: University of California Press, 1999), 233.

15. Plato, *Meno*, trans. W. K. C. Guthrie, in *The Collected Dialogues of Plato*, ed. Edith Hamilton and Huntington Cairns (Princeton: Princeton University Press, 1961), 381.

16. Jean Baudrillard, *The System of Objects*, trans. James Benedict (London: Verso, 1996), 121.

17. Hesiod, *The Works and Days, Theogony, The Shield of Herakles*, trans. Richard Lattimore (Ann Arbor: University of Michigan Press, 1991), 33–34.

18. Sigmund Freud, "The 'Uncanny,'" trans. Alix Strachey, in *Writings on Art and Literature* (Stanford: Stanford University Press, 1997), 193–229; Ernst Jentsch, "On the Psychology of the Uncanny," trans. Roy Sellars, *Angelaki* 2 (1995): 7–16.

19. Jentsch, "On the Psychology of the Uncanny," 12.

20. Freud, "'Uncanny,'" 195–201.

21. Ibid., 208–209.

22. Jean Piaget, *The Child's Conception of the World*, trans. Joan and Andrew Tomlinson (London: Routledge and Kegan Paul, 1929), 169–193.

23. Freud, "'Uncanny,'" 209.

24. John Cohen, *Human Robots in Myth and Science* (London: Allen and Unwin, 1966), 7.

25. Ibid., 106.

26. Ibid., 127.

27. Robert Plank, "Golem and the Robot," *Literature and Psychology* 15, 1 (1965): 12–27, 13. Per Schelde calls this theme "womb-envy." See Per Schelde, *Androids, Humanoids, and Other Science Fiction Monsters: Science and Soul in Science Fiction Films* (New York: New York University Press, 1993), 218–221. Another study that analyzes the sexuality theme in modern automata is Michel Carrouges, *Les machines célibataires* (Paris: Arcanes, 1954).

28. Huyssen, *After the Great Divide: Modernism, Mass Culture, Postmodernism* (Bloomington: Indiana University Press, 1986), 71.

29. Ibid.

30. Ibid., 227, n. 10.

31. Peter Gendolla makes a similar mistake in associating the appearance of uncanny automata in German literature of the early Romantic period with industrialization. See Gendolla, *Die lebenden Maschinen: Zur Geschichte des Maschinenmenschen bei Jean Paul, E. T. A. Hoffmann und Villiers de l'Isle Adam* (Marburg: Guttandin und Hoppe, 1980), 3–17.

32. See also Rodney A. Brooks, *Flesh and Machines: How Robots Will Change Us* (New York: Pantheon Books, 2002).

33. Linda M. Strauss, "Reflections in a Mechanical Mirror: Automata as Doubles and as Tools," *Knowledge and Society* 10 (1996): 179–209.

34. This, of course, is a simplified description of Locke's ideas. For a detailed discussion of the Locke-Leibniz disagreement, see Nicholas Jolley, *Leibniz and Locke: A Study of the New Essays on Human Understanding* (Oxford: Clarendon Press, 1984), 162–193. For Leibniz's critique of the tabula rasa idea, see Gottfried Wilhelm von Leibniz, *New Essays on Human Understanding*, trans. Peter Remnant and Jonathan Bennett (Cambridge: Cambridge University Press, 1981).

35. See for example the chapter "Do Dual Organizations Exist?" in Claude Lévi-Strauss, *Structural Anthropology*, vol. 1, trans. Claire Jacobson and Brooke Grundfest Schoepf (New York: Basic Books, 1963), 132–162. For an informative analysis of his ideas on reality formation, including as it relates to Kantian philosophy, see Ino Rossi, "The Unconscious in the Anthropology of Claude Lévi-Strauss," *American Anthropologist* 75, 1 (Feb. 1973): 20–48.

36. As Rossi shows in "The Unconscious in the Anthropology," Lévi-Strauss asserts such a biological determinism in a number of his works. See Lévi-Strauss, *Totemism*, trans. Rodney Needham (Boston: Beacon Press, 1962), 90; *Myth and Meaning* (New York: Schocken Books, 1979), 5–14; and *The Naked Man: Introduction to a Science of Mythology 4*, trans. John and Doreen Weightman (New York: Harper and Row, 1981), 678–680.

37. Contemporary work being done on the search for neurobiological roots of human cognition is very rich, but it seems problematic to tie binary thinking solely to biology. If one asserts that human beings think in a dualistic manner for biological reasons, then one would have to provide a biological explanation for how we can also go beyond the binary and comprehend notions of transitions, spectrums, and gray areas. The prevalence of dualism in human worldviews may be due simply to the fact that dividing things into two opposing categories is just the most rudimentary way of organizing things. For recent works on categories, the neurological connection, and theories of embodiment that are highly relevant to this topic, see George Lakoff, *Women, Fire, and Dangerous Things: What Categories Reveal about the Mind* (Chicago: University of Chicago Press, 1987); Mark Johnson, *The Body in the Mind: The Bodily Basis of Meaning, Imagination, and Reason* (Chicago: University of Chicago Press, 1987); James B. Ashbrook, ed., *Brain, Culture and the Human Spirit: Essays from an Emergent Evolutionary Perspective* (Lanham, Md.: University Press of America, 1993); and Antonio B. Damasio, *Descartes' Error: Emotion, Reason, and the Human Brain* (New York: Quill, 2000).

38. Mary Douglas, *Purity and Danger: An Analysis of Concepts of Pollution and Taboo* (London: Routledge, 1966), 163.

39. Ibid., 39–41.

40. Ibid., 55–57.

41. Ibid., 171.

42. Zakiya Hanafi, *The Monster in the Machine: Magic, Medicine, and the Marvelous in the Time of the Scientific Revolution* (Durham, N.C.: Duke University Press, 2000), 1–2.

43. Judith Shapiro, "Transsexualism: Reflections on the Persistence of Gender and Mutability of Sex," in Julia Epstein and Kristina Straub, eds., *Body Guards: The Cultural Politics of Gender Ambiguity* (New York: Routledge, 1991), 265–268.

44. See Aristotle, *Poetics*, in *Complete Works of Aristotle*, 2:2316–2340, and Mikhail Bakhtin, *Rabelais and His World*, trans. Hélène Iswolsky (Bloomington: Indiana University Press, 1984).

45. Douglas, *Purity and Danger*, 97.

46. Psychologist Paul Bloom, in commenting on an earlier work by Paul Rozin, offers a completely natural explanation of our attitude toward corpses, criticizing theories that are too conceptual. He claims that they disturb us because they are disgusting and physically dangerous, for the diseases they could cause, which is why we bury or cremate them. Although there is no doubt that the physical aspect of it is important, I find his singular approach unconvincing since it does not explain the elaborate rituals every human society has developed in the disposal process. If it were just a problem of sanitation, we would handle of corpses the way we do fecal matter, which is also disgusting and potentially harmful. But unlike fecal matter, which is gotten rid of as quickly and efficiently as possible, we treat a corpse with respect, take some care in its preparation for disposal, and even dwell on it for a time. See Paul Bloom, *Descartes' Baby: How the Science of Child Development Explains What Makes Us Human* (New York: Basic Books, 2004), 171–172.

47. Carlo Ginzburg, "Representation: The Word, the Idea, and the Thing," in

Wooden Eyes: Nine Reflections on Distance, trans. Martin Ryle and Kate Soper (New York: Columbia University Press, 2001), 63–78.

48. See also David Freedberg, *The Power of Images: Studies in the History and Theory of Response* (Chicago: University of Chicago Press, 1989), 213–215.

49. Ginzburg, "Representation," 64.

50. See Ernst Kantorowicz, *The King's Two Bodies: A Study in Medieval Political Theology* (Princeton: Princeton University Press, 1957).

51. Ginzburg, "Representation," 64–66 and 71–72.

52. Ralph E. Giesey, *The Royal Funeral Ceremony in Renaissance France* (Geneva: Librairie E. Droz, 1960), 148–149.

53. Ibid., 146.

54. All biblical quotations are from the King James Version.

55. See Alain Besançon, *The Forbidden Image: An Intellectual History of Iconoclasm*, trans. Jane Marie Todd (Chicago: University of Chicago Press, 2000).

56. Ernst Kris and Otto Kurz have explored this notion of the artist as a magician. They recount numerous stories from all over the world of art works being mistaken for living beings or actually coming alive, and analyze them in terms of earlier magical belief that equates the picture with the depicted, what is called "effigy magic": the belief that "a man's soul resides in his image, that those who possess this image also hold power over that person, and that all the pain inflicted on the image must be felt by the person it represents" (e.g., the voodoo doll). See Ernst Kris and Otto Kurz, *Legend, Myth, and Magic in the Image of the Artist: A Historical Experiment*, trans. Alastair Laing and Lottie M. Newman (New Haven: Yale University Press, 1979), 73. See also Max von Boehn's discussion of image magic in *Dolls and Puppets*, trans. Josephine Nicoll (Philadelphia: McKay, 1932), 57–67. Boehn claims that as late as the eighteenth century a man in Turin was executed under the charge that he tried to use image magic to kill a nobleman (60). See also the famous chapter "Pygmalion's Power" in E. H. Gombrich, *Art and Illusion: A Study in the Psychology of Pictorial Representation* (Princeton: Princeton University Press, 1960), 93–115.

57. See Marcel Mauss, *A General Theory of Magic*, trans. Robert Brain (London: Routledge, 2001), 31–55.

58. Victor Turner, *The Ritual Process: Structure and Anti-Structure* (Chicago: Aldine, 1969), 95.

59. Appian, *The Civil Wars*, trans. John Carter (London: Penguin Books, 1996), 149 (bk. 2, sec. 147).

60. For more on Roman funerals that includes mention of this episode, see Keith Hopkins, *Death and Renewal* (Cambridge: Cambridge University Press 1983), 217–226.

61. Gérard Walter, for instance, in his biography of Caesar, claims that the moving image was ordered by Marcus Antonius in great secrecy prior to the funeral. Gérard Walter, *Caesar: A Biography*, trans. Emma Craufurd (New York: Scribner's, 1952), 544.

62. Henri Bergson, *Laughter: An Essay on the Meaning of the Comic*, trans. Cloudesley Brereton and Fred Rothwell (Mineola, N.Y.: Dover Publications, 2005), 18.

63. See Chapter 7 for details.

64. See Frederick Burwick and Paul Douglass, *The Crisis in Modernism: Bergson and the Vitalist Controversy* (Cambridge: Cambridge University Press, 1992).

65. Daniel Dinello, *Technophobia! Science Fiction Visions of Posthuman Technology* (Austin: University of Texas Press, 2005), 74.

66. Edmund Burke, *A Philosophical Enquiry into the Origin of Our Ideas of the Sublime and Beautiful* (Oxford: Oxford University Press, 1990), see especially 119–125 (bk. 4, pts. 3–10). I am fully aware that the sublime is a historically contingent concept that must be understood in the context of eighteenth-century aesthetics and thought in the case of Burke. I am nevertheless using the idea in a somewhat ahistorical manner here since his notion of what causes sublime pleasure is particularly useful in understanding the power of the automaton. Burke himself presents what is now an implausible physiological explanation of this phenomenon, as the body's defensive measure against the negative effects of lethargy. He points to various disorders that result from a body being too much at rest, the solution to which is vigorous exercise. The same goes for the mind, which needs to experience violent emotions every once in a while to remain active. But rather than having the mind go through the trauma of real terror, the body allows the substitute of the sublime, which evokes similar extreme feelings but in a pleasurable and safe way. See 122–123.

67. Aristotle, *Poetics*, trans. R. Kassel, in *Complete Works of Aristotle*, 2:2318.

68. Ann Radcliffe, "On the Supernatural in Poetry," in E. J. Clery and Robert Miles, eds., *Gothic Documents: A Sourcebook 1700–1820* (Manchester: Manchester University Press, 2000), 168.

69. Kant's notion of the sublime deals mainly with the sense of vastness found in nature that seems to intimate notions of infinity and eternity. See Kant, *The Critique of Judgment*, trans. J. H. Bernard (New York: Prometheus Books, 2000), 101–150.

70. Henry Adams, *The Education of Henry Adams* (Oxford: Oxford University Press, 1999), 318.

71. Per Schelde makes the same link between scientists in science fiction films and shamans. See Schelde, *Androids, Humanoids*, 29–37.

72. Benvenuto Cellini, *Autobiography*, trans. George Bull (London: Penguin Books, 1998), 299. For an interesting mention of this story as it relates to the interest in mechanics in early modern Europe, see Horst Bredekamp, *The Lure of Antiquity and the Cult of the Machine*, trans. Allison Brown (Princeton: Wiener, 1995), 1–2.

73. Elizabeth King, "Perpetual Devotion: A Sixteenth-century Machine That Prays," in Jessica Riskin, ed., *Genesis Redux: Essays in the History and Philosophy of Artificial Life* (Chicago: University of Chicago Press, 2007), 274.

74. Villiers de l'Isle Adam, *Tomorrow's Eve*, trans. Robert Martin Adams (Urbana: University of Illionois Press, 1982); Herman Melville, "The Bell-Tower," in *Billy Budd and Other Stories* (New York: Penguin Books, 1986), 141–158; and Jeff VanderMeer, "Dradin, in Love," in *The City of Saints and Madmen* (Canton, Ohio: Prime Books, 2002), 15–72.

75. The article was originally published in Japanese in *Energy* 7, 4 (1970): 33–35; English translation by Karl F. MacDorman and Takashi Minato: www.androidscience

.com/theuncannyvalley/proceedings2005/uncannyvalley.html. Originally the Japanese word *shinwakan* was translated as "familiarity," but MacDorman recently changed it to "rapport" as the closer English expression of what Mori meant.

76. I am grateful to Karl MacDorman for permission to reproduce the graphs in Figures 3, 4, and 5.

77. MacDorman and Ishiguro, "Uncanny Advantage," 304; and David Hanson, "Exploring the Aesthetic Range for Humanoid Robots," www.androidscience.com/proceedings2006/6Hanson2006ExploringTheAesthetic.pdf.

78. For the male version, they used photographs of the science fiction writer Philip K. Dick, who wrote some of the most interesting android stories, including *Do Androids Dream of Electric Sheep* (1968), on which the film *Blade Runner* (1982) was based.

79. It may also be fruitful to explore the gender angle here as to why the peak and the dip are not so pronounced in the case of the female version.

80. MacDorman and Ishiguro, "Uncanny Advantage," 312–313.

81. E. T. A. Hoffmann, "Automata," trans. Major Alexander Ewing, in *The Best Tales of Hoffmann*, ed. E. F. Bleiler (London: Dover, 1967), 81; William Saroyan, *The Human Comedy* (New York: Dell, 1943), 151.

82. Gaby Wood, *Living Dolls: A Magical History of the Quest for Mechanical Life* (London: Faber and Faber, 2002), xvii.

83. On the contrary, scholars like Carlo Ginzburg, Michel Serres, and Kenneth Gross have shown that the purpose of still images of humans, like statues or effigies, is to shield us from the disturbing fact of death by providing us with stable, lasting representations in place of the decaying corpse. In addition to Ginzburg, "Representation," see Michel Serres, *Statues: Le second livre des foundations* (Paris: Éditions François Bourin, 1987), and Kenneth Gross, *The Dream of the Moving Statue* (Ithaca, N.Y.: Cornell University Press, 1992).

2. Between Magic and Mechanics

1. *The Works of Liudprand of Cremona*, trans. F. A. Wright (New York: Dutton, 1930), 236.

2. Norman Cantor, *Civilization of the Middle Ages* (New York: Harper Perennial, 1994), 225.

3. Liudprand, *Works of Liudprand*, 207–208. For more on the throne automata, see Gerard Brett, "The Automata in the Byzantine 'Throne of Solomon,'" *Speculum* 29 (1954): 477–487.

4. Arthur C. Clarke, *Profiles of the Future: An Inquiry into the Limits of the Possible* (New York: Harper and Row, 1973), 21. This "law" was added in the 1973 edition and does not appear in earlier versions of the book.

5. John Wilkins, "To the Reader," in *Mathematicall Magick*, in *The Mathematical and Philosophical Works of Right Reverend John Wilkins* (London: J. Nicholson, 1708), n.p.

6. Valerie I. J. Flint, *The Rise of Magic in Early Medieval Europe* (Princeton: Princeton University Press, 1991), 203–204.

7. For an analysis of such holy images see David Freedberg, *The Power of Images: Studies in the History and Theory of Response* (Chicago: University of Chicago Press, 1989), 99–135. See also Flint, *Rise of Magic*, 254–266.

8. For more on the concept of natural magic in the high Middle Ages, the best place is begin is, of course, Lynn Thorndike, *History of Magic and Experimental Science*, vol. 2 (New York: Columbia University Press, 1923); see the sections on William of Auvergne, 338–371, and on Albertus Magnus, 517–592. See also Richard Kieckhefer, *Magic in the Middle Ages* (Cambridge: Cambridge University Press, 1989), especially 116–150; and Nicholas H. Clulee, *John Dee's Natural Philosophy: Between Science and Religion* (London: Routledge, 1988), 133–135.

9. After Charles Haskins published his influential work on the era, *The Renaissance of the Twelfth Century* (Cambridge, Mass.: Harvard University, 1927), a substantial literature has appeared on the topic, including Charles R. Young ed., *The Twelfth-century Renaissance* (Huntington, N.Y.: Robert E. Krieger, 1977); and Robert L. Benson, Giles Constable, and Carol D. Lanham, eds., *Renaissance and Renewal in the Twelfth Century* (Toronto: University of Toronto Press, 1991). I particularly found enlightening Tina Stiefel's short but illuminating work, which emphasizes the central importance of Plato's *Timaeus* to the intellectual expansion of the period. See Tina Stiefel, *The Intellectual Revolution in Twelfth-century Europe* (London: Croom Helm, 1985).

10. On the importance of Arabic learning to the development of European natural magic, see Thorndike, *History of Magic and Experimental Science*, 2:66–93; Kieckhefer, *Magic in the Middle Ages*, 116–131; and William Eamon, *Science and the Secrets of Nature: Books of Secrets in Medieval and Early Modern Culture* (Princeton: Princeton University Press, 1994), 39–45.

11. Roger Bacon, *Letter Concerning the Marvelous Power of Art and of Nature and Concerning the Nullity of Magic*, trans. Tenney L. Davis (Easton, Penn.: Chemical, 1923), 26–32. That is also the case with Robert Grosseteste also for his advocacy of experimentation—see A. C. Crombie, *Robert Grossesteste and the Origins of Experimental Science, 1100–1700* (Oxford: Clarendon Press, 1953).

12. Thorndike, *History of Magic and Experimental Science*, 2:659–677.

13. Ibid., 2:813–821; Frances Yates, *Giordano Bruno and the Hermetic Tradition* (Chicago: University of Chicago Press, 1964), 49–57. For details on the Arabic origin and sources of the work, including that of image magic, see David Pingree, "Some of the Sources of the Ghayat al-Hakim," *Journal of the Warburg and Courtauld Institutes* 43 (1980): 1–15.

14. Thorndike, *History of Magic and Experimental Science*, 2:350–351.

15. Charles G. Nauert, *Agrippa and the Crisis of Renaissance Thought* (Urbana: University of Illinois Press, 1965), 120.

16. Two important articles on this topic are William Newman, "Technology and Alchemical Debate in the Late Middle Ages," *Isis* 80 (1989): 423–445; and William Eamon, "Technology as Magic in the Late Middle Ages and the Renaissance," *Janus* 70, 3–4 (1983): 170–199.

17. Lorraine Daston, "Marvelous Facts and Miraculous Evidence in Early Modern Europe," in James Chandler, Arnold I. Davidson, and Harry Harootunian,

eds., *Questions of Evidence: Proof, Practice, and Persuasion across the Disciplines* (Chicago: University of Chicago Press, 1991), 243–274.

18. Daston, "Marvelous Facts and Miraculous Evidence," 256.

19. Jean Gimpel, *The Medieval Machine: The Industrial Revolution of the Middle Ages* (New York: Penguin Books, 1977), 130–134.

20. Eamon, "Technology as Magic in the Late Middle Ages and the Renaissance," 176; and Lorraine Daston and Katharine Park, *Wonders and the Order of Nature, 1150–1750* (New York: Zone Books, 1998), 95–100.

21. For a full description of the wonders, as purchased by Philip the Good, see Miriam Sherwood, "Magic and Mechanics in Medieval Fiction," *Studies in Philology* 44, 4 (Oct. 1947): 567–592, 587–589.

22. For a detailed description and analysis of this episode see Scott Lightsey, *Manmade Marvels in Medieval Culture and Literature* (New York: Palgrave Macmillan, 2007), 27–53.

23. Elspeth Whitney, *Paradise Restored: The Mechanical Arts from Antiquity through the Thirteenth Century* (Philadelphia: American Philosophical Society, 1990), 3. See also Lynn White, *Medieval Technology and Social Change* (Oxford: Oxford University Press, 1962), 79–134; and Gimpel, *Medieval Machine*.

24. Eamon, *Science and the Secrets of Nature*, 82.

25. Whitney, *Paradise Restored*, 82–99.

26. Eamon, *Science and the Secrets of Nature*, 59 and 83.

27. Quoted in Carlo Ginzburg, "Representation: The Word, the Idea, and the Thing," in Ginzburg, *Wooden Eyes: Nine Reflections on Distance*, trans. Martin Ryle and Kate Soper (New York: Columbia University Press, 2001), 76.

28. Flint, *Rise of Magic*, 255–265.

29. Ginzburg, "Representation," 77–78.

30. Ibid., 77.

31. Several fine studies of automata in medieval fiction have been written, most recently by E. R. Truitt on their appearance in French literature: Truitt, "'Trei poëte, sages dotors, qui mout sorent di nigromance'": Knowledge and Automata in Twelfth-century French Literature," *Configurations* 12 (2004): 167–193. See also J. Douglas Bruce, "Human Automata in Classical Tradition and Medieval Romance," *Modern Philology* 10 (1912–1913): 511–526; Sherwood, "Magic and Mechanics in Medieval Fiction"; Penny Sullivan, "Medieval Automata: The 'Chambre de Beautés' in Benoît's *Roman de Troie*," *Romance Studies* 6 (Summer 1985): 1–20; Kevin LaGrandeur, "The Talking Brass Head as a Symbol of Dangerous Knowledge in *Friar Bacon* and in *Alphonsus, King of Aragon*," *English Studies* 80, 5 (Oct. 1999): 408–422.

32. See "The Journey of William of Rubruck," in Christopher Dawson, ed., *The Mongol Mission: Narratives and Letters of the Franciscan Missionaries in Mongolia and China in the Thirteenth and Fourteenth Centuries* (New York: Sheed and Ward, 1955), 89–220, on the automata: 175–176. For more on this interesting figure of William Buchier or Guillaume Boucher, see Leonardo Olschki, *Guillaume Boucher: A French Artist at the Court of the Khans* (New York: Greenwood Press, 1946), especially, on his automata, 89–98.

33. See Bruce, "Human Automata in Classical and Medieval Romance," 5; and Sherwood, "Magic and Mechanics in Medieval Fiction," 568.

34. See Sullivan, "Medieval Automata," 2–3. See also Truitt, "Trei poëte, sages dotors, qui mout sorent di nigromance," 167–169.

35. Sullivan, "Medieval Automata," 13. John Spago has also noted legends of automata and other fantastic devices associated with the figure of Virgil. See John Webster Spago, *Virgil the Necromancer: Studies in Virgilian Legends* (Cambridge, Mass.: Harvard University Press, 1934), especially 117–135.

36. See Bruce, "Human Automata in Classical and Medieval Romance," 11.

37. Edmund Spenser, *Spencer's Faerie Queene*, vol. 2, bks. 4–7 (Oxford: Clarendon Press, 1909), 166 (bk. 5 pt. 1).

38. Jessica Wolfe, *Humanism, Machinery, and Renaissance Literature* (Cambridge: Cambridge University Press, 2004), 203.

39. Truitt, "Trei poëte, sages dotors, qui mout sorent di nigromance," 172.

40. Jacques Le Goff, *The Medieval Imagination*, trans. Arthur Goldhammer (Chicago: University of Chicago Press, 1985), 36–37.

41. Ibid., 37.

42. Odoric, *The Travels of Friar Odoric*, trans. Henry Yule (Grand Rapids, Mich.: Eerdmans, 2002), 137.

43. Ibid., 137.

44. See Iain Macleod Higgins, *Writing East: The "Travels" of Sir John Mandeville* (Philadelphia: University of Pennsylvania Press, 1997), especially 156–169. See also Lightsey, *Manmade Marvels*, 137–158.

45. Mandeville, *Mandeville's Travels: Texts and Translations*, vol. 1, ed. Malcolm Letts (London: Hakluyt Society, 1953), 151. "The Book of John Mandeville" trans. unknown.

46. In the original Anglo-Norman French, "par artefice ou par nigromance," in Mandeville, *Mandeville's Travels*, 352.

47. On the sources of Chaucer's Squire's Tale see Kathryn L. Lynch, "East Meets West in Chaucer's Squire's and Franklin's Tales," *Speculum* 70, 3 (July 1995): 530–551. See also Lightsey's analysis of the Squire's Tale in the context of Chaucer's general attitude toward manmade marvels, *Manmade Marvels*, 74–80.

48. Geoffrey Chaucer, "The Canterbury Tales," in *The Complete Works of Chaucer*, ed. F. N. Robinson (Boston: Houghton Mifflin, 1933), 155 (5. 110).

49. Ibid., 156 (5.200).

50. Ibid., 158 (5.322).

51. Ibid., 155 (5.128–132).

52. William Shakespeare, *The Winter's Tale* (New York: Washington Square Press, 1998), 233 (5. 3. 130–139). For a discussion of this passage in the larger context of animation in literature see Kenneth Gross, *The Dream of the Moving Statue* (Ithaca: Cornell University Press, 1992), 100–109, and Scott Maisano, "Infinite Gesture: Automata and the Emotions in Descartes and Shakespeare," in Jessica Riskin, ed., *Genesis Redux: Essays in the History and Philosophy of Artificial Life* (Chicago: University of Chicago Press, 2007), 63–84.

53. For a concise account of Gerbert's life, see Harriet Pratt Lattin, editor's introduction to *The Letters of Gerbert with His Papal Privileges as Sylvester II* (New York: Columbia University Press, 1961), 1–20. For more details see Roland Allen, "Gerbert, Pope Silvester II," *English Historical Review* 27 (October 1892): 625–668.

54. William of Malmesbury, *Chronicle of the Kings of England*, trans. J. A. Giles (London: Henry G. Bohn, 1847), 173–174.
55. Ibid., 176.
56. Ibid., 181.
57. Allen, "Gerbert, Pope Silvester II," 666.
58. Ibid., 663.
59. The most complete account of the sources and transmissions of these talking head stories can be found in Arthur Dickson, *Valentine and Orson: A Study in Late Medieval Romance* (New York: Columbia University Press, 1929), 201–216.
60. I am deeply grateful to Arielle Saiber for providing the translation of this text from the Italian. The original text can be found in Matteo Corsini, *Rosaio della vita* (Florence: Societá Poligrafica Italiana, 1845), 15–16.
61. Dickson, *Valentine and Orson*, 214, n. 147.
62. *The Complete Works of John Gower*, vol. 2, ed. G. C. Macaulay (Grosse Pointe, Mich.: Scholarly Press, 1968), 307 (bk 4, l. 234–243).
63. *The Famous Historie of Fryer Bacon* (Edinburgh: O. Schulze, 1908), 111.
64. Ibid., 112.
65. See Robert Greene, *The Honourable History of Friar Bacon and Friar Bungay*, in Ashley Thorndike ed., *Minor Elizabethan Drama*, vol. 2 (London: Everyman's Library, 1958), 206.
66. Greene, "With seven years' tossing necromantic charms" in *Honourable History*, 206.
67. Allen, "Gerbert, Pope Sylvester II," 633.
68. Flint, *Rise of Magic*, 142–143.
69. For details on Albertus's scientific ideas see James A. Weishipl, ed., *Albertus Magnus and the Sciences: Commemorative Essays* (Toronto: Pontifical Institute of Mediaeval Studies, 1980), and Thorndike, *History of Magic and Experimental Science*, 2:521–592.
70. William R. Newman, *Promethean Ambitions: Alchemy and the Quest to Perfect Nature* (Chicago: University of Chicago Press, 2004), 49.
71. Lynn White has presented an interesting theory that the source of the talking head legends may be the description in Albertus's *De Meteoris* of a device called an aeolipiles or fire-blower. Originally appearing in works of ancient engineers like Hero and Vitruvius, it was a hollow metal vessel that was filled with water and heated. A blowing device then sent steam and water out of it through a hole and onto a fire, making the flames flare up. As verified by modern archaeologists through their findings in England, Germany, and other places, Albertus claimed that they were often made in the shape of a man, with the steam and water spraying out of its mouth or genitals. This report may have contributed to the legends, especially in connection with Albertus, but it is unlikely to be the source since the story attached to Gerbert is earlier and there are other more plausible origins of the talking head image. See White, *Medieval Technology and Social Change*, 90–91. For more on aeolipiles including archaeological findings, see W. L. Hildburgh, "Aeolipiles as Fire-Blowers," *Archaeologia* 94 (1951): 27–55.
72. See Crombie, *Robert Grosseteste*, especially 61–90 on his experimental

method; and 100–218 on Roger Bacon. On Bacon's praise of Grosseteste, see Thorndike, *History of Magic and Experimental Science*, 2:437. For the appearance of alchemical and magical ideas in Gower as they informed the Grosseteste story in his work, see George G. Fox, *The Mediaeval Sciences in the Works of John Gower* (New York: Haskell House, 1966), 114–155.

73. On his ideas on optics, see Crombie, *Robert Grossetests*, 91–127. See also Thorndike, *History of Magic and Experimental Science*, 2:440–443.

74. A. G. Molland, "Roger Bacon as Magician," *Traditio* 30 (1974): 442–469, 453.

75. For Grosseteste's attitude toward magic, see Thorndike, *History of Magic and Experimental Science*, 2:444–447, and Fox, *Mediaeval Sciences in the Works of John Gower*, 120–122; for Bacon's attitude toward magic, see Thorndike, *History of Magic and Experimental Science*, 2:659–674, also Newman, *Promethean Ambitions*, 117–118. For more on Bacon's scientific and magical ideas see Jeremiah Hackett, ed., *Roger Bacon and the Sciences: Commemorative Essays* (New York: Brill, 1977), especially Jeremiah Hackett, "Roger Bacon on Astronomy-Astrology: The Sources of the Scientia Experimentalis," 175–198.

76. See Dickson, *Valentine and Orson*, 201–205.

77. For details on the ancient Hermetic texts, including differing ideas on dating, see Frances Yates, *Giordano Bruno and the Hermetic Tradition* (Chicago: University of Chicago Press, 1964), especially 2–4, and more recently, Garth Fowden, *The Egyptian Hermes: A Historical Approach to the Late Pagan Mind* (Princeton: Princeton University Press, 1986), 68–74. See also *Hermetica*, trans. Brian Copenhaver (Cambridge: Cambridge University Press, 1992), lvii–lviii.

78. *Asclepius*, in Copenhaver, *Hermetica*, 81.

79. Ibid., 90.

80. Ibid., 90–91.

81. Augustine, *The City of God*, trans. Henry Bettenson (Harmondsworth, England: Penguin Books, 1984), 331.

82. Ibid., *City of God*, 332.

83. William of Auvergne quoted in Thorndike, *History of Magic and Experimental Science*, 2:350.

84. Kevin LaGrandeur, "Talking Brass Head as a Symbol of Dangerous Knowledge," 422.

85. Gabriel Naudé, *The History of Magick: By way of apology for all the wise men who have unjustly been reputed magicians, from the creation to the present age*, trans. J. Davies (London: J. Streater, 1657), 187; and Wilkins, *Mathematicall Magick*, 104. For an excellent discussion of this topic see Alexander Marr, "Understanding *Automata* in the Late Renaissance." *Journal de la Renaissance* 2 (2004): 205–221.

86. On the recovered texts on mechanics and their impact on the period, see E. R. Laird, "The Scope of Renaissance Mechanics," *Osiris* 2 (1986): 43–68; Wolfe, *Humanism, Machinery, and Renaissance Literature*, 29–57. For more on the impact of *Mechanical Problems* see Paul Lawrence Rose and Stillman Drake, "The Pseudo-Aristotelian *Questions of Mechanics* in Renaissance Culture," *Studies in the Renaissance* 18 (1971): 65–104. On Hero see Marie Boas, "Hero's *Pneumatica*: A Study of Its Transmission and Influence," *Isis* 119 (Feb. 1990): 38–48.

87. Wolfe, *Humanism, Machinery, and Renaissance Literature*, 37–50.

88. Pamela O. Long, "Power, Patronage, and the Authorship of Ars: From Mechanical Know-how to Mechanical Knowledge in the Last Scribal Age," *Isis* 88, 1 (Mar. 1997): 1–41. See also Eamon, "Technology as Magic in the Late Middle Ages and the Renaissance."

89. See Carlo M. Cipolla, *Clocks and Culture 1300–1700* (New York: Norton, 1977), 40–44; Gimpel, *Medieval Machine*, 147–170; and David S. Landes, *Revolution in Time: Clocks and the Making of the Modern World* (Cambridge, Mass.: Harvard University Press, 1983), 53–82.

90. See Cipolla, *Clocks and Culture*, 44; Landes, *Revolution in Time*, 83–84; and Derek de Solla Price, "Automata and the Origins of Mechanism and Mechanistic Philosophy," *Technology and Culture* 5, 1 (Winter 1964): 9–25, 18.

91. Alfred Chapuis and Edouard Gélis, *Automata, a Historical and Technological Study*, trans. Alec Reid (New York: Central Book, 1958), 49–66.

92. Daston and Park, *Wonders*, 100–106.

93. Ibid., 139–154.

94. Ibid., 260. See also the recent essay collection R. J. W. Evans and Alexander Marr, eds., *Curiosity and Wonder from the Renaissance to the Enlightenment* (Burlington, Vermont: Ashgate, 2006), especially Marr's discussion of the role of automata in collections of wonders, "*Gentille Curiosité:* Wonder-working and the Culture of Automata in the Late Renaissance," 149–170. The topic is also discussed in Patrick Mauriès, *Cabinets of Curiosities* (London: Thames and Hudson, 2002), 116–119, and Horst Bredekamp, *The Lure of Antiquity and the Cult of the Machine*, trans. Allison Brown (Princeton: Wiener, 1995), 37–62.

95. See R. J. W. Evans, *Rudolph II and His World: A Study in Intellectual History, 1576–1612* (Oxford: Clarendon Press, 1973), especially 196–242. See also Thomas DaCosta Kaufmann, "Remarks on the Collections of Rudolph II: The Kunstkammer as a Form of Representatio," *Art Journal* 38, 1 (Autumn 1978): 22–28.

96. Wolfe, *Humanism, Machinery, and Renaissance Literature*, 31–32.

97. For pictures of the lute player and the monk see José A. Garcia-Diego, *Juanelo Turriano, Charles V's Clockmaker: The Man and His Legend*, trans. Charles David Ley (Madrid: Editorial Castalia, 1986).

98. There is evidence that he did build automata for the clock-obsessed Charles, especially after the emperor's retirement to the monastery of Yuste in Extremadura in 1556, but Garcia-Diego has exposed the tenuous nature of the attribution of the extant automata to him (see *Juanelo Turriano*, 97–116). Elizabeth King has written a fascinating essay on the monk and her own search for its origins: "Perpetual Devotion: A Sixteenth-century Machine That Prays," in Riskin, *Genesis Redux*, 263–292.

99. Giovanni Paolo Lomazzo names Francis I as the French king who viewed the lion automaton. See Lomazzo, *A Tracte Containing the Artes of Curious Paintings*, trans. Richard Haydock (Amsterdam: Da Capo Press, 1969), 2 (bk. 2). For details on the lion automaton and an alternate idea on its presentation to Francis I, see Mark Elling Rosheim, *Leonardo's Lost Robots* (Berlin: Springer, 2006), 21–68.

100. Giorgio Vasari, *The Lives of Great Artists*, trans. Julia Conaway Bondanella and Peter Bondanella (Oxford: Oxford University Press, 1991), 293.

101. For details on the designs for the knight automaton, see Rosheim, *Leonardo's Lost Robots*, 69–112.

102. Silvio Bedini, "The Role of Automata in the History of Technology," *Technology and Culture* 5, 1 (Winter 1964): 26–42, 27–28; and Chapuis and Droz, *Automata*, 43–47.

103. Chapuis and Droz, *Automata*, 40.

104. Michel Eyquem de Montaigne, *The Complete Works of Montaigne: Essays, Travel Journal, Letters*, trans. Donald M. Frame (Stanford: Stanford University Press, 1943), 963. See also Bedini, "Role of Automata in the History of Technology," 26; and Chapuis and Droz, *Automata*, 42–43.

105. Frederika H. Jacobs, *The Living Image in Renaissance Art* (Cambridge: Cambridge University Press, 2005).

106. Lightsey provides a particularly illuminating analysis of how collections of wonders functioned in the courts of Europe as demonstrations of power in the late Middle Ages (see *Manmade Marvels*, 30–53).

107. Anthony Grafton, "The Devil as Automaton: Giovanni Fontana and the Meaning of a Fifteenth-century Machine," in Riskin, *Genesis Redux*, 46–62.

108. Yates, *Giordano Bruno and the Hermetic Tradition*, 12–19. See also Brian Copenhaver's important essay on this topic, "Hermes Trismegistus, Proclus, and the Question of Philosophy of Magic in the Renaissance," in Ingrid Merkel and Allen G. Debus, eds., *Hermeticism and the Renaissance: Intellectual History and the Occult in Early Modern Europe* (London: Folger Books, 1988), 79–106.

109. See Kieckhefer, *Magic in the Middle Ages*, 134, and Clulee, *John Dee's Natural Philosophy*, 105.

110. Yates, *Giordano Bruno and the Hermetic Tradition*, 398–403.

111. Nicholas Clulee sees Neoplatonic philosophy as the predominant inspiration for Renaissance philosophy and is somewhat critical of Yates's depiction of a coherent Hermetic worldview. See Clulee, *John Dee's Natural Philosophy*, 129–135.

112. Thorndike, like some of Agrippa's contemporary critics, views him as "a dabbler and a trifler," describing his work as little more than a compendium of other people's ideas with not much original contribution to the subject. Thorndike, *A History of Magic and Experimental Science*, vol. 5 (New York: Columbia University Press, 1941), 133–134. For a more appreciative assessment see Yates, *Giordano Bruno and the Hermetic Tradition*, 130–43; and Nauert, *Agrippa and the Crisis of Renaissance Thought*.

113. Aristotle, *Politics*, trans. Benjamin Jowett, in *The Complete Works of Aristotle*, vol. 2, ed. Jonathan Barnes (Princeton: Princeton University Press, 1984), 1989.

114. François Rabelais, *Gargantua and Pantagruel*, trans. J. M. Cohen (London: Penguin Books, 1955), 105. For the variations in the original, see François Rabelais, *Œuvres Complètes* (Paris: Éditions du Seuil, 1973), 118–119. See also Jean-Claude Beaune, "The Classical Age of Automata: An Impressionistic Survey from the Sixteenth to the Nineteenth Century," in Michel Feher, Ramona Naddaff, and Nadia Tazi, eds., *Fragments for a History of the Human Body*, pt. 1 (New York: Zone Books, 1989), 431.

115. Lucian Febvre asserts that Rabelais must have known about the first volume of *De Occulta Philosophia* that was published in 1531. Febvre, *The Problem of Unbelief*

in the Sixteenth Century: The Religion of Rabelais, trans. Beatrice Gottlieb (Cambridge, Mass.: Harvard University Press, 1982), 228. The full three-volume text, with the use of the word "automata" in the second volume, did not appear until 1533, and *Gargantua* was not published until a year after.

116. Rabelais, *Gargantua and Pantagruel*, 356.
117. Henry Cornelius Agrippa of Nettesheim, *Three Books of Occult Philosophy*, trans. James Freake (St. Paul, Minn.: Llewellyn, 2000), 233.
118. Ibid., 134.
119. Eamon, *Science and the Secrets of Nature*, 24.
120. For the passage in the sixteenth-century English translation, see Henry Cornelius Agrippa, *Three Books of Occult Philosophy, or of Magick*, trans. J. F. (Hastings, U.K.: Chthonios Books, 1986), 168.
121. Philostratus, *The Life of Apollonius of Tyana*, trans. F. C. Conybeare (London: Heinemann, 1912), 289–291 (bk. 3, part 27).
122. Cassiodorus, *Variae*, trans. S. J. B. Banish (Liverpool: Liverpool University Press, 1992), 20 (bk. 1, part 45).
123. Ibid., 20.
124. Newman, *Promethean Ambitions*, 164–237.
125. For more on these categories in John Dee's work, see Clulee, *John Dee's Natural Philosophy*, 154–162.
126. John Dee, *The Mathematicall Praeface to the Elements of Geometrie of Euclid of Megara* (New York: Science History, 1975), A.i.
127. See Clulee, *John Dee's Natural Philosophy*, 136 and 161; and Yates, *Giordano Bruno and the Hermetic Tradition*, 149.
128. Otto Mayr, *Authority, Liberty and Automatic Machinery in Early Modern Europe* (Baltimore: Johns Hopkins University Press, 1986), 9.
129. For details on the automata legends of Regiomontanus, see Ernst Zinner, *Regiomontanus: His Life and Work*, trans. Ezra Brown (Amsterdam: Elsevier Science, 1990), 135–140.
130. Ibid., 135.
131. Daston and Park, *Wonders*, 68–69 and 76–86.
132. Dee, *Mathematicall Praeface*, A.i.
133. For more on the scarab episode see J. Peter Zetterberg, "The Mistaking of 'the Mathematicks' for Magic in Tudor and Stuart England," *Sixteenth Century Journal* 12, 1 (Spring 1980): 81–108. On the accusation of witchcraft against Mary, see Clulee, *John Dee's Natural Philosophy*, 33–34.
134. Frances A. Yates, *The Occult Philosophy in the Elizabethan Age* (London: Routledge and Kegan Paul, 1979), 154–155.
135. Clulee, *John Dee's Natural Philosophy*, 203–230.
136. D. P. Walker, *Spiritual and Demonic Magic from Ficino to Campanella* (University Park: Pennsylvania State University Press, 2000), 46.
137. Walker, *Spiritual and Demonic Magic*, 50–51.
138. On Lomazzo and his theories on art and lifelikeness, see Jacobs, *Living Image in Renaissance Art*, 24–26.
139. Lomazzo, *Tracte Containing the Artes of Curious Paintinge*, 2 (bk. 2).

140. Ibid., 2.

141. John Baptista Porta, *Natural Magick*, trans. anonymous (New York: Basic Books, 1957), 385.

142. Ibid., 385–386.

143. See Chapuis and Droz, *Automata*, 16–17, and Dickson, *Valentine and Orson*, 191–192.

144. Quoted in Yates, *Giordano Bruno and the Hermetic Tradition*, 147–148.

145. For more on the bull of Phalaris, see Truesdell S. Brown, *Timaeus of Tauromenium* (Berkeley: University of California Press, 1958), 54–58.

146. For more on Campanella see Bernadino M. Bonansea, *Tommaso Campanella: Renaissance Pioneer of Modern Thought* (Washington D.C.: Catholic University of America Press, 1969).

147. Alexander Marr has pointed to the same magical-technological tension in other works of the period in which the list appears in different forms, including Martin del Rio's 1599 work on magic, the 1625 work by Gabriel Naudé, and, in England, a 1627 book by George Hakewill. For Marr's discussion of the Del Rio text see Marr, "*Gentile curosité*," 152–153, on the others see Marr, "Understanding *Automata*."

148. Daston and Park, *Wonders*, 205–208.

149. Miguel de Cervantes, *Don Quixote*, trans. John Ormsby (New York: Norton, 1981), 770.

150. Ibid., 775.

151. Leonora Cohen Rosenfield, *From Beast-machine to Man-Machine: The Theme of Animal Soul in French Letters from Descartes to La Mettrie* (New York: Oxford University Press, 1940), 31–33.

152. For more on Florentius Schuyl, see Rosenfield, *From Beast-machine to Man-Machine*, 245–249.

153. I am working from the French translation of Schuyl's introduction from Clerselier's 1664 edition of the treatise—*L'Homme de René Descartes, et un Traité de la Formation du Foetus du Mesme Autheur* (Paris: Chez Charles Angot, 1664), 421–422.

154. See Michael John Gorman, "Between the Demonic and the Miraculous: Athanasius Kircher and the Baroque Culture of Machines," in Daniel Stolzenberg, *The Great Art of Knowing: The Baroque Encyclopedia of Athansius Kircher* (Stanford: Stanford University Libraries, 2001), 59–70, 61. In the list in the works of John Wilkins discussed below, he also cites Coelius Rhodignius. See Wilkins, *Mathematicall Magick*, 104.

155. The source of the reference is given as Vassenaer in the French translation, which reads Wassenaer in the Latin original. I assume that the author is Nicolaes van Wassenaer, a Dutch historian whose twenty-one-volume history was published from 1622 to 1635. I have, unfortunately, been unable to track down the origin of the legend.

156. The major biography of John Wilkins is Barbara J. Shapiro, *John Wilkins, 1614–1672: An Intellectual Biography* (Berkeley: University of California Press, 1969).

157. Wilkins, *Mathematicall Magick*, 102.

158. On Kircher's interest in automata and other mechanical devices see Gorman, "Between the Demonic and the Miraculous." On that topic Gorman makes a similar point to mine in this chapter about the attraction of automata in the period coming from the viewer's uncertainty about how they were achieved.

159. Wilkins, *Mathematicall Magick*, 104.

160. Ibid., 112. As mentioned above, the Holy Roman Emperor during the time Regiomontanus was at Nuremberg was Friedrich III, but it is interesting that Wilkins names Charles V because, according to Ernst Zinner, the double image of an eagle that may have been the source of the legend of the flying automaton was first hung on the occasion of Charles V's visit to the city in 1541, twenty-eight years before Petrus Ramus published the account of the eagle and the fly. See Zinner, *Regiomontanus*, 135.

161. Denis Diderot and Jean le Rond D'Alembert, *Encyclopédie, ou dictionnaire raisonné des sciences, des arts et des métiers*, 3rd ed. (Geneva: Jean-Léonard Pellet, 1778): "Automate," 2:116–118; "Androide," 4:626–630.

162. Aulus Gellius, *The Attic Nights of Aulus Gellius* (London: Heinemann, 1927), 37.

163. Quoted in Manly Palmer Hall, *The Secret Teachings of All Ages: An Encyclopedic Outline of Masonic, Hermetic, Cabbalistic and Rosicrucian Symbolical Philosophy* (Los Angeles: Philosophical Research Society Press, 1947), cl. Because of the tantalizing obscurity of the origin of the word *android*, I was hoping to locate a medieval usage, but this quotation from the sixteenth-century work is the earliest reference I have been able to find. Marr also notes its appearance in Naudé's 1625 work ("Understanding *Automata*," 207, n. 11).

3. The Man-machine in the World-machine

1. *Mercure de France* (April 1978), 738.

2. Quoted in Alfred Chapuis and Edmond Droz, *Automata: A Historical and Technological Study*, trans. Alec Reid (New York: Central Book, 1958), 274. For the quotation in the original see André Doyon and Lucien Liaigre, *Jacques Vaucanson: Mécanicien de génie* (Paris: Presse Universitaires de France, 1966), 40.

3. Doyon and Liaigre, *Vaucanson*, 57–58.

4. Joseph Spence, *Letters from the Grand Tour* (Montreal: McGill-Queen's University Press, 1975), 413–414.

5. Johann Wolfgang von Goethe, "Annals; Or Day and Year Papers," in Goethe, *The Autobiography of Goethe*, trans. anonymous (London: George Bell, 1894), 320.

6. Chapuis and Droz, *Automata*, 234.

7. Chapuis and Droz, *Automata*, 279–284, and 289–295. For more on the Jaquet-Droz automata, including some beautiful photographs of them, see Roland Carrera, Dominique Loiseau, and Olivier Rous, *Androids: The Jaquet-Droz Automatons* (Lausanne: Scriptar, 1979).

8. Chapuis and Droz, *Automata*, 300–308.

9. On the Knaus automata, see Chapuis and Droz, *Automata*, 289–292. On speaking automata, see 320–324; and Thomas L. Hankins and Robert J. Silverman,

Instruments and the Imagination (Princeton: Princeton University Press), 186–198. Mical's speaking automata were possibly the inspiration for the fictional automaton in Allen Kurzweil's novel *A Case of Curiosities* (New York: Harcourt, Brace, Jovanovich, 1992). Given the fact that a number of speaking automata made their appearance in the 1770s and 1780s it is unlikely that another one would have caused the fuss that it does in this novel set on the eve of the French Revolution, but that is an overly fastidious nitpicking on the part of a historian. The novel does an excellent job of capturing the automaton craze in this period.

10. Richard D. Altick, *The Shows of London* (Cambridge, Mass.: Harvard University Press, 1978), 69–72.

11. Altick, *Shows of London*, 72–76. For more on Merlin see the catalogue of his works with several interesting essays on them—Anne French, *John Joseph Merlin: The Ingenious Mechanick* (London: Greater London Council, 1985).

12. See Otto Mayr, *Authority, Liberty and Automatic Machinery in Early Modern Europe* (Baltimore: Johns Hopkins University Press, 1986), 107–109, and Michel Foucault, *Discipline and Punish*, trans. Alan Sheridan (New York: Vintage Books, 1977), 136–156.

13. Doyon and Liaigre, *Vaucanson*, 185–203. See also Charles Coulston Gillispie, *Science and polity in France at the End of the Old Regime* (Princeton: Princeton University Press, 1980), 414–416.

14. Doyon and Liaigre, *Vaucanson*, 205–253. See also Gillispie, *Science and Polity in France*, 416–418.

15. Siegfried Giedion, *Mechanization Takes Command* (New York: Norton, 1948), 36. For details on Vaucanson's Aubenas enterprise, see Doyon and Liaigre, *Vaucanson*, 293–299; and Gillispie, *Science and Polity in France*, 418–420.

16. For details on this project, see Doyon and Liaigre, *Vaucanson*, 109–131.

17. Jeff Horn and Margaret C. Jacob, "Jean-Antoine Chaptal and the Cultural Roots of French Industrialization," *Technology and Culture* 39 (1998): 671–698, 674 n. 10.

18. Karl Marx, *Capital*, vol. 1, trans. Ben Fowkes (New York: Vintage Books, 1977), 503.

19. Immanuel Kant, *Critique of Practical Reason*, trans. Lewis White Beck (Chicago: University of Chicago Press, 1949), 206.

20. Thomas Carlyle, "Signs of the Times," in *Critical and Miscellaneous Essays*, ed. H. D. Traill (London: Chapman and Hall, 1899), 56–82, 65. Hermann von Helmoltz, "On the Interaction of Natural Forces," in *Popular Lectures on Scientific Subjects*, trans. E. Atkinson (New York: Longmans, Green, 1904), 137–171, 137–138.

21. Thomas Henry Huxley, "On the Hypothesis That Animals Are Automata, and Its History," in *Science and Culture, and Other Essays* (London: MacMillan, 1881), 199–245, 227.

22. Iwan Bloch, *The Sexual Life of Our Time in Its Relations to Modern Civilization*, trans. M. Eden Paul (New York: Allied Books, 1925), 648.

23. Villiers de l'Isle-Adam, *Tomorrow's Eve*, trans. Robert Martin Adams (Champaign: University of Illinois Press, 2001), 61.

24. Thomas Pynchon, *Mason and Dixon* (New York: Holt, 1997), 668.

25. Marquis de Condorcet, "Élogie de M. De Vaucanson," in *Œuvres completes de Condorcet*, vol. 2 (Paris: Chez Heinrichs, 1804), 437.

26. Quoted in Doyon and Liaigre, *Vaucanson*, 50, and for a full quotation of Desfontaines's praise of Vaucanson see 49–51.

27. Abbé Prevost, "Il ne faut point appréhender . . ." *Le Pour et contre* 14, 101 (1773): 212.

28. Doyon and Liaigre, *Vaucanson*, 41–42. See also Jacques de Vaucanson, *Le Mécanisme du Fluteur Automate, An Account of the Mechanism of an Automaton or Image Playing on the German-Flute*, trans. J. T. Desaguliers (Buren, The Netherlands: Uitgeverij Frits Knuf, 1979).

29. Quoted in Doyon and Liaigre, *Vaucanson*, 42.

30. Voltaire, "Discours en vers sur l'homme," in *The Complete Works of Voltaire*, vol.17, ed. W. A. Smeaton and Robert L. Walters (Oxford: Voltaire Foundation, Taylor Institution, 1991), 521. For Voltaire's recommendation to Frederick see Doyon and Liaigre, *Jacques Vaucanson*, 133–136; and Michael Cardy, "Technology as Play: The Case of Vaucanson," *Studies on Voltaire and the Eighteenth Century* 241 (1986): 109–123, 115–116.

31. Julien Offray de La Mettrie, *Man a Machine and Man a Plant*, trans. Richard A. Watson and Maya Rybalka (Indianapolis: Hackett, 1994), 69.

32. Denis Diderot, *Rameau's Nephew/D'Alembert's Dream*, trans. Leonard Tancock (London: Penguin Books, 1966), 211.

33. "Androide," in Denis Diderot and Jean le Rond D'Alembert, *Encyclopédie, ou dictionnaire raisonné des sciences, des arts, et des métiers*, 3rd ed., vol. 2 (Geneva: Jean-Léonard Pellet, 1778), 626–630.

34. "Automate," in Diderot and D'Alembert, *Encyclopédie*, 2:116–118.

35. Louis-Sebastien Mercier, *Panorama of Paris*, trans. Helen Simpson and Jeremy D. Popkin (University Park: Pennsylvania State University Press, 1999), 60.

36. Joan B. Landes, "The Anatomy of Artificial Life: An Eighteenth-century Perspective," in Jessica Riskin, ed., *Genesis Redux: Essays in the History and Philosophy of Artificial Life* (Chicago: University of Chicago Press, 2007), 96–116. See also Barbara Maria Stafford, *Artful Science: Enlightenment, Entertainment and the Eclipse of Visual Education* (Cambridge, Mass.: MIT press, 1994), 190–197; and "The Defecating Duck, or, the Ambiguous Origins of Artificial Life," *Critical Inquiry* 29, 4 (Summer 2003): 599–633.

37. Vaucanson, *Le Mécanisme du Fluteur Automate*, 22.

38. Ibid., 21.

39. Doyon and Liaigre, *Vaucanson*, 125–129.

40. Spence, *Letters from the Grand Tour*, 413.

41. E. J. Dijkterhuis, *The Mechanization of the World Picture: Pythagoras to Newton*, trans. C. Dikshoorn (Princeton: Princeton University Press, 1986).

42. For recent discussions see William J. Bouwsma, *The Waning of the Renaissance, 1550–1640* (New Haven: Yale University Press, 2000), and Theodore K. Rabb, *The Last Days of the Renaissance and the March to Modernity* (New York: Basic Books, 2006).

43. On the antienthusiasm attitude of seventeenth-century thinkers, see Steven Shapin and Simon Shaffer, *Leviathan and the Air-Pump: Hobbes, Boyle, and the Exper-

imental Life (Princeton: Princeton University Press, 1985), especially 283–331. See also Lorraine Daston and Katharine Park, *Wonders and the Order of Nature, 1150–1750* (New York: Zone Books, 1998), 334–346, and Michael Heyd, *"Be Sober and Reasonable": The Critique of Enthusiasm in the Seventeenth and Early Eighteenth Centuries* (Leiden: Brill, 1995).

44. Daston and Park, *Wonders*, 208–214.

45. Ibid., 210.

46. Robert Boyle, *A Free Enquiry into the Vulgarly Received Notion of Nature* (Cambridge: Cambridge University Press, 1996), 160.

47. Daston and Park, *Wonders*, 336–337.

48. Ibid., 361 and 329.

49. For alternate views see Anita Guerrini, "The Creativity of God and the Order of Nature: Anatomizing Monsters in the Early Eighteenth Century" in Charles T. Wolfe, ed., *Monsters and Philosophy* (London: College, 2005), 153–168; Margaret Ewalt, "Christianity, Coca, and Commerce in the Peruvian Mercury," *Studies in Eighteenth-Century Culture* 36 (Spring 2007): 187–212; and Paolo Bertucci, "Back from Wonderland: Jean Antoine Nollet's Italian Tour (1749)," in R. J. W. Evans and Alexander Marr, eds., *Curiosity and Wonder from the Renaissance to the Enlightenment* (Burlington, Ver.: Ashgate, 2006), 193–211.

50. Mary Terrall, *The Man Who Flattened the Earth: Maupertuis and the Sciences in the Enlightenment* (Chicago: University of Chicago Press, 2002), 204–205. In the same vein, see the discussion on Diderot's interest in monsters in Annie Ibrahim, "The Status of Anomalies in the Philosophy of Diderot," in Wolfe, *Monsters and Philosophy*, 169–186.

51. René Descartes, *The Philosophical Writings of Descartes*, vol. 1, trans. John Cottingham, Robert Stoothhoff, and Dugald Murdoch (Cambridge: Cambridge University Press, 1985), 115.

52. See Paolo Rossi, *Francis Bacon: From Magic to Science*, trans. Sacha Rabinovitch (London: Routledge and Kegan Paul, 1968), Betty Jo Teeter Dobbs, *The Janus Faces of Genius: The Role of Alchemy in Newton's Thought* (Cambridge: Cambridge University Press, 1991), and Lawrence Principe, *The Aspiring Adept: Robert Boyle and His Alchemical Quest* (Princeton: Princeton University Press, 1998).

53. There is a substantial literature on this topic now, the latest examples being William R. Newman, *Atoms and Alchemy: Chymistry and the Experimental Origins of the Scientific Revolutions* (Chicago: University of Chicago Press 2006); William R. Newman and Lawrence M. Principe, *Alchemy Tried in the Fire: Starkey, Boyle, and the Fate of Helmontian Chymistry* (Chicago: University of Chicago Press, 2002); and William Eamon, *Science and the Secrets of Nature: Books of Secrets in Medieval and Early Modern Culture* (Princeton: Princeton University Press, 1994). For older works see Frances Yates, "The Hermetic Tradition in the Renaissance," in Charles Singleton, ed., *Art, Science and History in the Renaissance* (Baltimore: Johns Hopkins University Press, 1967), 255–279; Charles Webster, *From Paracelsus to Newton: Magic and the Making of Modern Science* (Cambridge: Cambridge University Press, 1980). See also Brian P. Copenhaver, "Astrology and Magic," in Charles B. Schmitt, Quientin Skinner, Eckhard Kesseler, and Jill Kraye eds., *The Cambridge*

History of Renaissance Philosophy (Cambridge: Cambridge University Press, 1988), 264–300; Allen G. Debus, "Alchemy in the Age of Reason: The Chemical Philosophers of Early Eighteenth-century France," in Ingrid Merkel and Allen G. Debus, eds., *Hermeticism and the Renaissance* (Cranbury, New Jersey: Associated University Press, 1988), 231–250; and Karin Johannisson, "Magic, Science, and Institutionalization in the Seventeenth and Eighteenth Centuries," in Merkel and Debus, *Hermeticism and the Renaissance*, 251–261.

54. Keith Hutchinson, "What Happened to Occult Qualities in the Scientific Revolution?" *Isis* 73, 267 (June 1982): 233–253.

55. For comprehensive discussions of this topic see the recent works Newman, *Atoms and Alchemy*, and Eamon, *Science and the Secrets of Nature*.

56. Descartes, *Philosophical Writings*, 1:139.

57. For more details on Descartes' medical ideas, the best recent works are Dennis Des Chene, *Spirits and Clocks: Machine and Organism in Descartes* (Ithaca, N.Y.: Cornell University Press, 2001); and Gordon Barker and Katherine J. Morris, *Descartes' Dualism* (London: Routledge, 1996). For an excellent overview of the ideas in the context of Descartes' life, see Stephen Gaukroger, *Descartes: An Intellectual Biography* (Oxford: Oxford University Press, 1995), especially 269–299. See also Sergio Moravia, "From *Homme Machine* to *Homme Sensible*: Changing Eighteenth-century Models of Man's Image," *Journal of the History of Ideas* 39 (1978): 49–60; and Leonora Cohen Rosenfield, *From Beast-machine to Man-Machine* (New York: Oxford University Press, 1940).

58. Descartes, *Philosophical Writings*, 1:140–141.

59. See Gaukroger, *Descartes*, 271, and Rosenfield, *From Beast-machine to Man-Machine*, 19.

60. Aristotle, *Movement of Animals*, trans. A. S. L. Farquharson, in *Complete Works of Aristotle*, vol. 1, ed. Jonathan Barnes (Princeton: Princeton University Press, 1984), 1092.

61. Thomas Aquinas, "Psychology of Human Acts," in *Summa Theologiae*, vol. 17, trans. Thomas Gilby (Cambridge: Blackfriars, 1970), 129, (quest. 6, art. 2). Rosenfield speculates that Descartes may have come across this passage during his education at the Jesuit school in La Flèche, in Rosenfield, *From Beast-machine to Man-Machine*, 19–20.

62. Rosenfield, *From Beast-machine to Man-Machine*, 79–81.

63. Aristotle, *Movement of Animals*, 1092.

64. Sylvia Berryman, "The Imitation of Life in Ancient Greek Philosophy," Riskin, *Genesis Redux*.

65. Sylvia Berryman, "Ancient Automata and Mechanical Explanation," *Phronesis* 48, 4 (2003): 344–369, especially 362–366.

66. Heinrich von Staden, "Body and Machine: Interactions between Medicine, Mechanics, and Philosophy in Early Alexandria," in John Walsh and Thomas F. Reese, eds., *Alexandria and Alexandrianism* (Malibu: J. Paul Getty Museum, 1996), 85–106. In the third century CE, Athenaeus gave a detailed description of a grand procession organized in the city by Ptolemy II Philadelphos that included a great wagon featuring a twelve-feet-high statue of Nysa, the nurse of Dionysus, that stood

up, poured milk from a cup, and sat down again. See Athenaeus, *The Deipnosophists*, trans. Charles Burton Gulick (London: Heinemann, 1928), 401 (bk. 5, 198).

67. René Descartes, *The World and Other Writings*, trans. Stephen Gaukroger (Cambridge: Cambridge University Press, 1998), 99.

68. Ibid., 107.

69. See Gaukroger, *Descartes*, 63–64.

70. Chapuis and Droz, *Automata*, 40.

71. Aristotle, *Metaphysics*, trans. W. D. Ross, in *The Complete Works of Aristotle*, vol. 2, ed. Jonathan Barnes (Princeton: Princeton University Press, 1984), 1555. For a discussion of this passage and Hero see Berryman, "Ancient Automata," 361–362.

72. Dennis Des Chene has made a similar point about the "open" nature of the machine in Cartesian writings, in direct contrast to the secretive works of a Hermetic philosopher like Robert Fludd, connecting Descartes' mechanistic ideas with technological illustrations in Caus's engineering treatise. Des Chene, *Spirits and Clocks*, 99–100.

73. Descartes, *The Philosophical Writings of Descartes*, vol. 2, trans. John Cottingham, Robert Stoothhoff, and Dugald Murdoch (Cambridge: Cambridge University Press, 1984), 21.

74. Ibid., 1:205.

75. Ibid., 1:329–330.

76. René Descartes, *Oeuvres Inédites de Descartes*, trans. (Latin into French) Foucher de Careil (Paris: Auguste Durand, 1859), 35–37. See also Rosenfield, *From Beast-machine to Man-Machine*, 4.

77. Nicolas Joseph Poisson, *Commentaire ou Remarques sur La Méthode de René Descartes* (Vandosme, France: S. Hyp, 1670), 156.

78. Des Chene, *Spirits and Clocks*, 65–66.

79. See Anatole France, *Romance of the Queen Pédauque*, trans. anonymous (Garden City, N.Y.: Halcyon House, 1950), 83–84.

80. Gaukroger, *Descartes*, 1; Gaby Wood, *Living Dolls: A Magical History of the Quest for Mechanical Life* (London: Faber and Faber, 2002), 3–5.

81. Vigneul-Marveille, *Mélanges d'Histoire et de Litterature* (Paris: Claude Prudhomme, 1725), 134.

82. Gaukroger, *Descartes*, 294–295, and on Francine's death, see 353.

83. Julian Jaynes, "The Problem of Animate Motion in the Seventeenth Century," *Journal of the History of Ideas* 31 (1970): 119–234, 224.

84. Thomas Hobbes, *Leviathan* (New York: Norton, 1997), 9.

85. Boyle, *Free Enquiry*, 135.

86. Ibid., 127–128.

87. Gottfried Wilhelm Leibniz, *The Monadology and Other Philosophical Writings*, trans. Robert Latta (London: Oxford University Press, 1951), 254.

88. Ibid., 131–134.

89. Gottfried Wilhelm Leibniz, *Theodicy*, trans. E. M. Huggard (La Salle, Illinois: Open Court, 1985), 364–365; for a discussion of Leibniz on this issue, see Mayr, *Authority, Liberty and Automatic Machinery*, 70–73.

90. See Thomas S. Hall, *Ideas of Life and Matter*, 2 vols. (Chicago: University of Chicago Press, 1969), 1:218–348.

91. Quoted in Moravia, "From *Homme Machine* to *Homme Sensible*," 46.

92. Hall, *Ideas of Life and Matter*, 1:275–278.

93. See Hall. *Ideas of Life and Matter*, 1:321–325. For the relevant passages see Thomas Willis, *Two Discourse Concerning the Soul of Brutes*, trans. S. Pordage (Gainsville, Fl.: Scholars' Facsimiles and Reprints, 1971), 22–27.

94. See Hall, *Ideas of Life and Matter*, 1:342–348; also Zakiya Hanafi, *The Monster in the Machine: Magic, Medicine, and the Marvelous in the Time of the Scientific Revolution* (Durham, N.C.: Duke University Press, 2000), 129–133; and Dennis Des Chene, "Abstracting from the Soul: The Mechanics of Locomotion," in Riskin, *Genesis Redux*, 89–91. On the nature of fever, see Giovanni Alfonso Borelli, *On the Movement of Animals*, trans. Paul Maquet (Berlin: Springer-Verlag, 1989), 426–427. This criticism by Borelli was somewhat unfair since Descartes made careful observations of animal carcasses in butcher shops and also conducted dissections himself.

95. Borelli, *On the Movement of Animals*, 36.

96. Ibid., 282.

97. Ibid., 283.

98. Hall, *Ideas on Life and Matter*, 1:343–344.

99. Borelli, *On the Movement of Animals*, 319.

100. Ibid., 320.

101. Ibid., 398.

102. See Theodore M. Brown, "The College of Physicians and the Acceptance of Iatromechanism in England, 1665–1695," *Bulletin of the History of Medicine* 44, 1 (1970): 12–30. See also Michael Hawkins, "'A Great and Difficult Thing': Understanding and Explaining the Human Machine in Restoration England," in Iwan Rhys Morus, ed., *Bodies/Machine* (Oxford: Berg, 2002), 15–38. Emily Booth has also written an excellent study of the mechanistic works of Walter Charleton: *"A Subtle and Mysterious Machine": The Medical World of Walter Charleton (1619–1707)* (Dordrecht: Springer, 2005).

103. Quoted in Mayr, *Authority, Liberty and Automatic Machinery*, 90.

104. Robert Hooke, preface to *Micrographia or Some Physiological Descriptions of Minute Bodies Made by Magnifying Glasses with Observations and Inquiries Thereupon* (New York: Dover, 1961), n.p.

105. See Rosenfield, *From Beast-machine to Man-Machine*, 28–70. On Le Cat and Vaucanson see Doyon and Liaigre, *Vaucanson*, 109–131.

106. On La Forge, see Rosenfield, *From Beast-machine to Man-Machine*, 243–245.

107. Louis de La Forge, *Traité de l'Espirit de l'Homme* (Hildesheim: Olms, 1984), 329–331.

108. Pierre Sylvain Régis, *Cours entire de philosophie* (New York: Johnson Reprint, 1970), 504–505. For more on Régis, see Rosenfield, *From Beast-machine to Man-Machine*, 257–261.

109. David-Renaud Boullier, *Essai philosophique sur L'ame des bêtes* (Paris: Fayard, 1985), 205.

110. Denis Diderot, *Select Essays from the Encyclopedia*, trans. anonymous (London: Samuel Leacroft, 1772), 153–154.

111. La Mettrie, *Man a Machine*, 59. See also the elaboration of his ideas in his *Man a Plant*, 77–92.

112. La Mettrie, *Man a Machine*, 69.

113. Aram Vartanian, *La Mettrie's L'Homme Machine: A Study in the Origins of an Idea* (Princeton: Princeton University Press, 1960), 16. See also Vartanian, *Science and Humanism in the French Enlightenment* (Charlottesville, Va.: Rookwood Press, 1999), 45–87.

114. La Mettrie, *Man a Machine*, 32.

115. Vartanian, *La Mettrie's L'Homme Machine*, 20.

116. Kathleen Wellman, *La Mettrie: Medicine, Philosophy, and Enlightenment* (Durham, N.C.: Duke University Press, 1992), 181–186.

117. There were mechanistic materialists during this period, most notably Thomas Hobbes, but he and other figures whom the historians Margaret Jacob and Jonathan Israel have identified as belonging to the "radical" Enlightenment were few and most of them were on the margins of intellectual life until the mid-eighteenth century, not least because many faced political persecution and professional isolation for denying the existence of the transcendent soul. See Margaret C. Jacob, *The Radical Enlightenment: Pantheists, Freemasons and Republicans* (London: Allen and Unwin, 1981); and Jonathan I. Israel, *Radical Enlightenment: Philosophy and the Making of Modernity 1650–1750* (Oxford: Oxford University Press, 2001).

118. Descartes, *Philosophical Writings*, 1:139; Leibniz, *Monadology*, 254; and Willis, *Two Discourse Concerning the Soul of Brutes*, 24.

119. Cotton Mather, *The Christian Philosopher* (Urbana: University of Illinois Press, 1994), 237.

120. Livy, *The Early History of Rome*, trans. Aubrey de Sélincourt (Harmondsworth, England: Penguin Books, 1960), 141–142.

121. Mary Douglas, *Purity and Danger: An Analysis of Concepts of Pollution and Taboo* (London: Routledge, 1966), 116.

122. For a discussion on this point in the context of Renaissance political discourse see Paul Archambault, "The Analogy of the 'Body' in Renaissance Political Literature," *Bibliotheque d'Humanisme et Renaissance* 29, 1 (1967): 21–53. See also Jacques Le Goff, "Head or the Heart? The Political Use of Body Metaphors in the Middle Ages," trans. Patricia Ranum, in Michel Feher, Ramona Naddaff, and Nadia Tazi, eds., *Fragments for a History of the Human Body*, pt. 3 (New York: Zone Books, 1989), 13–26.

123. Quoted in Christopher Hill, "William Harvey and the Idea of Monarchy," in Charles Webster, ed., *The Intellectual Revolution of the Seventeenth Century* (London: Routledge and Kegan Paul, 1974), 160–181.

124. Michel Foucault, *Discipline and Punish: The Birth of the Prison*, trans. Alan Sheridan (New York: Vintage Books, 1995), 136.

125. Hobbes, *Leviathan*, 9.

126. Ibid.

127. Quoted in Mayr, *Authority, Liberty and Automatic Machinery*, 108.

128. Etienne Bonnot de Condillac, *A Treatise on Systems*, in *Philosophical Writings of Etienne Bonnot Abbé de Condillac*, eds. and trans. Franklin Philip and Harlan Lane (Hillsdale, N.J.: Erlbaum, 1982), 140.

129. Condillac, "A Treatise on Sensations," in *Philosophical Writings*, 155.

130. Mayr, *Authority, Liberty and Automatic Machinery*, 102–144.

131. Ibid., 107.

132. Lionel Rothkrug, *Opposition to Louis XIV: The Political and Social Origins of the French Enlightenment* (Princeton: Princeton University Press, 1965), 105. For a more detailed discussion of the political use of this idea in the reign of the Sun King see Jean-Marie Apostolidès, *Le roi-machine: Spectacle et politiques au temps de Louis XIV* (Paris: Editions de Minuit, 1981).

133. Rothkrug, *Opposition to Louis XIV*, 129. For the continuity of the idea in the eighteenth century see George Armstrong Kelley, *Mortal Politics in Eighteenth-century France* (Waterloo: University of Waterloo Press, 1986), especially 32–34.

134. Jacob, *Radical Enlightenment*, 31.

135. Foucault, *Discipline and Punish*, 135–169.

136. Wellman, *La Mettrie*, 181.

137. Mayr makes an overly determined a connection between the mechanistic analogy and the absolutist state. It leads him to draw the wrong conclusion that any thinker engaged in the mechanistic description of the state must be a supporter of autocracy, as when he notes such language in Rousseau's works. *Authority, Liberty and Automatic Machinery*, 109–111.

138. Henry Fielding, *Tom Jones* (London: Penguin Books, 1994), 231.

139. For more on this theme see Rosenfeld, *From Beast-machine to Man-Machine*, especially 50–63 and 154–179, George Boas, *The Happy Beast in French Thought of the Seventeenth Century* (Baltimore: Johns Hopkins University Press, 1933), and Wallace Shugg, "The Cartesian Beast-machine in English Literature (1663–1750)," *Journal of the History of Ideas* 29 (1968): 279–292. For a more recent work on the subject in English literature see Erica Fudge, *Perceiving Animals: Humans and Animals in Early Modern English Culture* (New York: St. Martin's Press, 2000).

140. Rosenfeld, *From Beast-machine to Man-Machine*, 156–157. For more on La Fontaine's criticism of Descartes see Russell Ganim, "Scientific Verses: Subversion of Cartesian Theory and Practice in the 'Discours à Madame de La Sablière,'" in Anne L. Birberick, ed., *Refiguring La Fontaine: Tercentenary Essays* (Charlottesville, Virginia: Rookwood Press, 1996), 101–125.

141. Jean de la Bruyère, *Characters*, trans. Henri van Laun (London: Oxford University Press, 1963), 309–310.

142. Cyrano de Bergerac, *Voyages to the Moon and the Sun*, trans. Richard Aldington (New York: Orion Press, 1962), 74–152.

143. Jonathan Swift, *Gulliver's Travels* (New York: Norton and Norton, 2002), 80.

144. Ibid., 86.

145. Voltaire, "Micromégas," in *Micromégas and Other Short Fictions*, trans. Theo Cuffee (London: Penguin Books, 2002), 29.

146. For discussions on the topic, see Leo Braudy, "*Fanny Hill* and Materialism," *Eighteenth-Century Studies* 4, 1 (1970): 21–40. See also Douglas J. Stewart, "Pornography, Obscenity, and Capitalism," *Antioch Review* 35, 4 (1977): 389–398.

147. John Cleland, *Fanny Hill or, Memoirs of a Woman of Pleasure* (New York: Modern Library, 2001), 186–187.

148. Ibid., 96.

149. Boyle, *Free Enquiry*, 13.

150. "Automate," in Diderot and D'Alembert, *Encyclopédie*, 116.

151. In the introductory essay "The Life of Dr. Robert Hooke," in *The Posthumous Works of Robert Hooke*, (New York: Johnson Reprint, 1969), Hooke is quoted as saying "I contrived and made many trials about the Art of flying in the Air, and moving very swift on the Land and Water, of which I shew'd several designs to Dr. Wilkins then Warden of Wadham College, and at the same time made a Module, which by the help of Springs and Wings, rais'd and sustain'd itself in the Air" (iv).

152. Charles Hutton, *A Mathematical and Philosophical Dictionary* (London: J. Johnson and G. G. and J. Robinson, 1796), 176.

153. Eamon, *Science and the Secrets of Nature*, 319–350.

154. Francis Bacon, *New Atlantis*, in *Francis Bacon: A Critical Edition of the Major Works*, ed. Brian Vickers (Oxford: Oxford University Press, 1996), 471.

155. Ibid., 485–486.

156. Shapin and Shaffer, *Leviathan and the Air-Pump*.

157. The two most comprehensive accounts of public science are, for Britain, Larry Stewart, *The Rise of Public Science: Rhetoric, Technology, and Natural Philosophy in Newtonian Britain, 1660–1750* (Cambridge: Cambridge University Press, 1992), and for France, Geoffrey V. Sutton, *Science for a Polite Society: Gender, Culture, and the Demonstration of Enlightenment* (Boulder, Colo.: Westview Press, 1995). See also the important essay on this topic, Simon Schaffer, "Natural Philosophy and Public Spectacle in the Eighteenth Century," *History of Science* 21, 51 (Mar. 1983): 1–43; Anita Guerrini, "Anatomists and Entrepreneurs in Early Eighteenth-century London," *Journal of the History of Medicine* 59, 2 (2004): 219–239; and Margaret C. Jacob, *Scientific Culture and the Making of the Industrial West* (Oxford: Oxford University Press, 1997). On electrical experiments see the authoritative study J. L. Heilbron, *Electricity in the Seventeenth and Eighteenth Centuries: A Study of Early Modern Physics* (Berkeley: University of California Press, 1979), especially 229–305.

158. Roland Barthes, *Sade/Fourier/Loyola*, trans. Richard Miller (Berkeley: University of California Press, 1989), 152.

159. Descartes, *Philosophical Writings*, 2:405.

160. Frances Burney, *Evelina* (New York: Norton, 1998), 63. For more on the Burney-Merlin connection see Altick, *Shows of London*, 70–71, and French, *John Joseph Merlin*, 17–29. Also, for an interesting discussion of the role of the automaton in Burney's works see Deidre Shauna Lynch, *The Economy of Character: Novels, Market Culture, and the Business of Inner Meaning* (Chicago: University of Chicago Press, 1998), 192–206.

161. Burney, *Evelina*, 64.

162. La Bruyère, *Characters*, 109.

4. From the Man-machine to the Automaton-man

1. Letter to François-Joseph de Conzieé, comte des Charmettes, 17 January 1742, in *Correspondance complète de Jean-Jacques Rousseau*, vol. 1, ed. R. A. Leigh (Geneva: Institut et Musée Voltaire, 1965), 139 (no. 43).

2. Jean-Jacques Rousseau, *Julie, or the New Heloise*, trans. Philip Stewart and Jean Vaché (Hanover, N. H.: University Press of New England, 1997), 192.

3. For his break with the Encyclopedists and his attitude toward them, see Graeme Garrard, *Rousseau's Counter-Enlightenment: A Republican Critique of the Philosophes* (Albany: State University of New York Press, 2003), 29–35, 83–101.

4. Robert Boyle, *A Free Enquiry into the Vulgarly Received Notion of Nature* (Cambridge: Cambridge University Press, 1996), 13.

5. Ibid., 13.

6. See Thomas S. Hall, *Ideas of Life and Matter*, 2 vols. (Chicago: University of Chicago Press, 1969), 1:396–398.

7. See Hall, *Ideas of Life and Matter*, 1:351–366; and Lester S. King, "Stahl and Hoffmann: A Study in Eighteenth-century Animism," *Journal of the History of Medicine* 19, 2 (April 1964): 118–130. For a concise discussion of the role of Glisson and Stahl in the mechanicism-vitalism debate, see Sergio Moravia, "From *Homme Machine* to *Homme Sensible*: Changing Eighteenth-century Models of Man's Image," *Journal of the History of Ideas* 39 (1978): 49–60, 49–51.

8. See Hall, *Ideas of Life and Matter*, 1:399–408.

9. The best recent works that deal with issues of sensibility and the vital force in eighteenth-century medicine are Peter Hanns Reill, *Vitalizing Nature in the Enlightenment* (Berkeley: University of California Press, 2005), Elizabeth Williams, *A Cultural History of Medical Vitalism in Enlightenment Montpellier* (Burlington, Vermont: Ashgate, 2003), and Anne C. Vila, *Enlightenment and Pathology: Sensibility in the Literature and Medicine of Eighteenth-century France* (Baltimore: Johns Hopkins University Press, 1998).

10. Quoted in Anita Guerrini, "James Keill, George Cheyne, and Newtonian Physiology, 1690–1740," *Journal of the History of Biology* 18, 2 (1985): 247–266, 263.

11. See Theodore M. Brown, "From Mechanism to Vitalism in Eighteenth-century English Physiology," *Journal of the History of Biology* 7, 2 (Fall 1974): 179–216. See also Hall, *Ideas of Life and Matter*, 2:68–73, 91–99, and 107–118.

12. See Anita Guerrini, "Isaac Newton, George Cheyne and the 'Principia Medicinae,'" in Roger French and Andrew Wear, eds., *The Medical Revolution of the Seventeenth Century* (Cambridge: Cambridge University Press, 1989), 222–245; and Guerrini, "James Keill, George Cheyne, and Newtonian Physiology."

13. Isaac Newton, *Newton's Philosophy of Nature: Selections from His Writings* (New York: Hafner Press, 1953), 145 (query 24 of the *Opticks*).

14. See the most comprehensive work on those figures, Williams, *Cultural History of Medical Vitalism*.

15. See Reill, *Vitalizing Nature*, 33–70.

16. George Buffon, "Initial Discourse," in *From Natural History to the History of Nature*, ed. and trans. John Lyon and Philip R. Sloane (Notre Dame, Ind.: University of Notre Dame Press, 1981). 127.

17. Quoted in Moravia, "From *Homme Machine* to *Homme Sensible*," 46.

18. For a detailed discussion of the differences see Williams, *Cultural History of Medical Vitalism*, 146–184.

19. See Denis Diderot, "D'Alembert's Dream," in *Rameau's Nephew / D'Alembert's Dream*, trans. Leonard Tancock (London: Penguin Books, 1966), 158–159.

20. Reill, *Vitalizing Nature*, 5; on vitalism in opposition to mechanicism and animism, see 139–142.

21. *The Works of Robert Whytt, M.D.* (Edinburgh: T. Becket and P. A. De Hondt, 1768), 322.

22. Denis Diderot, "From 'Elements of Physiology,'" trans. Jean Stewart and Jonathan Kemp, in *Diderot, Interpreter of Nature: Selected Writings* (New York: International Publishers, 1936), 135–136.

23. Williams, *Cultural History of Medical Vitalism*, 80–105.

24. François Boissier de Sauvages de la Croix, *Nosologie méthodique ou distribution des maladies en classe, en genres et en especes*, vol. 1, trans. (Latin to French) Eouvion (Lyon: Jean-Marie Bruyset, 1772), 170.

25. See Williams, *Cultural History of Medical Vitalism*, 95–101.

26. Vila, *Enlightenment and Pathology*, 111–151.

27. The scholarship on eighteenth-century culture and literature of sensibility is enormous, but the works I have found useful for this study are Janet Todd, *Sensibility: An Introduction* (London: Methuen, 1986); John Mullan, *Sentiment and Sociability: The Language of Feeling in the Eighteenth Century* (Oxford: Clarendon Press, 1988); G. J. Barker-Benfield, *The Culture of Sensibility: Sex and Society in Eighteenth-century Britain* (Chicago: University of Chicago Press, 1992); and Markman Ellis, *The Politics of Sensibility: Race, Gender and Commerce in the Sentimental Novel* (Cambridge: Cambridge University Press, 1996). On the impact of sensibility on the sciences see Jessica Riskin, *Science in the Age of Sensibility* (Chicago: University of Chicago Press, 2002).

28. Robert Darnton, "Readers Respond to Rousseau: The Fabrication of Romantic Sensitivity," in *The Great Cat Massacre and Other Episodes in French Cultural History* (New York: Vintage Books, 1985), 242. On the Rousseau cult, see Garrard, *Rousseau's Counter-Enlightenment*, 35–40.

29. Margaret C. Jacob, *The Radical Enlightenment: Pantheists, Freemasons and Republicans* (London: Allen and Unwin, 1981); Jonathan I. Israel, *Radical Enlightenment: Philosophy and the Making of Modernity 1650–1750* (Oxford: Oxford University Press, 2001).

30. Jacob, *Radical Enlightenment*, 22–27, and 152–154.

31. Immanuel Kant, *The Critique of Judgment*, trans. James Creed Meredith (Oxford: Oxford University Press, 1952), 222–223.

32. David Hume, *Dialogues Concerning Natural Religion* (London: Penguin Books, 1990), 87.

33. Paul-Henri Thiry d'Holbach, *Le Bon sens ou idées naturelles opposées aux idées surnaturelles* (Paris: Editions Rationalistes, 1971), 40.

34. The original title of the Mercier novel is *L'An deux mille quatre cent quarante, rêve s'il en fût jamais* (The Year Two Thousand Four Hundred Forty, a Dream If Ever There Was One). See Robert Darnton, *The Forbidden Best-sellers of Prerevolutionary France* (New York: Norton, 1996), 115–136.

35. For more on Bordeu, see Williams, *Cultural History of Medical Vitalism*, 147–162.

36. Louis Sebastien Mercier, *Memoirs of the Year 2500*, trans. W. Hooper (Clifton, New Jersey: Augustus M. Kelley, 1973), 262.

37. Lavoisier is commonly associated with the scientific mechanicism of the late

Enlightenment, but Peter Reill has demonstrated the important role of vitalist ideas in his works. See Reill, *Vitalizing Nature*, 101–114.

38. Simon Schaffer, "Enlightened Automata," in William Clark, Jan Golinski, and Simon Schaffer, eds., *The Sciences in Enlightened Europe* (Chicago: University of Chicago Press, 1999), 126–165.

39. Quoted in Schaffer, "Enlightened Automata," 129.

40. Ibid. Cynthia Koepp has also shown the tension in Diderot's attitude toward laborers as it appears in the *Encyclopédie*. See Cynthia J. Koepp, "Alphabetical Order: Work in Diderot's *Encyclopédie*," in Steven Laurence Kaplan and Cynthia J. Koepp, eds., *Work in France: Representations, Meaning, Organization, and Practice* (Ithaca, N.Y.: Cornell University Press, 1986), 229–257.

41. Mark Seltzer, *Bodies and Machines* (New York: Routledge, 1992), 109.

42. For discussions on George Cheyne and the influence of his medical ideas on the works of Richardson and others, see Barker-Benfield, *Culture of Sensibility*, 6–36; and Carol Houlihan Flynn, "Running Out of Matter: The Body Exercised in Eighteenth-century Fiction," in G. S. Rousseau, ed., *The Language of Psyche: Mind and Body in Enlightenment Thought* (Berkeley: University of California Press, 1990), 147–185.

43. Samuel Richardson, *Pamela* (London: Penguin Books, 1985), 394.

44. D'Holbach, *Le Bon sens*, 40.

45. Samuel Richardson, *Clarissa* (London: Penguin Books, 1985), 658 (letter 202). This notion of the man-machine as a manipulated person has precedence in the Renaissance when, as Jessica Wolfe has shown, the resurgence of interest in machines and mechanics led to the introduction of mechanical language in the courtly discourse of the sixteenth and early seventeenth centuries in which the idea of "machination" came to denote actions in the social and political sphere. See Wolfe, *Humanism, Machinery, and Renaissance Literature* (Cambridge: Cambridge University Press, 2004), 56–124.

46. Jean de la Bruyère, *Characters*, trans. Henri van Laun (London: Oxford University Press, 1963), 309–310.

47. Voltaire, "The Story of the Good Brahmin," in *Micromégas and Other Short Fictions*, trans. Theo Cuffee (London: Penguin Books, 2002), 73.

48. Jean-Jacques Rousseau, *Emile or On Education*, trans. Allan Bloom (New York: Basic Books, 1979), 61–62.

49. Ibid., 364.

50. Ibid., 118.

51. Ibid.

52. D'Holbach, *Le Bon sens*, 4.

53. Denis Diderot, *The Nun*, trans. Leonard Tancock (London: Penguin Books, 1972), 51. For more on the role of sensibility in this novel see Vila, *Enlightenment and Pathology*, 162–180.

54. *Jacques le fataliste et son maître*, written between 1755 and 1784, unpublished during his lifetime.

55. Denis Diderot, *Jacques the Fatalist*, trans. Michael Henry (London: Penguin Books, 1986), 40.

56. Ibid., 74.

57. Ibid., 235.

58. Mercier, *Memoirs of the Year 2500*, 23.
59. Ibid.
60. Ibid., 87.
61. Ibid., 201.
62. Anne Vila again provides a detailed and illuminating analysis of the role of sensibility in libertine literature, especially in the works of Laclos and De Sade (*Enlightenment and Pathology*, 258–292).
63. Choderlos de Laclos, *Les Liaisons Dangereuses*, trans. Richard Aldington (New York: Knopf, 1992), 209, and 223–224. See also Vila, *Enlightenment and Pathology*, 283–285.
64. Roland Barthes, *Sade/Fourier/Loyola*, trans. Richard Miller (Berkeley: University of California Press, 1989), 152–153. For another analysis of the machine image in de Sade's works see Vera Lee, "The Sade Machine," *Studies on Voltaire and the Eighteenth Century* 98 (1972): 207–208.
65. Susan Sontag, "The Pornographic Imagination," in *Styles of Radical Will* (New York: Dell, 1966), 52.
66. Marquis de Sade, *Juliette* (New York: Grove Press, 1968), 147. Both David B. Morris and Anne Vila have shown the impact of vitalist physiology on de Sade's works. Morris, "The Marquis de Sade and the Discourses of Pain: Literature and Medicine in the Revolution," in Rousseau, ed., *The Language of Psyche: Mind and Body in Enlightenment Thought* (Berkeley: University of California Press, 1990): 291–330; and Vila, *Enlightenment and Pathology*, 264–269, 286–292.
67. Thomas Paine, "Rights of Man," in *Common Sense, Rights of Man, and Other Essential Writings of Thomas Paine* (New York: Signet Classics, 2003), 281.
68. Quoted in Moncure Daniel Conway, *The Life of Thomas Paine* (New York: Putnam, 1892), 111.
69. The classic study of this topic is Robert Darnton, *Mesmerism and the End of the Enlightenment in France* (Cambridge, Mass.: Harvard University Press, 1968). See also Riskin, *Science in the Age of Sensibility*, 189–225; and Simon Schaffer, "Self Evidence," *Critical Inquiry* 18 (Winter 1992): 327–362.
70. Quoted in Schaffer, "Self Evidence," 351.
71. Ibid., 353.
72. Alison Winter, *Mesmerized: Powers of Mind in Victorian Britain* (Chicago: University of Chicago Press, 1998), 37–38, 62–64, and 120–121.
73. Sigmund Freud, *The Psychopathology of Everyday Life*, in *The Standard Edition of the Complete Psychological Works of Sigmund Freud*, vol. 6, trans. Alan Tyson (London: Hogarth Press, 1960), 177.
74. Ibid.
75. Ibid., 178.
76. Erich Fromm, *Escape from Freedom* (New York: Holt, 1969), 183–204. Jacques Lacan, *Four Fundamental Concepts of Psycho-Analysis*, trans. Alan Sheridan (London: Hogarth Press, 1977), 53–64.
77. Thomas Henry Huxley, "On the Hypothesis That Animals Are Automata, and Its History," in *Science and Culture, and Other Essays* (London: MacMillan, 1881), 239–240.

78. Huxley, "On the Hypothesis That Animals Are Automata," 234–236, 241–242.

79. Immanuel Kant, *Critique of Practical Reason*, trans. Lewis White Beck (Chicago: University of Chicago Press, 1949), 206.

80. William James, "Are We Automata?" in *Essays in Psychology* (Cambridge, Mass.: Harvard University Press, 1983), 46.

81. Ibid., 59.

82. For a different response to the Huxley article by another contemporary, see William Carpenter, "The Limits of Human Automatism," in *Nature and Man: Essays Scientific and Philosophical* (New York: D. Appleton, 1889), 284–315. For a discussion of the debates on Huxley's article, see M. Norton Wise, "The Gender of Automata in Victorian Britain," in Riskin, *Genesis Redux*, 161–195, especially 184–190.

83. Percy Bysshe Shelley, "Queen Mab," in *Major Works*, ed. Zachary Leader and Michael O'Neill (Oxford: Oxford University Press, 2003), 29.

84. Stendhal, *The Red and the Black*, trans. Roger Gard (London: Penguin Books, 2002), 57, 80.

85. Ibid., 124.

86. Ibid., 498.

87. Alexandre Dumas, *Twenty Years After*, trans. anonymous (Oxford: Oxford University Press, 1998), 25.

88. Foucault, *Discipline and Punish*, 135–136.

89. Dumas, *Twenty Years After*, 56–57.

90. Ibid., 46.

91. Henry Adams, *Democracy: An American Novel* (New York: Modern Library, 2003), 8.

92. Ibid., 50.

93. Ibid., 50–51.

94. On Weeks's museum see Arthur W. J. G. Ord-Hume, *Clockwork Music: An Illustrated History of Mechanical Musical Instruments from the Music Box to the Pianola, from Automaton Lady Virginal Players to Orchestrion* (New York: Crown, 1973), 22; and Richard D. Altick, *The Shows of London* (Cambridge, Mass.: Harvard University Press, 1978), 350–353. On the Kaufmann automaton see Myles W. Jackson, *Harmonious Triads: Physicists, Musicians, and Instrument Makers in Nineteenth-century Germany* (Cambridge, Mass.: MIT Press, 2006), 89–93. On Faber's "Euphonia," see Ord-Hume, *Clockwork Music*, 22, 44–45, and 49, and Altick, *Shows of London*, 353–356.

95. Christian Bailly, *Automata: The Golden Age, 1848–1914* (London: Hale, 1987).

96. See Chapter 5, note 9.

97. The best introduction to the chess-player and to writings on it is the recent work of popular history—Tom Standage, *The Turk: The Life and Times of the Famous Eighteenth-century Chess-playing Machine* (New York: Walker, 2002). For a scholarly purpose, however, the most useful book, which not only provides an excellent narrative of the object's history but also includes many of the most important writings in the appendix, including Carl Gottlieb von Windisch's "Inanimate Reason" and Edgar Allen Poe's "Maelzel's Chess-Player," is Gerald M. Levitt, *The Turk: the Chess Automaton* (Jefferson, North Carolina: McFarland, 2000). See also Schaffer, "Enlightened Automata," 154–163.

98. Philip Thicknesse, "The Speaking Figure, and the Automaton Chess-player, Exposed and Detected," in Levitt, *Turk*, 202.

99. Ibid., 202.

100. Carl Gottlieb von Windisch, "Inanimate Reason; or a Circumstantial Account of That Astounding Piece of Mechanism, M. De Kempelen's Chess-player," in Levitt, *Turk*, 197.

101. Levitt, in Levitt, *Turk*, 25–29.

102. Ibid., 39–42.

103. Louis Dutens, "Letter to *Gentleman's* Magazine," in Levitt, *Turk*, 192.

104. Quoted in Standage, *Turk*, 46.

105. Charles Babbage, *Passages from the Life of a Philosopher* (New York: Kelley, 1969), 465.

106. Babbage's brief description of the Turk is from two notes that were found in his copy of Windisch's "Inanimate Reason." Their contents are reproduced in "De Kempelen's Automaton Chess-Player," *Notes and Queries* 12 (February 25, 1922): 155–156. See also Standage, *Turk*, 140.

107. Windisch, "Inanimate Reason," 194.

108. Robert Willis, "An Attempt to Analyse the Automaton Chess Player of Mr. de Kempelen," in Levitt, *Turk*, 216–220.

109. See Jacques-François Mouret, "The Automaton Chess Player," in Levitt, *Turk*, 221–222.

110. Ambrose Bierce, "Moxon's Master," in *The Complete Short Stories of Ambrose Bierce* (Lincoln: University of Nebraska Press, 1970), 89–97. The automaton in Hermann Melville's story "The Bell-Tower" (1855) also kills its inventor, but it is revealed at the end that the whole thing may have been an accident.

111. Walter Benjamin, "Theses on the Philosophy of History," in *Illuminations* (New York: Schocken Books, 1968), 253.

112. Slavoj Žižek, *The Puppet and the Dwarf: The Perverse Core of Christianity* (Cambridge, Mass.: MIT Press, 2003), 3.

113. Arturo Pérez-Reverte, *The Flanders Panel*, trans. Margaret Jull Costa (Orlando, Fl.: Harcourt, 1994), and Robert Löhr, *The Chess Machine*, trans. Anthea Bell (New York: Penguin Press, 2007). In reality, the cabinet was big enough so that Kempelen did not have to find a dwarf for the purpose, and Kempelen is portrayed in the novel as a ruthless fraud out to perpetrate the notion that he created a playing machine, which contradicts the view we get of him in the Windisch letters. But since Löhr admits to this discrepancy in the afterword to the novel, this is, once again, merely the fastidious nitpicking of a historian.

114. Zakiya Hanafi, *The Monster in the Machine: Magic, Medicine, and the Marvelous in the Time of the Scientific Revolution* (Durham, N.C.: Duke University Press, 2000), 53–96.

5. The Uncanny Automaton

1. Christopher Clark, *Iron Kingdom: The Rise and Downfall of Prussia, 1600–1947* (Cambridge, Mass.: Harvard University Press, 2006), 274.

2. Ibid., 275.

3. For most Americans today his name might ring a bell, as Harry Houdini (1874–1926) adopted the name of the famous Frenchman when he embarked on his own career in magic. Houdini, born Erich Weiss, was an Austro-Hungarian Jew from Budapest who immigrated with his family to the United States in 1878.

4. See Jean-Eugène Robert-Houdin, *Memoirs of Robert-Houdin: Ambassador, Author, and Conjurer*, trans. Lascelles Wraxall (London: Laurie, 1942), 19–39.

5. "Travail d'automate" in the original, though it is translated as "mechanical work" by Wraxall; Robert-Houdin, *Memoirs of Robert-Houdin*, 34. For the original French text see Jean-Eugène Robert-Houdin, *Confidences et revelations: Comment on devient sorcier* (Geneva: Editions Slatkine, 1980), 24.

6. Robert-Houdin, *Memoirs of Robert-Houdin*, 40–42.

7. See Richard D. Altick, *The Shows of London* (Cambridge, Mass.: Harvard University Press, 1978), 362; Christian Bailly, *Automata: The Golden Age, 1848–1914* (London: Hale, 1987), 20–22; and Barbara Maria Stafford and Frances Terpak, *Devices of Wonder: From the World in a Box to Images on a Screen* (Los Angeles: Getty, 2001), 272–273.

8. Linda M. Strauss, "Reflections in a Mechanical Mirror: Automata as Doubles and as Tools," *Knowledge and Society* 10 (1996): 179–209, 199. See also Altick, *Shows of London*, 504, and Harry M. Geduld and Ronald Gottesman, eds., *Robots, Robots, Robots* (Boston: New York Graphic Society, 1978), 24.

9. One might object to this generalization considering the use of myriad machinery in the scientific research into musical and acoustical theories in nineteenth-century Germany as detailed in Myles W. Jackson's superb book *Harmonious Triads: Physicists, Musicians, and Instrument Makers in Nineteenth-century Germany* (Cambridge, Mass.: MIT Press, 2006). A careful reading of the book, however, reveals that with few exceptions, like Johann Gottfried Kaufmann and his son Friedrich Kaufmann's Bellonion, an automatic music-playing machine, and Friedrich Kaufmann's automaton trumpeter, none of the machines discussed can be labeled as automata, in the sense of machines in the shape of living beings or fully self-moving devices. The vast majority of the objects discussed in the book were manually operated instruments and machines that go beyond the definition of automata. But for a discussion of nineteenth-century German automata in the context of scientific theories over the organic versus the natural, see 75–110.

10. For general discussions of this topic see Peter Gendolla, *Die lebenden Machinen: Zur Geschichte der Maschinenmenschen bei Jean Paul, E. T. A. Hoffmann und Villiers de l'Isle Adam* (Marburg: Guttandin und Hoppe, 1980); and Leselotte Sauer, "Romantic Automata" in Gerhart Hoffmesister, ed., *European Romanticism: Literary Cross-currents, Modes, and Models* (Detroit: Wayne State University Press, 1990), 287–306.

11. See Immanuel Kant, "Metaphysical Foundations of Natural Science," in *Philosophy of Material Nature*, trans. James W. Ellington (Indianapolis: Hackett, 1985), 76–94.

12. For a concise overview of Blumenbach's medical ideas, see Thomas S. Hall, *Ideas of Life and Matter*, 2 vols. (Chicago: University of Chicago Press, 1969),

2:99–105. See also Peter Hanns Reill, *Vitalizing Nature in the Enlightenment* (Berkeley: University of California Press, 2005), 144–147.

13. For this reason Timothy Lenoir labels him a vital materialist; Lenoir, "Kant, Blumenbach, and Vital Materialism," *Isis* 71 (1980): 77–108.

14. The most comprehensive recent work on *Naturphilosophie* is Robert J. Richards, *The Romantic Conception of Life: Science and Philosophy in the Age of Goethe* (Chicago: University of Chicago Press, 2002). See also Andrew Cunningham and Nicholas Jardine, eds., *Romanticism and the Sciences* (Cambridge: Cambridge University Press, 1990). On the issues of the transition from the late Enlightenment to Romantic *Naturphilosophie*, I found particularly enlightening the last chapter of Reill, *Vitalizing Nature*, 199–236. Another useful study of this issue is Nicholas Jardine, "Inner History; Or, How to End Enlightenment," in William Clark, Jan Golinski, and Simon Schaffer, eds., *The Sciences in Enlightened Europe* (Chicago: University of Chicago Press, 1999), 477–494.

15. See Robert Stern, introduction to Friedrich Wilhelm Joseph von Schelling, *Ideas for a Philosophy of Nature*, trans. Errol E. Harris and Peter Heath (Cambridge: Cambridge University Press, 1988), x–xi.

16. Reill, *Vitalizing Nature*, 159–198.

17. For a concise and lucid critique of such a position see Frederick C. Beiser, *The Romantic Imperative: The Concept of Early German Romanticism* (Cambridge, Mass.: Harvard University Press, 2003), 43–55.

18. Reill, *Vitalizing Nature*, 8–9.

19. For an eloquent description of this situation and its relation to the development of *Naturphilosophie*, see Reill, *Vitalizing Nature*, 199–203.

20. Reill, *Vitalizing Nature*, 67, 107–108.

21. Goethe, "Polarity," in *The Collected Works*, vol. 12, *Scientific Studies*, ed. and trans. Douglas Miller (New York: Suhrkamp, 1988), 156. For more on Goethe's scientific ideas see Thomas S. Hall, *Ideas of Life and Matter*, 2 vols. (Chicago: University of Chicago Press, 1969), 2:42–45; David Seamon and Arthur Zajong eds., *Goethe's Way of Science* (Albany: State University of New York Press, 1998); Peter Hanns Reill, "*Bildung, Urtyp* and Polarity: Goethe and Eighteenth-century Physiology," *Goethe Yearbook* 3 (1986): 139–148; Frederick Armine, Francis J. Zucker, and Harvey Wheeler, eds., *Goethe and the Sciences: A Reappraisal* (Boston: Reidel, 1987), and most recently Richards, *Romantic Conception of Life*, 407–508.

22. Evellen Richards, "'Metaphorical Mystifications': the Romantic Gestation of Nature in British Biology," in Cunningham and Jardine, *Romanticism and the Sciences*, 130–143, 131. See also H. A. M. Snelders, "Romanticism and Naturphilosophie and the Inorganic Sciences 1797–1840: An Introductory Survey," *Studies in Romanticism* 9 (Summer 1970): 193–215.

23. Quoted in S. R. Morgan, "Schelling and the Origins of His *Naturphilosophie*," in Cunningham and Jardine, *Romanticism and the Sciences*, 25–37, 34. For more on Schelling's scientific ideas, see Richards, *Romantic Conception of Life*, 289–306.

24. Morgan, "Schelling and the Origins of His *Naturphilosophie*," 35.

25. Hans Christian Oersted, "The Spiritual in the Material," in *The Soul in Nature*, trans. Leonora and Joanna B. Horner (London: Dawson of Pall Mall, 1966), 7. For more on Oersted as a *Naturphilosophen* see H. A. M. Snelders, "Oersted's Discovery of Electromagnetism," in Cunningham and Jardine, *Romanticism and the Sciences*, 228–240.

26. Elizabeth Williams, *A Cultural History of Medical Vitalism in Enlightenment Montpellier* (Burlington, Vermont: Ashgate, 2003), 146–184.

27. See Denis Diderot, "D'Alembert's Dream," in *Rameau's Nephew/D'Alembert's Dream*, trans. Leonard Tancock (London: Penguin Books, 1966), 158–159.

28. There is a large body of scholarly works dealing with this issue, but the ones I found most useful for the purpose of this study were M. H. Abrams, *Natural Supernaturalism: Tradition and Revolution in Romantic Literature* (New York: Norton, 1971); Rosemary Jackson, *Fantasy: The Literature of Subversion* (London: Routledge, 1981); Tobin Siebers, *The Romantic Fantastic* (Ithaca, N.Y.: Cornell University Press, 1984); and José B. Monleon, *A Specter Is Haunting Europe: A Sociohistorical Approach to the Fantastic* (Princeton: Princeton University Press, 1990).

29. See Chris Baldick, *In Frankenstein's Shadow: Myth, Monstrosity, and Nineteenth-century Writing* (Oxford: Clarendon Press, 1987), 10–29; and Anne K. Mellor, *Mary Shelley: Her Life, Her Fiction, Her Monsters* (New York: Routledge, 1989), 70–88.

30. Tzvetan Todorov, *The Fantastic: A Structural Approach to a Literary Genre*, trans. Richard Howard (Ithaca, N.Y.: Cornell University Press, 1973), 24–40. Howard translates Todorov's word "étrange" as "uncanny." I have taken Tili Boon Cuillé's helpful suggestion to use the closer term "strange" to avoid confusion with the predominant way I use the term "uncanny" throughout this book, as detailed in Chapter 1.

31. William Wordsworth, *The Prelude, 1799, 1805, 1850* (New York: Norton, 1979), 262–264; in the 1805 version, bk. 7, 681–692.

32. E. T. A. Hoffmann, *The Life and Opinion of the Tomcat Murr*, trans. Anathea Bell (London: Penguin Books, 1999), 32.

33. Ibid., 301.

34. Ibid., 125–126.

35. Gendolla, *Die lebenden Machinen*, 4.

36. Ibid., 3.

37. William Blake, "Milton," in *The Portable Blake* (New York: Penguin Books, 1974), 412.

38. On the Industrial Revolution in Germany see David S. Landes, *The Unbound Prometheus: Technological Change and Industrial Development in Western Europe from 1750 to the Present* (Cambridge: Cambridge University Press, 1969), 124–192.

39. Baldick, *In Frankenstein's Shadow*, 44–45, and 54–55. Andreas Huyssen makes the same association of all nineteenth-century automaton literature with anxieties caused by the Industrial Revolution, as he was informed by Gendolla's book. See Andreas Huyssen, *After the Great Divide: Modernism, Mass Culture, Postmodernism* (Bloomington: Indiana University Press, 1986), 69–70.

40. For a good introduction to his works see Jean Paul, *Jean Paul: A Reader*, ed. Timothy J. Casey and trans. Erika Casey (Baltimore: Johns Hopkins University Press, 1992).

41. Jean Paul, "Menschen sind Maschinen der Engel," in *Sämtliche Werke*, pt. 2, vol. 1, ed. Eduard Berend (Munich: Carl Hanser Verlag, 1974), 1028–1031, 1028. I am grateful to Steve Rowan for his help with the translation.

42. Ibid., 1029.

43. Ibid.

44. For another interesting discussion of this passage see Adelheid Voskuhl, "Motions and Passions: Music-playing Women Automata and the Culture of Affect in Late Eighteenth-century Germany," in Jessica Riskin, ed., *Genesis Redux: Essays in the History and Philosophy of Artificial Life* (Chicago: University of Chicago Press, 2007), 293–320, especially 298–303.

45. Jean Paul, "Unterthängiste Vorstellung unser, der sämtlicher Spieler und redenden Damen in Europa entgegen und wider die Einführung der Kempelischen Spiel-und Sprechmaschinen," in *Werke*, pt. 1, vol.1, ed. Nobert Miller (Munich: Carl Hanser Verlag, 1976), 275–292.

46. Jean Paul, "Machinenmann nebst seinen Eigenschaften," in *Werke*, pt. 1, vol. 1, 544–551.

47. Jean Paul, "Personalien vom Bedienten- und Maschinenmann" in *Werke*, vol. 4, ed. Nobert Miller (Munich: Carl Hanser Verlag, 1963), 901–907.

48. Ibid., 903.

49. Ibid.

50. Jean Paul, "Machinenmann nebst seinen Eigenschaften," 551.

51. Such satirical images of the ruler as an automaton can be found in later French works as well—see Charles Nodier, "Voyages pittoresque et industriel dans le Paraguay-Roux et la Palingénésie austral," in *Contes* (Paris: Éditions Garnier Frères, 1961), 459; and Albert Robida, *The Twentieth Century*, trans. Philippe Willems (Middletown, Conn.: Wesleyan University Press, 2004), 269.

52. See Richards, *Romantic Conception of Life*, 92–93, or for a more comprehensive account, Peter Firchow, translator's introduction to Friedrich Schlegel, *Lucinde and the Fragments* (Minneapolis: University of Minnesota Press, 1971), 3–39.

53. Schlegel, *Lucinde*, 66.

54. Ibid., 66–68.

55. Ibid., 67.

56. Ibid.

57. Ibid.

58. Georg Büchner, *Leonce and Lena*, in *Complete Plays and Prose*, trans. Carl Richard Mueller (New York: Hill and Wang, 1963), 102.

59. Ibid.

60. Büchner, *The Hessian Courier*, in *Complete Plays and Prose*, 176–177.

61. Baldick, *In Frankenstein's Shadow*, 10–29; and Mellor, *Mary Shelley*, 70–88.

62. See Gotthilf Heinrich von Schubert, *Ansichten von der Nachtseite der Naturwissenschaft* (Darmstadt: Wissenschaftliche Buchgesellschaft, 1967). For Hoffmann's knowledge of his works see James M. McGalthery, *Mysticism and Sexuality, E. T. A. Hoffmann*, pt. 1, *Hoffmann and His Sources* (Las Vegas: Peter Lang, 1981), 160–161.

63. For a discussion of Schubert's ideas in the context of *Naturphilosophie* see Nicholas A. Rupke, "Caves, Fossils and the History of the Earth," in Cunningham and Jardine, *Romanticism and the Sciences*, 254–256.

64. E. T. A. Hoffmann, "Automata," in *The Best Tales of Hoffmann* (New York: Dover, 1967), 81. In my quotations from this translation, I have modified the title to the singular, as in the original, used the original German names of the characters, and slightly changed some sentences. For more on the story, see James M. McGlathery, *Mysticism and Sexuality, E. T. A. Hoffmann*, pt. 2, *Interpretations of the Tales* (New York: Peter Lang, 1985), 103–104; and Horst S. Daemmrich, *The Shattered Self: E. T. A. Hoffmann's Tragic Vision* (Detroit: Wayne State University Press, 1973), 75–76.

65. Hoffmann, "Automata," 87.

66. Ibid., 99.

67. In Myles Jackson's analysis of this story, Ludwig's diatribe against machines is quoted at length to emphasize the tale's pro-organic, anti-mechanistic stance. Although I largely agree with Jackson's interpretation, the story's mysterious ending subverts the simple organic-versus-mechanical dichotomy to reveal the uncanny consequence of the collapse of the polarity, which is a possibility well within the worldview of *Naturphilosophie*. See Jackson, *Harmonious Triads*, 79–81.

68. Heinrich von Kleist, "On the Marionette Theater," trans. Roman Paska, in Michel Feher, Ramona Naddaff, and Nadia Tazi, eds., *Fragments for a History of the Human Body* (New York: Zone, 1989), 415–420.

69. S. S. Prawer, "Hoffmann's Uncanny Guest: A Reading of *Der Sandmann*," *German Life and Letters* 18, 4 (July 1965): 297–308, 305. See also McGlathery, *Mysticism and Sexuality*, pt. 2, 57–59; and Daemmrich, *Shattered Self*, 47–51.

70. E. T. A. Hoffmann, "The Sandman," in *Tales of Hoffmann*, trans. R. J. Hollingdale (London: Penguin Books, 1982), 91–92.

71. Sigmund Freud, "The 'Uncanny,'" trans. Alix Strachey, in *Writings on Art and Literature* (Stanford: Stanford University Press, 1997), 206.

72. Hoffmann, "Sandman," 116.

73. Ibid., 119.

74. Ibid., 120.

75. Ibid., 121–122.

76. Ibid., 103.

77. Johann Wolfgang von Goethe, "The Annals; or Day and Year Papers," trans. Charles Nisbet, in *The Autobiography of Goethe*, bks. 14–20 (London: George Bell, 1894), 316.

78. For details on his life see Grete de Francesco, *The Power of the Charlatan*, trans. Miriam Beard (New Haven: Yale University Press, 1939), 250–267.

79. Francesco, *Power of the Charlatan*, 252.

80. Goethe, "Annals," 317.

81. Ibid., 331. See also Francesco, *Power of the Charlatan*, 251–252.

82. Goethe, "Annals," 331–332.

83. For an introduction to his life and works see Roland Hoermann, *Achim von Arnim* (Boston: Twayne, 1984).

84. Ludwig Achim von Arnim, "Armut, Reichtum, Schuld und Busse der Gräfin Dolores," in *Die Erzählungen und Romane* (Leipzig: Insel Verlag, 1982), 294.

85. Arnim, "Armut, Reichtum, Schuld und Busse der Gräfin Dolores," 295.

86. Altick, *Shows of London*, 353.

87. Arnim, "Armut, Reichtum, Schuld und Busse der Gräfin Dolores," 309–310.

88. See Johann Wolfgang von Goethe, "The Sorcerer's Apprentice," in *Goethe, the Lyricist*, ed. and trans. Edwin H. Zeydel (Chapel Hill: University of North Carolina Press, 1955), 105, and *The Collected Works*, vol. 2, *Faust I and II*, ed. and trans. Stuart Atkins (Princeton: Princeton University Press, 1984), 175–177.

89. Ludwig Achim von Arnim, "Isabella of Egypt," in *Novellas of 1812*, trans. Bruce Duncan (Lewiston, Me.: Edwin Mellen Press, 1997), 1–95.

90. Hoffmann, "Nutcracker and the King of Mice," in *Best Tales of Hoffmann*, 139–140.

91. Jane (Webb) Loudon, *The Mummy! A Tale of the Twenty-second Century* (Ann Arbor: University of Michigan Press, 1994).

92. Mary Shelley, *Frankenstein, Or the Modern Prometheus* (Chicago: University of Chicago Press, 1974), 6.

93. Mellor, *Mary Shelley*, 89–114. For more on Humphry Davy's scientific ideas see Christopher Lawrence, "The Power and the Glory: Humphry Davy and Romanticism," in Cunningham and Jardine, *Romanticism and the Sciences*, 213–227.

94. Shelley, *Frankenstein*, 35.

95. Mellor, *Mary Shelley*, 104–106.

96. Shelley, *Frankenstein*, 32–35.

97. Baldick, *In Frankenstein's Shadow*, 33–35.

98. Shelley, *Frankenstein*, 53.

99. S. S. Prawer, *Frankenstein's Island: England and the English in the Writings of Heinrich Heine* (Cambridge: Cambridge University Press, 1986), 42–82.

100. Heinrich Heine, "Concerning the History of Religion and Philosophy in Germany," trans. Max Knight, in *Heinrich Heine: Selected Works*, ed. Helen M. Mustard (New York: Random House, 1973), 325.

101. Heine, "Concerning the History of Religion and Philosophy in Germany," 365–366.

102. Prawer, *Frankenstein's Island*, 166.

103. Hermann von Pückler-Muskau, *A Regency Visitor: The English Tour of Prince Pückler-Muskau Described in His Letters, 1826–1828*, trans. Sarah Austin (New York: Dutton, 1958), 50. The reference in the letter to the English actor T. P. Cooke playing the monster makes it certain that the dramatization was Richard Brinsley Peake's *Presumption: Or the Fate of Frankenstein*, which was first performed in 1823. See Baldick, *In Frankenstein's Shadow*, 58–59. For greater details on the early dramatization of the novel, see Steven Early Forry, *Hideous Progenies: Dramatizations of Frankenstein from Mary Shelley to the Present* (Philadelphia: University of Pennsylvania Press, 1990), 3–42; for the text of the Peake play, 135–160.

104. Prawer, *Frankenstein's Island*, 284.

105. Heine, "Concerning the History of Religion and Philosophy in Germany," 366.

6. The Living Machines of the Industrial Age

1. Samuel Butler, "Darwin among the Machines," in *A First Year in Canterbury Settlement and Other Early Essays* (London: Jonathan Cape, 1923), 209.
2. Ibid., 212–213.
3. Ibid.
4. Samuel Butler, *Erewhon* (New York: Penguin Books, 1970), 81.
5. Ibid., 97.
6. Butler, "The Mechanical Creation," in *First Year in Canterbury Settlement*, 231–237.
7. Thomas Carlyle, "Signs of the Times," in *Critical and Miscellaneous Essays*, ed. H. D. Traill (London: Chapman and Hall, 1899), 59.
8. Frances Burney, *Evelina* (New York: Norton, 1998), 64.
9. Samuel Johnson, *Rambler* 83 (January 1, 1751), in Robert Lynam, ed., *The British Essayists* (London: J. F. Dove, 1827), 345–349, 347.
10. Ibid., 347–348.
11. Quoted in Edward Baines, *History of the Cotton Manufacture in Great Britain* (New York: Kelley, 1966), 229–230.
12. See David S. Landes, *The Unbound Prometheus: Technological Change and Industrial Development in Western Europe from 1750 to the Present* (Chicago: Chicago University Press, 1969), 86.
13. David Brewster, *Letters on Natural Magic* (London: J. Murray, 1832), 285–286.
14. Hermann von Helmoltz, "On the Interaction of Natural Forces," in *Popular Lectures on Scientific Subjects*, trans. E. Atkinson (New York: Longmans, Green, 1904), 138.
15. For Babbage's discussion of the Jacquard loom, see Charles Babbage, *Passages from the Life of a Philosopher* (New York: Kelley, 1969), 116–117.
16. Ibid., 17–18.
17. Ibid., 365–366.
18. Ibid., 462.
19. For a more detailed analysis of the silver lady figure see Simon Schaffer, "Babbage's Dancer and the Impresarios of Mechanism," in Francis Spufford and Jenny Uglow, eds., *Cultural Babbage: Technology, Time and Invention* (London: Faber and Faber, 1996), 53–80. See also Doron Swade, "'It Will Not Slice a Pineapple': Babbage, Miracles and Machines" in the same volume, 34–51; and on his more comprehensive work, Doron Swade, *The Difference Engine: Charles Babbage and the Quest to Build the First Computer* (New York: Viking, 2001).
20. See Frederick Gregory, *Scientific Materialism in Nineteenth-century Germany* (Dordrecht: Reidel, 1977).
21. See Thomas S. Kuhn, "Energy Conservation as an Example of Simultaneous Discovery," in *Essential Tension: Selected Studies in Scientific Tradition and Change* (Chicago: University of Chicago Press, 1977), 66–104; and Anson Rabinbach, *The Human Motor: Energy, Fatigue and the Origins of Modernity* (Berkeley: University of California Press, 1992), 45–68.

22. Helmoltz, "On the Interaction of Natural Forces," 162.

23. For details on thermodynamic physiology of Helmholtz and other materialist scientists of the era see Robert Brain and M. Norton Wise, "Muscles and Engines: Indicator Diagrams and Helmoltz's Graphic Methods," in Mario Biagioli, ed., *The Science Studies Reader* (New York: Routledge, 1999), 51–66. See also Rabinbach, *Human Motor*, 45–68.

24. Crosbie Smith and M. Norton Wise, *Energy and Empire: A Biographical Study of Lord Kelvin* (Cambridge: Cambridge University Press, 1989), 256.

25. See Rabinbach, *Human Motor*, 84–119.

26. Etienne-Jules Marey, *Animal Mechanism: A Treatise on Terrestrial and Aërial Locomotions*, trans. anonymous (New York: Appleton, 1893), 1.

27. Rabinbach, *Human Motor*, 47.

28. Quoted in Schaffer, "Babbage's Dancer," 63–64. See also Simon Schaffer, "Babbage's Intelligence: Calculating Engines and the Factory System," *Critical Inquiry* 91 (Fall 1994): 203–227.

29. Benjamin Disraeli, *Coningsby or the New Generation* (London: Peter Davies, 1927), 164.

30. Disraeli, *Coningsby*, 164. For an analysis of this passage in terms of gender issues, see M. Norton Wise, "The Gender of Automata in Victorian Britain," in Jessica Riskin, ed., *Genesis Redux: Essays in the History and Philosophy of Artificial Life* (Chicago: University of Chicago Press, 2007), 163–195, 172–173.

31. Herman Melville, "The Paradise of Bachelors and the Tartarus of Maids," in *Billy Budd and Other Stories* (New York: Penguin Books, 1986), 284.

32. Ibid., 277–278.

33. Adam Smith, *The Wealth of Nations* (New York: Modern Library, 2000), 4–9.

34. Charles Dickens, *Hard Times* (New York: Penguin Books, 1995), 28.

35. Ibid., 68.

36. Elizabeth Gaskell, *North and South* (London: Penguin Books, 1970), 166.

37. Karl Marx, *Capital*, vol. 1, trans. Ben Fowkes (New York: Vintage, 1977), 549.

38. Ibid., 503.

39. Ibid., 481.

40. For an excellent and concise overview of debates on modernism see Torbjörn Wandel, "Too Late for Modernity," *Journal of Historical Sociology* 18, 3 (2005): 255–268. For a work dealing directly with the role of technology in modernism, see Tim Armstrong, *Modernism, Technology and the Body* (Cambridge: Cambridge University Press, 1998). Among the many classic works dealing generally with modernism that I found particularly useful are Marshall Berman, *All That Is Solid Melts into Air: The Experience of Modernity* (New York: Penguin Books, 1988), and Matei Calinescu, *Five Faces of Modernity: Modernism, Avant-garde, Decadence, Kitsch, Postmodernism* (Durham, N.C.: Duke University Press, 1987).

41. Wandel, "Too Late for Modernity," 262.

42. For some general works on the machine symbolism in nineteenth-century culture see, for the British context, Herbert L. Sussman, *Victorians and the Machine* (Cambridge, Mass.: Harvard University Press, 1968); for the French, Jacques Noiray, *Le romancier et la machine: L'image de la machine dans le roman français*

(*1850–1900*), vol. 1, *L'univers de Zola* (Paris: Librarie José Corti, 1981), *Le romancier et la machine: L'image de la machine dans le roman français (1850–1900)*, vol. 2, *Jules Verne—Villiers de l'Isle Adam* (Paris: Librarie José Corti, 1982), and Kai Mikkonen, *The Plot Machine: The French Novel and the Bachelor Machines in the Electric Years (1880–1914)* (Amsterdam: Rodopi, 2001). See also Christoph Asendorf, *Batteries of Life: On the History of Things and Their Perception in Modernity*, trans. Don Reneau (Berkeley: University of California Press, 1993), and M. Norton Wise, "Architectures for Steam," in Pater Gallison and Emily Thompson, eds., *The Architecture of Science* (Cambridge, Mass.: MIT Press, 1999), 107–140.

43. For another analysis of this picture see Alison Winter, *Mesmerized: Powers of Mind in Victorian England* (Chicago: University of Chicago Press, 1998), 16–18.

44. H. G. Wells, "The Lord of the Dynamos," in *Selected Short Stories* (New York: Penguin Books, 1979), 184.

45. Ibid., 186.

46. Ibid., 192.

47. Henry Adams, *The Education of Henry Adams* (Oxford: Oxford University Press, 1999), 318.

48. Dolf Sternberger, *Panorama of the Nineteenth Century*, trans. Joachim Neugroschel (New York: Urizen, 1977), 17.

49. Sternberger, *Panorama of the Nineteenth Century*, 17–18.

50. J. K. Huysmans, *Against the Grain (À Rebours)*, trans. anonymous (New York: Dover, 1969), 22.

51. Huysmans, *Against the Grain*, 22–23. For an interesting discussion of the "eroticism of machines" see Asendorf, *Batteries of Life*, 105–111, and the more comprehensive work on the topic, Michel Carrouges, *Les machines célibataires* (Paris: Arcanes, 1954). Also on the cultural impact of railway see Wolfgang Schivelbusch, *The Railway Journey: The Industrialization of Time and Space in the Nineteenth Century* (Berkeley: University of California Press, 1977).

52. Émile Zola, *La Bête Humaine*, trans. Leonard Tancock (London: Penguin Books, 1977), 155.

53. Ibid., 300.

54. Ibid., 365.

55. Ibid., 366. For more on the role of machines in Zola's works see Noiray, *Le romancier et la machine*, vol. 1, *L'univers de Zola*. For a more concise study, Geoff Wollen, "Zola's Thermodynamic Vitalism," *Romance Studies* 6 (1985): 48–62.

56. Gerhart Hauptmann, "Lineman Thiel," in *Lineman Thiel and Other Tales*, trans. Stanley Radcliffe (London: Angel Books, 1989), 68.

57. Hauptmann, "Lineman Thiel," 72.

58. Jeffrey Herf, *Reactionary Modernism: Technology, Culture, and Politics in Weimar and the Third Reich* (Cambridge: Cambridge University Press, 1984), 1–17.

59. For a detailed discussion of Edison's doll see Gaby Wood, *Living Dolls: A Magical History of the Quest for Mechanical Life* (London: Faber and Faber, 2002), 107–125.

60. Villiers de l'Isle-Adam, *Tomorrow's Eve*, trans. Robert Martin Adams (Champaign: University of Illinois Press, 2001), 44. For more on Villiers and *L'Eve*

future, see Mikkonen, *Plot Machine*, 149–183; Jacques Noiray, *Le romancier et la machine: L'image de la machine dans le roman français (1850–1900)*, 243–385; John Anzalone, ed., *Jeering Dreamers: Essays on L'Eve Future* (Amsterdam: Rodopi, 1996); Annette Michelson, "On the Eve of the Future: The Reasonable Facsimile and the Philosophical Toy," *October* 29 (Summer 1984): 3–20; and Carrouges, *Les machines célibataires*, 145–163.

61. Villiers, *Tomorrow's Eve*, 65.
62. Ibid., 85.
63. Ibid., 198.
64. Ibid., 64.
65. Ibid., 211.
66. There are a lot of interesting issues of gender in this novel and other automaton works of the period that I cannot adequately discuss in this study, though I explored some aspects of it in Minsoo Kang, "Building the Sex Machine: The Subversive Potential of the Female Robot," *Intertexts* 9, 1 (Spring 2005): 5–22. I hope to deal with this topic in greater detail in a future work.
67. For an introduction to Alfred Jarry and his works, see Roger Shattuck, *The Banquet Years: Origin of the Avant-Garde in France, 1885 to World War I* (New York: Vintage, 1968), 187–251. On *Le Surmâle*, see Linda Klieger Stillman, *Alfred Jarry* (Boston: Twayne, 1983), 100–104; Keith Beaumont, *Alfred Jarry: A Critical and Biographical Study* (Leicester: Leicester University Press, 1984), 243–257; and Carrouges, *Les machines célibataires*, 93–127.
68. Alfred Jarry, *The Supermale*, trans. Ralph Gladstone and Barbara Wright (Cambridge, Mass.: Exact Change, 1999), 7.
69. Rabinbach, *Human Motor*, 19–42.
70. Ibid., 143–144.
71. Jarry, *Supermale*, 49–71.
72. Ibid., 58–59.
73. Ibid., 124.
74. Ibid., 134.
75. Ibid., 136.
76. Ibid., 39.
77. Ibid., 39–40.
78. See Frederick Burwick and Paul Douglass, *Crisis in Modernism: Bergson and the Vitalist Controversy* (Cambridge, Mass.: Cambridge University Press, 1992).
79. For more on this poem and other Marinetti poetry featuring machinery, see Shirley Vinall, "The Emergence of Machine Imagery in Marinetti's Poetry," *Romance Studies* 6 (1985): 78–95.
80. F. T. Marinetti, "To My Pegasus," in *Selected Poems and Related Prose*, trans. Elizabeth R. Napier and Barbara R. Studholme (New Haven: Yale University Press, 2002), 38.
81. On Marinetti's interest in aviation and its influence on his works, see Jeffrey T. Schnapp, "Propeller Talk," *Modernism/Modernity* 1,3 (1994): 153–178.
82. F. T. Marinetti, "The Pope's Monoplane," in *Selected Poems and Related Prose*, 43.

83. F. T. Marinetti, "Technical Manifesto of Futurist Literature," in *Let's Murder the Moonshine: Selected Writings*, trans. R. W. Flint and Arthur A. Coppotelli (Los Angeles: Sun and Moon Classics, 1991), 94.

84. F. T. Marinetti, "Multiplied Man and the Reign of the Machine," in *Let's Murder the Moonshine*, 98–99.

85. Marinetti, "Multiplied Man and the Reign of the Machine," 99. It should be no surprise that Marinetti read the works of Alfred Jarry, with whom he was on friendly terms. See Günter Berghaus, *The Genesis of Futurism: Marinetti's Early Career and Writings, 1899–1909* (Leeds, U.K.: Society for Italian Studies, 1995), 28–29, and 58–59.

86. For a detailed study of those changes, see Stephen Kern, *The Culture of Time and Space, 1880–1918* (Cambridge, Mass.: Harvard University Press, 1983). See also Schivelbusch, *The Railway Journey*. ·

87. Schnapp, "Propeller Talk," 161.

88. For details see Berghaus, *Genesis of Futurism*, 78–92. On Futurist drama, including manifestoes and plays, see Michael Kirby, *Futurist Performance* (New York: Dutton, 1971), and the comprehensive Günter Berghaus, *Italian Futurist Theatre, 1909–1944* (Oxford: Clarendon Press, 1998).

89. F. T. Marinetti, *Poupées électriques* (Paris: E. Sansot, 1909), 38.

90. Ibid., 134–135.

91. Ibid., 132.

92. Mary's ultimate effect on John exemplifies Marinetti and the general Futurist attitude toward woman as a debilitating force for the creative man. For a concise discussion of the misogynistic view, see Caroline Tisdall and Angelo Bozzolla, *Futurism* (New York: Thames and Hudson, 1977), 153–163. For a nuanced discussion of this issue see Walter L. Adamson, *Embattled Avant-Gardes: Modernism's Resistance to Commodity Culture in Europe* (Berkeley: University of California Press, 2007), 99–105.

93. Adamson, *Embattled Avant-Gardes*, 79.

94. For some works that deal with this issue in detail, see Cinzia Blum, *The Other Modernism: Marinetti's Futurist Fiction of Power* (Berkeley: University of California Press, 1996), 55–78; Alice Yaeger Kaplan, *Reproductions of Banality: Fascism, Literature and French Intellectual Life* (Minneapolis: University of Minnesota Press, 1986), 76–92; and Barbara Spackman, "Mafarka and Son: Marinetti's Homophobic Economics," *Modernism/Modernity* 1,3 (1994): 89–107.

95. F. T. Marinetti, *Mafarka the Futurist*, trans. Carol Diethes and Steven Cox (London: Middlesex University Press, 1997), 145–146.

96. Ibid., 189.

97. Ibid., 197.

98. Ibid., 188.

99. An important element of his posthumous recognition is Michel Foucault's 1963 book on his works, Foucault's only full-length study of a single literary figure. See Foucault, *Death and the Labyrinth: The World of Raymond Roussel*, trans. Charles Ruas (London: Athlone Press, 1987). See also Mark Ford, *Raymond Roussel and the Republic of Dreams* (Ithaca: Cornell University Press, 2000), 216–239; and Carrouges, *Les machines célibataires*, 61–92.

100. Raymond Roussel, *Impressions of Africa*, trans. Lindy Foord and Rayner Heppenstall (London: Calder, 2001), 152–157.
101. Ibid., 35–37.
102. Ibid., 36.
103. Ibid., 45.
104. Ibid., 94.
105. Ibid., 140–150.
106. Raymond Roussel, *Locus Solus*, trans. Rupert Copeland Cuningham (Berkeley: University of California Press, 1970), 26–36.
107. Ibid., 51–87.
108. Ibid., 118.
109. Ford, *Raymond Roussel*, 17–18; and Foucault, *Death and the Labyrinth*, 77–79.
110. Roussel, *Impressions of Africa*, 143.
111. Roussel, *Locus Solus*, 26.
112. Alain Robbe-Grillet, "Enigma and Transparency in Raymond Roussel," in *For a New Novel: Essays on Fiction*, trans. Richard Howard (Evanston: Northwestern University Press, 1965), 81.
113. Ford, *Raymond Roussel*, 116; see also Ruth Brandon, *Surreal Lives: The Surrealists, 1917–1945* (New York: Grove Press, 1999), 48–51.
114. Quoted in Dawn Ades, Neil Cox, and David Hopkins, *Marcel Duchamp* (New York: Thames and Hudson, 1999), 106. For a discussion of the influence of scientific and technological ideas on the construction of the installation, see Linda Dalrymple Henderson, "Etherial Bride and Mechanical Bachelors: Science and Allegory in Marcel Duchamp's 'Large Glass,'" *Configurations* 4, 1 (Winter 1996): 91–120.
115. Michel Carrouges adopted the term *bachelor machines* for his 1954 study of machine symbolism in modernist literature. See *Les machines célibataires*, 27–59.
116. For an interesting discussion of automata and their influence on modern art, including Duchamp's works, see Jack Burnham, *Beyond Modern Sculpture: The Effects of Science and Technology on the Sculpture of this Century* (New York: Braziller, 1967), 185–217. See also the chapter on machine art in Harry M. Geduld and Ronald Gottesman, eds., *Robots, Robots, Robots* (Boston: New York Graphic Society, 1978), 78–83.

7. The Revolt of the Robots

1. G. K. Chesterton, "The Invisible Man," in *The Annotated Innocence of Father Brown* (Mineola, N.Y.: Dover, 1998), 102–120.
2. Ibid., 111–112.
3. Ibid., 116.
4. Ibid., 118.
5. See Martin Gardner, note to Chesterton, "Invisible Man," 107, no. 5.
6. Aristotle, *Politics*, trans. Benjamin Jowett, in *The Complete Works of Aristotle*, vol. 2, ed. Jonathan Barnes (Princeton: Princeton University Press, 1984), 1989.
7. For works that deal with technological warfare of World War I, see Hubert C. Johnson, *Breakthrough! Tactics, Technology, and the Search for Victory on the Western*

Front in World War I (Novato, Calif.: Presidio Press, 1996); Stephen Kern, *The Culture of Time and Space, 1880–1918* (Cambridge, Mass.: Harvard University Press, 1983), 287–312; and Modris Eksteins, *Rites of Spring: The Great War and the Birth of the Modern Age* (New York: Anchor Books, 1989).

8. Erich Maria Remarque, *All Quiet on the Western Front*, trans. A. W. Wheen (New York: Fawcett Crest, 1958), 115–116.

9. Henri Barbusse, *Under Fire*, trans. Robin Buss (New York: Penguin Books, 2003), 18.

10. Ibid., 305.

11. Anson Rabinbach, *The Human Motor: Energy, Fatigue and the Origins of Modernity* (Berkeley: University of California Press, 1992), 238–258. For more on Taylorism, see Daniel Nelson, *Frederick W. Taylor and the Rise of Scientific Management* (Madison: University of Wisconsin Press, 1980).

12. Rabinbach, *Human Motor*, 259–270. See also Charles S. Maier, "Between Taylorism and Technocracy: European Ideologies and the Vision of Industrial Productivity in the 1920s," *Journal of Contemporary History* 5, 2 (1970): 27–61; and George G. Humphreys, *Taylorism in France: The Impact of Scientific Management on Factory Relations and Society* (New York: Garland, 1986).

13. See Richard Stites, *Revolutionary Dreams: Utopian Vision and Experimental Life in the Russian Revolution* (Oxford: Oxford University Press, 1989), 30–36.

14. Stites, *Revolutionary Dreams*, 41–46; and James Schmidt and James Miller, "Aspects of Technology in Marx and Rousseau," in Teresa de Laurentis, Andreas Huyssen, and Kathleen Woodward, eds., *The Technological Imagination: Theories and Fictions* (Madison, Wisc.: Coda Press), 85–94.

15. See Stites, *Revolutionary Dreams*, 49–55. For more on this interesting figure see Kendall E. Bailes, "Alexei Gastev and the Soviet Controversy over Taylorism, 1918–1924," *Soviet Studies* 29, 3 (July 1977): 373–394; Kurt Johansson, *Aleksej Gastev: Proletarian Bard of the Machine Age* (Stockholm: Almquist and Wiksell International, 1983); and Rolf Hellebust, "Aleksei Gastev and the Metallization of the Revolutionary Body," *Slavic Review* 56, 3 (Fall 1997): 500–514.

16. I am grateful to Gregory Palatnik for providing me with this translation from the Russian. The poem in the original can be found in Johansson, *Aleksej Gastev*, 136.

17. See Yevgeny Zamyatin, trans. Mirra Ginsburg, *We* (New York: Madison, 1972). On the work in its historical context, see Stites, *Revolutionary Dreams*, 187–189. For an enlightening discussion of Gastev and Zamyatin works in relation to each other see Patricia Garden, "Utopia and Anti-Utopia: Aleksei Gastev and Evgeny Zamyatin," *Russian Review* 46, 1 (Jan. 1987): 1–18.

18. Alison Winter, *Mesmerized: Powers of Mind in Victorian Britain* (Chicago: University of Chicago Press, 1998), 281–284.

19. See Iwan Rhys Morus, *Frankenstein's Children: Electricity, Exhibition, and Experiment in Early-nineteenth-century London* (Princeton: Princeton University Press, 1998), 231–255; and Andreas Killen, *Berlin Electropolis: Shock, Nerves, and German Modernity* (Berkeley: University of California Press, 2006), 48–80.

20. Killen, *Berlin Electropolis*, 72, see also 48–49 and 77–78.

21. See Paul Roazen, introduction to Victor Tausk, *Sexuality, War, and Schizophrenia: Collected Psychoanalytic Papers*, trans. Eric Mosbacher et al. (New Brunswick: Transaction, 1991), 1–28.

22. Victor Tausk, "On the Origin of the 'Influencing Machine' in Schizophrenia," in *Sexuality, War, and Schizophrenia*, 186.

23. Tausk, "On the Origin of the 'Influencing Machine' in Schizophrenia," 195.

24. Andreas Killen quotes the Berlin neurologist Albert Eulenberg, who also had an occasion to describe the device as an "uncanny machine." See Killen, *Berlin Electropolis*, 77.

25. For extended analyses of German expressionist cinema in historical context see Siegfried Kracauer, *From Caligari to Hitler: A Psychological History of the German Film* (Princeton: Princeton University Press, 1969), especially 61–87; and Lotte H. Eisner, *The Haunted Screen: Expressionism in the German Cinema and the Influence of Max Reinhardt*, trans. Roger Greaves (Berkeley: University of California Press, 1969).

26. Frank Whitford, "The Many Faces of George Grosz," in George Grosz, *The Berlin of George Grosz: Drawings, Watercolours and Prints 1912–1930* (New Haven: Yale University Press, 1997), 10–12. See also Serge Sabarsky, ed., *George Grosz: The Berlin Years* (New York: Rizzoli, 1985).

27. See Grosz, *Berlin of George Grosz*, 122.

28. George Grosz, "Zu Meinen Neuen Bildern: On My New Paintings," in *Berlin of George Grosz*, 36.

29. Leni Riefenstahl, *Memoir*, (New York: St. Martin's Press, 1992), 174.

30. Ibid., 174.

31. Ibid., 175; for the fate of Riefenstahl's only copy of the print after World War II, see 324. It has apparently been found since then, as it has been shown in several recent film festivals.

32. Jeffrey Herf, *Reactionary Modernism: Technology, Culture, and Politics in Weimar and the Third Reich* (Cambridge: Cambridge University Press, 1984), 1–17.

33. For an argument of this nature, see Zygmunt Bauman, *Modernity and the Holocaust* (Ithaca, N.Y.: Cornell University Press, 1989), especially 83–116.

34. Original title *If This Is a Man*. Primo Levi, *Survival in Auschwitz: The Nazi Assault on Humanity*, trans. Stuart Woolf (London: Collier Macmillan, 1961), 45.

35. Elie Wiesel, *Night*, trans. Stella Rodway (New York: Bantam Books, 1960), 81.

36. See Michael Heim, translator's introduction in Karel Čapek, *The White Plague* (New York: Theatre Communications Group, 1988).

37. William E. Harkins, *Karel Čapek* (New York: Columbia University Press, 1962), 84.

38. Karel Čapek, *R. U. R. (Rossum's Universal Robots)*, trans. Claudia Novack-Jones, in *Toward the Radical Center: A Karel Čapek Reader*, ed. Peter Kussi (Highland Park, N.J.: Catbird Press, 1990), 38–39.

39. Ibid., 50.

40. Ibid., 51–69.
41. Ibid., 96.
42. Ibid., 100.
43. For more on the play, see Harkins, *Karel Čapek*, 84–95; Ivan Klíma, *Karel Čapek: Life and Work*, trans. Norma Comrada (North Haven, Conn.: Catbird Press, 2002), 72–84; and Barbara Bengels, "'Read History': Dehumanization in Karel Čapek's *R.U.R.*," in Thomas P. Dunn and Richard D. Erlich, eds., *The Mechanical God: Machines in Science Fiction* (Westport, Conn.: Greenwood Press, 1982), 13–17.
44. See Harkins, *Karel Čapek*, 94.
45. Klíma, *Karel Čapek*, 244, no. 17.
46. For more on Rolland, the Russian Revolution, and the debate with Henri Barbusse, see William Thomas Starr, *Romain Rolland and a World at War* (Evanston, Ill.: Northwestern University Press, 1956), 137–151, and David James Fisher, *Romain Rolland and the Politics of Intellectual Engagement* (Berkeley: University of California Press, 1988), 90–111.
47. Romain Rolland, *The Revolt of the Machines*, trans. William A. Drake (Ithaca: Dagon Press, 1932), 11.
48. Ibid., 11–12.
49. Ibid., 20.
50. Ibid., 26.
51. Ibid.
52. Ibid., 51.
53. Ibid., 55.
54. Ibid., 56–57.
55. For details on Vasari and *The Anguish of the Machines*, see Günter Berghaus, *Italian Futurist Theatre, 1909–1944* (Oxford: Clarendon Press, 1998), 494–515; on his reaction to *R.U.R.* see 503.
56. Ruggero Vasari, *The Anguish of the Machines*, trans. Anna Lawton, in Harry M. Geduld and Ronald Gottesman, eds., *Robots, Robots, Robots* (Boston: New York Graphic Society, 1978), 62.
57. Ibid., 64–65.
58. Ibid., 65.
59. See Michael Minden and Holger Bachmann, *Fritz Lang's Metropolis: Cinematic Visions of Technology and Fear* (Rochester, N.Y.: Camden House, 2000), 4.
60. Thea von Harbou, *Metropolis*, trans. anonymous (Rockville, MD: Sense of Wonder Press, 2001), 12.
61. For the description of the scene in the film, see Fritz Lang, *Metropolis: A Film by Fritz Lang* (New York: Simon and Schuster, 1973), 32.
62. Harbou, *Metropolis*, 23.
63. Ibid., 24.
64. Ibid., 52.
65. Ibid., 33–34, and 59–60. For the film scene, see Lang, *Metropolis*, 51–52.
66. Ibid., 65.
67. Ibid., 177–178.

68. Ibid., 156.

69. See Peter S. Fisher, *Fantasy and Politics: Visions of the Future in the Weimar Republic* (Madison: University of Wisconsin Press, 1991) and I. F. Clarke, *Voices Prophesying War: Future Wars 1763–3749* (Oxford: Oxford University Press, 1992), 131–160.

70. See Kracauer, *From Caligari to Hitler*, 162–164; Peter Gay, *Weimar Culture: The Outsider as Insider* (New York: Norton, 2001), 141–142; Roger Dadoun, "*Metropolis*: Mother-city—'Mittler'—Hitler," in Constance Penley, Elisabeth Lyon, Lynn Spigel, and Janet Bergstrom, eds., *Close Encounters: Film, Feminism, and Science Fiction* (Minneapolis: University of Minnesota Press, 1991), 133–159; Andreas Huyssen, "The Vamp and the Machine," in *After the Great Divide: Modernism, Mass Culture, Postmodernism* (Bloomington: Indiana University Press, 1986), 65–81; and R. L. Rutsky, "The Mediation of Technology and Gender: *Metropolis*, Nazism, Modernism" in Minden and Bachmann, *Fritz Lang's Metropolis*, 217–245.

71. For the discussion of nationalist fantasies of the period see Fisher, *Fantasy and Politics*, 21–157, and for a discussion of *Metropolis* in this context, 126–142.

72. For more nuanced interpretations of the film, see Ludmilla Jordanova, "Science, Machines, and Gender," 173–195, and Andrew Webber, "Canning the Uncanny: The Construction of Visual Desire in *Metropolis*," 251–269, both in Minden and Bachmann, *Fritz Lang's Metropolis*.

73. See Eric D. Weitz, *Weimar Germany: Promise and Tragedy* (Princeton: Princeton University Press, 2007), especially 139–168.

74. Herf, *Reactionary Modernism*, 18–19.

75. Huyssen, "Vamp and the Machine," 81.

76. Huyssen, "Mass Culture as Woman: Modernism's Other," in *After the Great Divide*, 44–62.

77. For Lang's alternate ending, see Robert Bloch, "The Master and *Metropolis*," in Danny Peary, ed., *Omni's Screen Flights / Screen Fantasies: The Future According to Science Fiction Cinema* (Garden City, N.Y.: Doubleday, 1984), 88.

78. Jean Baudrillard, *The System of Objects*, trans. James Benedict (London: Verso, 1996), 121.

Conclusion

1. Ernst Jünger, *The Glass Bees*, trans. Louise Bogan and Elizabeth Mayer (New York: New York Review Books, 2000). See also Stanislaw Lem, *The Cyberiad*, trans. Michael Kandel (San Diego: Harvest Books, 1974) and *Mortal Engines* (San Diego: Harvest Books, 1977); and John Sladek, *The Complete Roderick* (Woodstock, New York: Overlook Press, 1980) and *Tick-Tock* (London: Gollancz, 1980).

2. Bruce Clark, "From Thermodynamics to Virtuality," in Bruce Clarke and Linda Dalrymple Henderson, eds., *From Energy to Information: Representation in Science and Technology, Art, and Literature* (Stanford: Stanford University Press, 2002), 17–33.

3. See Norbert Wiener, *The Human Use of Human Beings: Cybernetics and Society* (Boston: Da Capo Press, 1950). There is a significant literature on this topic; for

a good general introduction see N. Katherine Hayles, *How We Became Posthuman: Virtual Bodies in Cybernetics, Literature, and Informatics* (Chicago: University of Chicago Press, 1999), especially 50–112.

4. The word *cyborg* is a combination of "cybernetic organism" and denotes a being that consists of some organic, natural parts and some artificial. For the origin of the word, see Manfred E. Clynes and Nathan S. Kline, "Cyborgs and Space," in Chris Hables Gray, ed., *The Cyborg Handbook* (New York: Routledge, 1995), 29–33.

5. Donna J. Haraway, "A Cyborg Manifesto: Science, Technology, and Socialist-feminism in the Late Twentieth Century," in *Simians, Cyborgs, and Women: The Reinvention of Nature* (New York: Routledge, 1991), 149–181, originally published in a slightly different version as "Manifesto for Cyborgs: Science, Technology, and Socialist Feminism in the 1980s," *Socialist Review* 80 (1985): 65–108. The most comprehensive work on cyborg theory is Gray, *Cyborg Handbook*. See also Andy Clark, *Natural-born Cyborgs: Minds, Technologies, and the Future of Human Intelligence* (Oxford: Oxford University Press, 2003).

6. Kevin Warwick, *March of the Machines: The Breakthrough in Artificial Intelligence* (Urbana: University of Illinois Press, 2004), 280.

7. Hans Moravec, *Mind Children: The Future of Robot and Human Intelligence* (Cambridge, Mass.: Harvard University Press, 1988).

8. Philip K. Dick, "Second Variety," in *Robots, Androids, and Mechanical Oddities: The Science Fiction of Philip K. Dick* (Carbondale: Southern Illinois University Press, 1984), 54.

9. These laws first appeared in the stories collected in his 1950 book *I, Robot*. See also Isaac Asimov, "Runaround," in *Robot Visions* (New York: ROC, 1991), 113–134, and his other stories and essays on robots in his *Robot Dreams* (New York: Ace Books, 1986).

10. Isaac Asimov, "The Bicentennial Man," in *Robot Visions*, 245–290.

11. See Ray Kurzweil, *The Singularity Is Near: When Humans Transcend Biology* (New York: Penguin Books, 2005). See also his earlier work *The Age of Spiritual Machines: When Computers Exceed Human Intelligence* (New York: Penguin Books, 1999). Cyberpunk writers like William Gibson and Bruce Sterling have also explored the dystopian implications of such technological possibilities. For an excellent example see the short story by Walter Jon Williams "Daddy's World," in Kim Stanley Robinson, ed., *Nebula Awards Showcase 2002* (New York: ROC, 2002), 31–59.

12. In addition to the already cited works by Moravec, Warwick, Kurzweil, and Hayles on contemporary robotics and cybernetics, I particularly found useful Rodney A. Brooks, *Flesh and Machines: How Robots Will Change Us* (New York: Pantheon Books, 2002), Joel Garreau, *Radical Evolution: The Promise and Perils of Enhancing Our Minds, Our Bodies—and What It Means to Be Human* (New York: Broadway Books, 2005), and Lee Gutkind, *Almost Human: Making Robots Think* (New York: Norton, 2006).

13. Haraway, "Cyborg Manifesto," 154.

14. Ibid., 177.

15. Some recent examples are Roger B. Rollin, "*Deus in Machina:* Popular Culture's Myth of the Machine," *Journal of American Culture* 2 (Summer 1979): 297–308; Mark Crispin Miller, *Boxed In: The Culture of TV* (Evanston, Ill.: Northwestern University Press, 1988), 297–303; Peter Wollen, *Raiding the Icebox: Reflections on Twentieth-century Culture* (Bloomington: Indiana University Press, 1993), 41–47; and J. P. Telotte, *Replications: A Robotic History of the Science Fiction Film* (Urbana: University of Illinois Press, 1995), 29–53.

16. Claude Lévi-Strauss, *Totemism*, trans. Rodney Needham (Boston: Beacon Press, 1989).

17. See, for instance, Chris Hables Gray, *Cyborg Citizen: Politics in the Posthuman Age* (New York: Routledge, 2002); Francis Fukuyama, *Our Posthuman Future: Consequences of the Biotechnological Revolution* (New York: Picador, 2003); James Hughes, *Citizen Cyborg: Why Democratic Societies Must Respond to the Redesigned Human of the Future* (Cambridge, Mass.: Westview Press, 2004); and Byron L. Sherwin, *Golems among Us: How a Jewish Legend Can Help Us Navigate the Biotech Century* (Chicago: Dee, 2004).

18. Keith E. Stanovich, *The Robot's Rebellion: Finding Meaning in the Age of Darwin* (Chicago; University of Chicago Press, 2004), xii.

19. Gilles Deleuze and Félix Guattari, *Anti-Oedipus: Capitalism and Schizophrenia*, trans. Robert Hurley, Mark Seem, and Helen R. Lane (Minneapolis: University of Minnesota Press, 1983).

Acknowledgments

My first acknowledgment goes to all the wonderful professors I studied with at UCLA during my two phases as a graduate student, especially Mario Biagioli, Carlo Ginzburg, Michael Heim, Margaret Jacob, Anne Mellor, Ted Porter, Peter Reill, Mary Terrall, Norton Wise, and Robert Wohl. I am also grateful to the scholars I have met over the last years who shared my enthusiasm for automata and who provided me with a great deal of insight into the subject through our conversations. They include Bruce Clarke, Linda Dalrymple, Scott Lightsey, Scott Maisano, Elly Truitt, Kevin LaGrandeur, Arielle Saiber, John Tresch, and Adelheid Voskuhl. Members of the Eighteenth-Century Salon group at Washington University in St. Louis gave me extremely helpful comments on several of the chapters. I am especially grateful to Tili Boon Cuillé, Matt Erlin, Emily Guignon, Rebecca Messbarger, Annie Smart, and Jim Tierney. Thanks to my colleagues at the University of Missouri–St. Louis for their support during the time I finished this book, especially Deborah Cohen, Priscilla Dowden-White, Kevin Fernlund, Louis Gerteis, Jack Gillingham, Winston Hsieh, Andrew Hurley, Dick Mitchell, Adell Patton, Steve Rowan, Carlos Schwantes, and Laura Westhoff. Also to Ann Downer-Hazell and Vanessa Hayes, and Lindsay Waters and Phoebe Kosman at Harvard University Press. And my last acknowledgment goes to friends who continue to provide me with moral support, intellectual discussion, and much of my life's enjoyment. Thanks to John Dalton, Ron Davidson, Heinz Insu Fenkl, Kathy Gentile, Frank Grady, Kevin Lambert, Ken Little, Janet McGarry, Amy Pawl, Courtenay Raia, Robert Rosenstone, and Amy Woodson-Boulton.

Index

120 Days of Sodom (Sade), 165
2001: A Space Odyssey, 298

À Rebours (Huysmans), 239
Académie Royale des Sciences, 110, 141, 142, 228
Adams, Henry, 43, 172–174, 237–238
Agrippa von Nettesheim, Heinrich Cornelius, 5, 9, 44, 58, 59, 68, 79, 84–89, 90, 92, 93, 94, 97, 100, 116, 139, 219
Albertus Magnus, 9, 44, 58, 69, 70–75, 77, 78, 93, 94, 97, 98, 99, 219
Alchemy, 58, 74–75, 115, 116, 214–215
All Quiet on the Western Front (Remarque), 269
Allen, Roland, 70
American, The (James), 4–5
Android, the word, 99, 328n163
Anguish of the Machines, The (Vasari), 268
Animal Automatism, 111, 116–118, 122, 126–128, 129–130, 136–138
Animal Mechanism (Marey), 231
Animism, 23–24, 113, 116, 167
Anomalies, categorical, 29–38, 42, 45–46
Ansichten von der Nachtseite der Naturwissenschaft (Schubert), 207
Antimagnétisme, L' (Paulet), 167
Anti-Oedipus: Capitalism and Schizophrenia (Deleuze and Guattari), 308
Antonius, Marcus, 36–37
Apollinaire, Guillaume, 262
Apollonius of Tyana, 87
Appian, 36–37
Archimedes of Syracuse, 80

Archytas of Tarentum, the flying dove of, 16, 87, 93, 94, 97, 98, 100, 140, 176
"Are We Automata?" (James), 169–170
Aristotle, 17, 19, 21, 32, 58, 84, 86, 93, 97, 117–118, 120, 124, 132, 139, 148, 266
Arkwright, Richard, 107
Armut, Reichtum, Schuld und Buße der Gräfin Dolores (Arnim), 216–217
Arnim, Ludwig Achim von, 10, 27, 188, 193, 216–217
Artificial Intelligence, 12, 298–299, 300
Asclepius, 75–77, 83, 87, 92
Asimov, Isaac, 11, 302
Astrology, 58, 59, 67–68, 69, 71–72, 93, 115, 116
Attempt to Analyse the Automaton Chess Player of Mr. de Kempelen, An (Willis), 180
Attic Nights (Aulus Gellius), 16, 87
Augustine of Hippo, 76–77
Aulus Gellius, 16, 87, 99
Austrian Succession, war of, 154
Auswahl aus des Teufels Papieren (Jean Paul), 201
"Automate, Die." *See* "Automaton, The"
Automatic Theater (Hero of Alexandria), 16
"Automaton, The," (Hoffmann), 51, 207–210
Automaton, definitions of, 7–8, 18–19, 61, 84–85, 112, 135–136, 139–140, 148, 159

Babbage, Charles, 144, 176, 179, 228, 231, 298
Bacon, Francis, 115, 140–141
Bacon, Roger, 9, 58, 59, 69, 71–75, 78, 98

Baglivi, Giorgio, 125, 128, 151, 152
Baillet, Adrien, 123
Bailley, Christian, 175
Bakhtin, Mikhail, 32
Baldick, Chris, 199, 219
Barbusse, Henri, 269–270, 282
Barthes, Roland, 165
Barthez, Paul Joseph, 150, 151, 193
Battleship Potemkin (Eisenstein), 277
Baudrillard, Jean, 20–21, 296, 307–308
Bayle, François, 129
Beckford, William, 194
Beireis, Gottfried Christoph, 104, 214–216
"Bell Tower, The" (Melville), 27, 46
Bellini, Lorenzo, 128
Bellmer, Hans, 262
Benjamin, Walter, 180–181
Benno II, Cardinal and Bishop of Osnabrück, 69–70
Benoît de Sainte-Maure, 64
Bergerac, Cyrano de, 137
Bergson, Henri, 38–39, 250
Bernard of Clairvaux, 58
Berryman, Sylvia, 118
Bête humaine, La (Zola) 240–241
Bettini, Maurizio, 20
"Bicentennial Man, The" (Asimov), 302
Bierce, Ambrose, 27, 180
Bildungstrieb, 189, 191
Blade Runner, 11, 298
Blake, William, 199
Bloch, Iwan, 108
Blumenbach, Johann Friedrich, 189
Boccioni, Umberto, 286
Body-State Analogy, 132–134
Boerhaave, Herman, 149
Boethius, Anicius Manlius Severinus, 87–89, 90
Bois-Reymond, Emil du, 230
Bolshevik Revolution, 10, 268, 270–272
Bon sens, Le (D'Holbach), 156, 159–160, 162
Bonaparte, Napoleon, 178
Bonaventure d'Argonne. *See* Vigneul-Marville
Bordeu, Théophile de, 150, 151, 157, 193
Borelli, Giovanni, 125, 126–128, 131, 140, 151, 152
Boullier, David Renaud, 129–130
Boyle, Robert, 6, 113, 115, 124, 125, 139, 140, 141, 148
Brahe, Tycho, 81
"Brazen Android, The" (O'Connor), 27
Brentano, Clemens, 193, 216
Brewster, David, 144, 227–228
Briefe über den Schachspieler des von Kempelen (Windisch), 179–180
Brocklesby, Richard, 149
"Bronze Horseman, The" (Pushkin), 2–3
Brooks, Rodney, 27
Bruno, Giordano, 60

Buchier, William (Guillaume de), 64
Büchner, Georg, 188, 204–206
Büchner, Ludwig, 230
Buffon, Georges-Louis Leclerc, 150–151, 189, 191
Burke, Edmund, 41, 196, 317n66
Burney, Frances, 143–144, 226
Bush, George W., 4
Butler, Samuel, 223–225, 226, 233, 243, 263, 267, 300, 301
Butor, Michel, 257

Cabinet of Dr. Caligari, The, 273
Caelius Rhodiginus, Ludovico, 97
Caesar, Gaius Julius, 36–37
Campanella, Tommaso, 9, 59, 60, 80, 94–95
Canterbury Tales, The, 67–68
Čapek, Josef, 279
Čapek, Karel, 10, 268, 279–282, 296
Capital (Marx), 234
Caractères, Les (La Bruyère), 137, 144–145, 160
Cardano, Girolamo, 98, 113
Cardigan, Pat, 299
Carlyle, Thomas, 108, 175, 225, 267
Cartwright, Edmund, 227
Cassiodorus, Magnus Aurelius, 87–89, 93
Castle, George, 128
Castle of Otranto, The (Walpole), 194
Categories, conceptual and binary, 29–38, 45–46, 63, 100, 102, 114, 315n37
Catharsis, 32
Cellini, Benvenuto, 45
Cervantes, Miguel de, 95
Chaptal, Jean-Antoine, 108
Chapuis, Alfred, 11
Charcot, Jean Martin, 272–273
Charles I of England, 133
Charles V, Holy Roman Emperor, 81, 98
Charles VI of France, 33
Charleton, Walter, 128
Chaucer, Geoffrey, 67–68, 71
Chess Machine, The (Löhr), 181
Chess-Player or the Turk, Wolfgang von Kemplen's, 149, 174, 175–183, 201, 207, 217, 228
Chesterton, G. K., 264–266
Cheyne, George, 149, 150, 159
Chirac, Pierre, 129
Christian Philosopher, The (Mather), 131
Chronicles of the Kings of England (William of Malmesbury), 69
Chymical Galenist, The (Castle), 128
Cicero, Marcus Tullius, 37
City of God (Augustine of Hippo), 76–77
Clarissa (Richardson), 153, 160
Clark, Christopher, 185–186
Clarke, Arthur C., 56, 60
Clausius, Rudolf, 230
Cleland, John, 138–139, 165

INDEX   367

Clerselier, Claude, 96–97, 129
Clocks, 80, 90, 118, 127, 128, 130, 137, 139, 140, 266
Cockburn, William, 149
Cohen, John, 22, 24–25
Colbert, Jean-Baptiste, 135
Cole, William, 128
Commentaire ou Remarques de la Méthode de René Descartes (Poisson), 122
Communist Manifesto (Marx), 281
Conan Doyle, Arthur, 4, 195
Condillac, Étienne Bonnot de, 134
Condorcet, Nicolas de, 107, 109
Confessio Amantis (Gower), 71
Coningsby (Disraeli), 232
Constantinople, wonders of, 55–56, 64
Contes d'Hoffmann (Offenbach), 211
Copernius, Nicolaus, 96
Corpus Hermeticum, 83–84
Corsini, Matteo, 70–71, 78, 79
Cours entier de philosophie (Régis), 129
Cox, James, 106, 175
Creative Evolution (Bergson), 250
Crébillon, Prospero de, 153
Critique of Pure Reason (Kant), 29, 156, 170
Ctesibius, 90
Cullen, William, 149
Cybernetics, 12, 297–298, 299
"Cyborg Manifesto." *See* "Manifesto for Cyborgs."
Cyborgs and Cyborg Theory, 299, 302–304, 360n4

Daedalus, the moving statues of, 14, 17, 19–20, 35, 86, 90, 93, 94, 97, 98, 266
D'Alembert, Jean le Rond, 99, 110, 157
D'Alembert's Dream (Diderot), 110
"Darwin among the Machines" (Butler), 223–224
Darwin, Charles, and Darwinian Evolution, 218, 223, 225, 226, 243, 248, 249–250
Darwin, Erasmus, 218
Daston, Lorraine, 60, 95, 113, 114
"Daum Marries Her Pedantic Automaton 'George' in May 1920. John Heartfield Is Very Glad of It" (Grosz), 276
Davy, Humphry, 218
Dawkins, Richard, 307
Day the Earth Stood Still, The, 298
De Caus, Salomon, 81–82, 120
Dee, John, 5, 9, 59, 60, 80, 89–92, 94, 116
Deep Blue, IBM computer, 176
Dehumanization, 232, 233–235, 263, 266, 267–268, 275–276, 277–278, 288–290
Deleuze, Gilles, 308
Democracy, An American Novel (Adams), 172–174
Des Chene, Denis, 122
Desaguliers, John Theophilus, 141, 142
Descartes, Francine, 123

Descartes, René, 6, 96–97, 111, 115, 116–124, 125, 131, 133, 134, 135, 137, 138, 142–143, 151, 152, 155, 156, 174, 182, 246
Desfontaines, Pierre, 109
D'Holbach, Paul–Henri Thiry, 9, 151, 156, 159–160, 162
Dialogues Concerning Natural Religion (Hume), 156
Dick, Philip K., 11, 46, 298, 301
Dickens, Charles, 233–234, 240
Dickson, Arthur, 75
Diderot, Denis, 9, 99, 110, 147, 151, 152, 158, 159, 162–164, 193
Difference Engine, Charles Babbage's, 179, 229
Dijksterhuis, E. J., 112
Dinello, Daniel, 40
Diodorus Siculus, 20
Dionysian and Apollonian impulses (Nietzsche), 32, 250
Discourse on the Method (Descartes), 112, 115, 116–117
Disraeli, Benjamin, 232
Don Quixote (Cervantes), 95.
Douglas, Mary, 30–32, 42, 132
Dracula (Stroker), 274
"Dradin, in Love" (VanderMeer), 46
Dualism, 117, 123, 124, 131, 145, 155
Duchamp, Marcel, 262
Duck Automaton, Vaucanson's, 104, 107, 108, 140, 182, 214
Dumas, Alexandre, 171–172
Dutens, Louis, 177, 179
"Dynamo and the Virgin, The" (Adams), 43, 237–238

Edison, Thomas, 44, 109, 243, 272
Education of Henry Adams, The. See "Dynamo and the Virgin, The."
Edward II of England, 33
Eerie Peak, 50–52
Effigies, 32–34
"Ehefrau als bloßem Holze" (Jean Paul), 27
Eisenstein, Sergei, 277
Electricity and Electromagnetism, 141–142, 192–193, 218–219, 225, 244, 245, 249, 255, 272–272
Elements of Physiology (Diderot), 152
Elizabeth I of England, 91
Emerald Table, 83
Emile: or, On Education (Rousseau), 161–162
Encyclopédie, ou dictionnaire raisonné des sciences, des arts et des métiers (Diderot and D'Alembert eds.), 99, 100, 110, 114, 130, 140, 151, 157, 158
Engel, Ernst, 238, 243, 253, 276
Enslen, Karl, 185, 187
"Enthusiatic" Approach to Knowledge, 95–96, 113

Erasistratus, 119
Erewhon (Butler), 224–225
Escape from Freedom (Fromm), 168
Essai philosophique sur l'ame des bêtes (Boullier), 129
Essay on the History of Civil Society (Ferguson), 157
Essay on the Vital and Other Involuntary Motions of Animals (Whytt), 152
Euclid of Alexandria, 80, 89
Eve future, L'. *See Tomorrow's Eve*
Evelina (Burney), 143–144, 226
Evolution. *See* Darwin and Darwinian Evolution.
Experimental Science, 74–75, 100, 116, 140–143
Expressionist Cinema, 273–275

Faber, Joseph, 175
Faerie Queen, The (Spencer), 65
Famous Historie of Fryer Bacon, The, 71–72
Fanny Hill, or the Memoirs of a Woman of Pleasure (Cleland), 138, 165
Fantastic Literature, 194–196
Faraday, Michael, 272
Fast, Cheap and Out of Control, 27
Faust (Goethe), 218
Ferguson, Adam, 157, 231
Ficino, Marsilio, 58, 79–80, 92
Fielding, Henry, 136
Fisher, Peter S., 293
Flanders Panel, The (Pérez-Reverte), 181
Fleury, André-Hercule de, 106, 154
Fontana, Giovanni, 80, 82, 83
Fontenelle, Bernard le Bovier, 110
Forbidden Planet, The, 298
Foucault, Michel, 133, 135, 171–172
France, Anatole, 122
Francine, Descartes's automaton, 122–123
Francini, Tommaso and Alessandro, 81, 119, 123
Francis I of France, 33, 45
Frankenstein, or the Modern Prometheus (Shelley), 44, 195, 199, 218–220, 221–222
Franklin, Benjamin, 167, 178
Frederick II of Prussia, 106, 110, 134, 185–186
Frederick V, Elector Palatine, 81
Frederick William II of Prussia, 186
Free Enquiry into the Vulgarly Received Notion of Nature, A, (Boyle), 113, 124
Freud, Sigmund, 22–24, 25, 167–168, 212, 214, 250
Fromm, Erich, 168
Futurism, 243, 251–251, 286, 287–288

Galen of Pergamum, 118, 124
Galvani, Luigi, 218–219

Gargantua and Pantagruel (Rabelais), 84–85
Gaskell, Elizabeth, 234
Gastev, Alexei, 271
Gaukroger, Stephen, 123
Gendolla, Peter, 199
Gerbert of Aurillac (Pope Sylvester II), 9, 69–70, 71–75, 77, 78
Gibson, William, 11, 298, 299
Giesey, Ralph, 32–34
Ginzburg, Carlo, 32–35, 63
Giovio, Paolo, 99–100
Glass Bees, The (Jünger), 298
Glisson, Francis, 149
Godwin, William, 194
Goebbels, Josef, 277
Goethe, Johann Wolfgang von, 104, 189, 191–192, 214–216, 218
Goldsmith, Oliver, 153
Golem, The, 274
"Golem and the Robot, The" (Plank), 26
Goodall, Charles, 128
Gothic Literature. *See* Fantastic Literature
Gower, John, 71, 78
Graffigny, Françoise de, 153
Grafton, Anthony, 83
Gray, Stephen, 141, 142
Greene, Robert, 72
Gregory VII, Pope, 70
Grimm, Friedrich Melchior von, 107, 176, 179
Grosseteste, Robert, 9, 58, 69, 71–75, 78
Grosz, George, 262, 276
Guattari, Félix, 308
Gulliver's Travels (Swift), 137–138

Hales, Stephen, 129, 150
Haller, Albrecht von, 149
Hanafi, Zakiya, 31, 184
Hands of Orlac, The, 274
Hanson, David, 47
Haraway, Donna, 299, 303–304
Harbou, Thea von, 288–296
Hard Times (Dickens), 233–234
Hartmann, Edouard von, 250
Harvey, William, 128, 133
Hauksbee, Francis, 141
Hauptmann, Gerhart, 241–242
Heine, Heinrich, 220–222
Helm, Brigitte, 294
Helmholtz, Hermann von, 108, 144, 175, 228, 230–231
Henry IV of France, 81
Henry IV, Holy Roman Emperor, 70
Hephaestus (Vulcan), the tripods of, 15, 17, 84, 86, 90, 93, 97
Herder, Johann Gottfried, 189
Herf, Jeffrey, 242, 277, 294
Hermes (Mercury) Trismegistus, 83, 87, 90, 93

INDEX

Hermetic Magic, 59, 75–77, 79–80, 83–84, 96, 100, 115, 116, 140, 151, 189
Hero of Alexandria, 14, 16, 19, 80, 81, 82, 88, 90, 98, 119, 120
Herodian, Roman Emperor, 33
Hesiod, 21
Hessian Courier, The (Büchner), 206
Histoire de Juliette, L' (Sade), 165
Histoire naturelle (Buffon) 150–151
History of the Imagination, 12
Hobbes, Thomas, 115, 123, 133–134, 141
Hoffmann, E. T. A., 10, 27, 46, 51, 188, 193, 197–198, 206–214, 222
Holocaust, 277–278, 289
Homer, 15, 17, 84, 86, 139
Homme machine, L'. See *Man, a Machine*
Honourable History of Friar Bacon and Friar Bungey, The (Greene), 72
Hooke, Robert, 125, 128, 140
Hound of the Baskervilles (Conan Doyle), 195
Hugh of St. Victor, 62
Human Comedy, The (Saroyan), 51–52
Human Robots in Myth and Science (Cohen), 22, 24–25
Humboldt, Alexander von, 189
Hume, David, 156
Hunter, John, 149, 150
Hutchinson, Keith, 116
Hutton, Charles, 140
Huxley, T. H., 108, 168–169, 170, 175, 307
Huysmans, J. K., 239
Huyssen, Andreas, 26, 294–295
Hydraulic Automata, 81–82, 119–121

I, Robot, 302
Iamblichus Chalcidensis, 79
Idolatry, 34–35, 58, 62–63, 75–77, 98
Iliad (Homer), 14, 15
Image Magic. See Theurgy
Impressions of Africa (Roussel), 257, 258–259, 260–261, 262
Inanimate Reason. See *Briefe über den Schachspieler des von Kempelen.*
Industry and Industrial Revolution, 26, 38–39, 108, 157, 198–199, 220–222, 225, 228, 231–237, 266–268
Influencing Machine, 272–273
"Invisible Man, The" (Chesterton), 264–266
Iron Man, 267
"Isabella of Egypt" (Arnim), 27, 218
Ishiguro, Hiroshi, 47, 50–51
Israel, Jonathan, 155

Jacob, Margaret, 135, 155
Jacquard, Joseph-Marie, 228
Jacques the Fatalist (Diderot), 163–164
James I of England (James VI of Scotland), 91
James, William, 169–170, 307
Jane Eyre (Bronte), 4
Jaquet-Droz, Pierre and Henri-Louis, the automata of, 104, 108, 121, 140, 174, 199, 202
Jarry, Alfred, 10, 235, 243, 246–250, 252, 257, 261, 262, 267, 268
Jayne, Julian, 123
Jean Paul (Friedrich Richter), 10, 27, 109, 188, 200–203
Jentsch, Ernst, 22–23, 25
John of Salisbury, 58
Johnson, Samuel, 226
Jordan, Michael, 4
Joseph II of Austria, 178
Julia, Domna, 87
Jünger, Ernst, 46, 242, 298
Jurassic Park, 44
Justine, ou les Malheurs de la Vertu (Sade), 165, 166

Kant, Immanuel, 28, 29, 108, 156, 169, 186, 188, 189, 307
Kantorowicz, Ernst, 33
Kasparov, Gary, 176
Kaufmann, Friedrich, 175
Keill, James, 129, 150
Kelley, Edward, 91
Kempelen, Wolfgang von, 106, 140, 174, 175–183, 199, 201, 207, 217, 227, 228
Kepler, Johannes, 81
Keyser, Conrad, 80
King, Elizabeth, 45
King, Stephen, 3–4, 195
King Kong, 44
King's Two Bodies, The (Kantorowicz), 33
Kircher, Athanasius, 97, 98
Kleist, Heinrich von, 210–211
Knaben Wunderhorn, Des (Arnim and Brentano eds.), 216
Kratzenstein, Christian Gottlieb, 106
Kurzweil, Ray, 303

La Bruyère, Jean de, 137, 144–145, 160, 162
La Fontaine, Jean de, 136
La Forge, Louis de, 129
La Mettrie, Julien de, 26, 110, 112, 130–131, 136, 151, 174
Lacan, Jacques, 168
Laclos, Choderlos de, 159, 165
LaGrandeur, Kevin, 77
Lagrange, Joseph-Louis, 157
Lamy, François, 129
Lang, Fritz, 10, 268, 288–296
Laplace, Pierre-Simon, 157
Large Glass or The Bride Stripped Bare by Her Bachelors, Even (Duchamp), 262
"Laughter" (Bergson), 38–39
Lavoisier, Antoine Laurent, 157, 167, 191

Le Cat, Claude Nicolas, 107, 129
Le Goff, Jacques, 66
Léger, Fernand, 262
Leibniz, Gottfried Wilhelm, 6, 28, 108, 124, 131, 135, 155, 169, 188, 298, 307
Lem, Stanislaw, 298
Leonardo da Vinci, 81, 93
Leonce and Lena (Büchner), 204–206
Leschot, Jean, 106
Letters on Natural Magic (Brewster), 227–228
Leviathan (Hobbes), 123, 133–134
Levi, Primo, 278
Lévi-Strauss, Claude, 29, 306
Lewis, Matthew, 194
Liaisons dangereuses (Laclos), 165
"Liberal Education; And Where to Find It, A" (Huxley), 4
Life and Opinions of the Tomcat Murr, The (Hoffmann), 197–198
Liminal, 28, 35–36, 44, 65–66, 78, 114, 183
"Lineman Thiel" (Hauptmann), 241–242
Liuprand of Cremona, 55–56, 64
Livy, Titus, 37, 132
Locke, John, 28, 155
Locus Solus (Roussel), 257, 259–260
Loew, Judah, 81
Löhr, Robert, 181
Lomazzo, Giovannio Paolo, 92–93, 95
Long, Pamela, 80
"Lord of Dynamos, The" (Wells), 237
Louis I of Bavaria, 104
Louis IX of France, 64
Louis XIV of France, 135, 154
Louis XV of France, 106, 107, 129
Louis XVI of France, 106, 166, 183
Lucinde (Schlegel), 203–204
Ludwig, Karl, 230

MacDorman, Karl, 47, 50–51
"Machinenmann nebst seinen Eigenschaften" (Jean Paul), 109
MacKenzie, Henry, 153
Maelzel, Johann Nepomuk, 178–179, 180
"Maelzel's Chess-Player" (Poe), 178
Mafarka the Futurist (Marinetti), 254, 256–257, 288
Magia e grazia (Campanella), 94–95
Magician, automaton maker as, 35–36, 43–44
Maillardet, Henri, 104
"Maître Zacharias" (Verne), 27
Man, a Machine (La Mettrie), 26, 110, 112, 130–131
Man of Feeling (MacKenzie), 153
Mandeville, John, 67, 114, 120
"Manifesto for Cyborgs" (Haraway), 299, 303–304
March of Intellect, The (Seymour), 235–237

March of the Machines: The Breakthrough in Artificial Intelligence (Warwick), 300
Marey, Étienne-Jules, 231
Maria Theresa, 176, 178
Marie Antoinette, 104
Marinetti, Filippo Tommaso, 10, 235, 243, 246, 250–257, 261, 262, 267, 268, 286, 288
Marivaux, Pierre de, 153
Marx, Karl, 108, 234, 281
Mary I of England, 91
Maskelyne, John Nevil, 187
Mason & Dixon (Pynchon), 109
Materialism, 130–131, 134, 151, 155, 193, 230
Mathematical and Philosophical Dictionary (Hutton), 140
Mathematical Magic. *See* Natural Magic
Mathematicall Magick (Wilkins), 56, 97–99
Mather, Cotton, 131
Matrix, The, 11, 298
Maupertuis, Pierre-Louis, 114, 138
Mauss, Marcel, 35, 44
Maxwell, James Clerk, 272
Mayr, Otto, 134–135
McCulloch, Warren, 299
Mead, Richard, 149
"Mechanical Creation, The" (Butler), 225
Mechanical Problems (Strato), 80
Mechanics, the study of, 62, 74–75, 80–83, 85–91, 97, 100, 116, 140
Mechanistic Philosophy, 95–96, 100, 102, 112–116, 142–143, 148, 155–156, 158–159, 182–184, 186, 188, 194, 230, 266
Mechanistic Physiology, 96–97, 116–132, 142–143, 151, 152–153, 158, 165, 231
Mechanistic Political Theory, 133–136, 142–143, 158–159
Medici, Cosimo and Lorenzo di, 79, 83
Medieval Industrial Revolution, 62
Meditation on First Philosophy (Descartes), 121
Melville, Herman, 27, 45, 232–333
Mémoire descriptif (Vaucanson), 110, 111, 142
Memoires of the Year 2440 (Mercier), 156–157, 164–165
Mendelssohn, Moses, 203
Menenius Agrippa, 132
Meno, (Plato), 20
"Menschen sind Maschinen der Engel" (Jean Paul), 200–201
Mercier, Louis-Sebastien, 9, 110, 156–157, 159, 164–165
Merlin, John Joseph, 106, 143, 144, 229
Mesmer, Franz Anton, 167
Metaphysical Foundations of Natural Science (Kant), 188

Metropolis (Lang and Harbou), 10, 26, 27, 44, 268, 288–296
Mical, Abbé, 104
Micrographia (Hooke), 128–129
"Micromégas" (Voltaire), 138
Milton (Blake), 199
Mind Children (Moravec), 300–301
Modernism, 235, 243, 250, 262–263
Moleschott, Jacob, 230
Monadologie (Leibniz), 124
Mongolian Empire, wonders of, 64, 66–68
Montaigne, Michel Eyquem de, 82
Montana, Arnika, 217
Montefeltro, Frederico and Guidobaldo da, 80
Montpellier University of Medicine, vitalistic physiologists of, 150, 151, 152, 156–157
Moravec, Hans, 300–301
Mori, Masahiro, 47–49, 317–318n75
Mouret, Jacques–François, 180
"Movement of Animals" (Aristotle), 117–118
"Moxon's Master" (Bierce), 27, 180
"Multiplied Man and the Reign of the Machine" (Marinetti) 252–253
Mummy! A Tale of the Twenty–second Century, The (Webb), 218
Museum of Westward Expansion (St. Louis, Missouri), 45

Natural Magic, 6, 58–60, 71–75, 85–95, 102
Natural Magick (Porta), 93–94
Naturphilosophie, 189–194, 207, 219, 222, 225, 229, 263
Naudé, Gabriel, 78
Nazi Regime, 242, 276–278, 293
Necromancy, 65, 69, 102, 196
Neumann, John von, 299
New Atlantis, The (Bacon), 141
Newton, Isaac, 115, 129, 135, 149–150, 155, 157
Nicephorus II Phocas, Byzantine Emperor, 55
Nicolai, Christian Friedrich, 111
Nietzsche, Friedrich, 32, 250–251
Night (Wiesel), 278
Nodier, Charles, 27
Nollet, Jean–Antoine, 141, 142
North and South (Gaskell), 234
Nosferatu: A Symphony of Horror, 273–274
Nosologia methodica (Sauvages), 152
Nouvelle Heloise, La (Rousseau), 147, 153
"Nutcracker and the King of Mice, The" (Hoffmann), 218

Occulta Philosophia, De (Agrippa), 7, 84–89
O'Connor, William Douglas, 27
Odoric of Pordenone, 66

Oersted, Hans Christian, 189, 192–193, 272
Offenbach, Jacques, 211
Oken, Lorenz, 189
On the Economy of Machinery and Manufactures (Babbage), 228, 231
"On the Hypothesis That Animals are Automata, and Its History" (Huxley), 108, 168–169
"On the Interaction of Natural Forces" (Helmholtz), 108, 228
"On the Marionette Theater" (Kleist), 210–211
On the Movement of Animals (Borelli), 126–128
"On the Origin of the 'Influencing Machine' in Schizophrenia" (Tausk), 273
On the Origin of Species (Darwin), 223
"On the Psychology of the Uncanny" (Jentsch), 22–23
On the Soul (Aristotle), 19–20
"On the Supernatural in Poetry" (Radcliffe), 41
Opticks (Newton), 150
Otto I, Holy Roman Emperor, 55, 69

Paine, Thomas, 159, 166
Palingenesien (Jean Paul), 202
Pamela (Richardson), 153, 159, 168
Pappus of Alexandria, 80
Paracelsus (Theophrastus Bombastus von Hohenheim), 44, 219
"Paradise of Bachelors and the Tartarus of Maids, The" (Melville), 232–233
Park, Katharine, 95, 113, 114
Pascal, Blaise, 298
Passages from the Life of a Philosopher (Babbage), 228, 229
Passions of the Soul, The (Descartes), 121
Paulet, Jacques, 167
Péleringe de Charlemagne, 64
Père Goriot (Balzac), 4
Pereira, Gómez, 117
Péret, Benjamin, 262
Péréz–Reverte, Arturo, 181
Perrault, Claude, 129
"Personalien vom Bedienten–und Maschinen Mann" (Jean Paul), 27, 202–203
Pertinax, Roman Emperor, 34
Peter of Abelard, 58
Phalaris, the bronze bull of, 94
Philip the Good, Duke of Burgundy, 62, 80
Philodor, François–André Danican, 178
Philosophical Enquiry into the Origins of Our Ideas on the Sublime and the Beautiful (Burke), 41, 317n66
Philosophie dans le boudoir, La (Sade), 165
Philosophy of Manufactures (Ure), 231
Philostratus the Athenian, 87
Piaget, Jean, 23

Picabia, Francis, 262
Picatrix, The, 59
Pindar, 16
Pitcairne, Archibald, 129, 149, 150
Plato, 20, 21, 58, 79, 98, 132
Plotinus, 79
Pneumatics, (Hero of Alexandria), 16, 81
Poe, Edgar Allen, 176
Poisson, Nicolas, 122
Polarity, 191–192
Polinière, Pierre, 141
Politics, (Aristotle), 17
"Pope's Monoplane, The" (Marinetti), 251
Pornography, automaton and, 108, 138–139, 165–166
Porphyry of Tyre, 79
Porta, Giambattista della, 80, 93–95, 97, 100, 140
Pottier de la Hestroye, Jean, 135
Poupée électriques (Marinetti), 254–255
Power, automaton and the theme of, 3–5, 25, 43–44, 266–267
Prawer, S. S., 211, 221
Preludes, The (Wordsworth), 196, 216
Preternatural, medieval concept of, 60–61, 66, 100, 102, 115, 184
Prévost, Antoine François, 109, 153
Price, Richard, 154–155
Priestly, Joseph, 155
Principia Mathematica (Newton), 150
Principles of Philosophy (Descartes), 121
Proclus Lycaeus, 79
Psychopathology of Everyday Life (Freud), 167–168
Public Science, 116, 140–143
Pückler–Muskau, Hermann von, 221–222
Puppet and the Dwarf, The (Žižek), 181
Pygmalion, 16
Pynchon, Thomas, 109

Queen Mab (Shelley), 170–171

Rabelais, François, 84–85
Rabinbach, Anson, 231, 247
Radcliffe, Ann, 41, 194
Radical Enlightenment, 155, 158–159
Ramus, Petrus, 90
Rankine, William, 230
Reactionary Modernism, 242, 277
Rechsteiner, Johann–Bartholomé, 104
Red and the Black, The (Stendhal), 171
Regiomontanus (Johannes Müller of Königsberg), 90, 97, 98, 140, 176
Régis, Pierre Sylvain, 129
Reill, Peter, 151
Religieuse, La (*The Nun*, Diderot), 162–163
Remarque, Erich Maria, 269
"Representation: The Word, the Idea, the Thing" (Ginzburg), 32–34

Revolt of the Machines, The (Rolland), 10, 268, 283–286
Richard II of England, 62
Richardson, Samuel, 153, 159, 160, 168
Riefenstahl, Leni, 276–277
Rights of Man (Paine), 166
Robbe–Grillet, Alain, 257, 261
Robert, Count of Artois, 61–62
Robert–Houdin, Jean–Eugène, 186–187
Robespierre, Maximilien, 166, 183
Robocop, 267
Robot, the word, 8, 279
Robot's Rebellion, The (Stanovich), 307
Rohault, Jacques, 129
Rolland, Romain, 268, 282–286, 296
Roman de Troi (Benoît), 64
Romantic School, The (Heine), 220–221
Romantic Science and Worldview. *See* Naturphilosophie
Rosario della vita (Corsini), 70–71
Rostand, Edmond, 262
Rothkrug, Lionel, 135
Rôtisserie de la Reine Pédauque, La, (France), 122
Rousseau, Jean–Jacques, 9, 146–147, 153, 159, 161–162
Roussel, Raymond, 10, 235, 243, 257–262, 262, 267, 268
Royal Society, 56, 97, 141
Rudolph II, Holy Roman Emperor, 81
R.U.R: Rossum's Universal Robots (Čapek), 10, 268, 279–282, 286

Sade, Donatien Alphonse François de, 159, 165–166
Saint–Germain–en–Lay Automata, 81, 119–121, 123, 129
"Sandman, The" (Hoffmann), 3, 27, 45, 211–214, 222
Sant'Elia, Antonio, 286
Saroyan, William, 51–52
Sauvages, François Boissiers de, 150, 152
Scaliger, Julius Caesar, 140
"Scandal in Bohemia, A" (Conan Doyle), 4
Schaffer, Simon, 141, 157
Schelling, Friedrich, 189, 192, 203–204
Schlegel, August Wilhelm, 194, 221
Schlegel, Friedrich, 194, 203–204, 221
Schnapp, Jeffrey T. 253–254
Schopenhauer, Arthur, 250
Schubert, Gotthilf Heinrich von, 207
Schuyl, Florentius, 96–97
Schwarzenegger, Arnold, 46
Scientific Revolution, 95–96, 100
Search for Truth, The (Descartes), 142–143
"Second Variety" (Dick), 301
Selfish Gene, The (Dawkins), 307
Self–Mover, 19, 85, 148, 311–312n7
Seligmann, Kurt, 262

Seltzer, Mark, 7, 159
Sentimentality, 148, 153–154, 158
Septimius Severus, Roman Emperor, 33–34, 87
Seutonius Tranquillus, Gaius, 37
Seven Years War, 154
Sexual Life of Our Time and Its Relations to Modern Civilization (Bloch), 108
Seymour, Robert, 235–237
Shakespeare, William, 36, 68
Shannon, Claude, 299
Shapin, Steven, 141
Shelley, Mary, 43–44, 195, 198, 199, 218–220
Shelley, Percy Bysshe, 170–171, 218
"Signs of the Times" (Carlyle), 108, 225
Sladek, John, 298
Slavery, automaton and the theme of, 19–21, 266, 296, 307–308
Smith, Adam, 231, 233
Smith, Crosbie, 231
Smith, Will, 302
Smithsonian Institution, 45, 81
Sombart, Werner, 242
Sontag, Susan, 165
"Sorcerer's Apprentice" (Goethe), 218
"Speaking Figure, and the Automaton Chess-Player, Exposed and Detected" (Thicknesse), 176
Spence, Joseph, 104, 111, 144
Spencer, Edmund, 65
Spengler, Oswald, 242
Spinoza, Baruch, 155
Staden, Heinrich von, 119
Stahl, Georg-Ernst, 149
Stahltier, Das (Zielke), 277
Stanovich, Keith E., 307
Star Wars, 39–40
Star Wars—Episode II: Attack of the Clones, 40
Steam Engine, 199, 225, 226, 229–231, 238, 266
Stendhal, 171
Stephenson, Neal, 299
Sterling, Bruce, 299
"Story of a Good Brahim, The" (Voltaire), 160–161
Strasburg Cathedral, the astronomical clock of, 80, 139
Strauss, Linda, 28
Stroker, Bram, 274
Student of Prague, The, 274
Sublime, automaton and the, 41–47
Suger of St. Denis, 58, 91
Sullivan, Penny, 64
Supermale, The (Jarry), 246–250, 260
Superman, 248–250, 252–253, 256–257, 262–263
Surrealism, 243, 262
Survival in Auschwitz (Levi), 278

Swift, Jonathan, 137, 224
Symbolism, 243

Tableau de Paris, Le (Mercier), 110
Talos, 15, 19, 21
Talus, 65
Tausk, Victor, 273
Taylor, William Cooke, 231–232
Taylor, Frederick Winslow, and Taylorist Regime, 10, 38–39, 268, 270, 291, 296
Tchaikovsky, Pyotr Ilych, 218
"Technical Manifesto of Futurist Literature" (Marinetti), 254
Technophobia, 77, 220–222, 233–235, 272–279, 294–295
Terminator, The, 11, 46–47, 267, 298
Terrall, Mary, 114
Thermodynamics, 225, 230, 243, 245, 249–250, 253, 255, 266
"Thesis on the Philosophy of History" (Benjamin), 180–181
Theurgy, 59, 68, 75–77, 93, 98, 316n55
Thicknesse, Philip, 176
Thomas Aquinas, 60, 71, 78–79, 93, 98, 118
Thompson (Lord Kelvin), William, 230, 231
Three Musketeers, The (Dumas), 171
Tieck, Ludwig, 193
Timaeus, 58
"To My Pegasus" (Marinetti), 251
Todorov, Tzvetan, 195, 207
Toland, John, 155
Tom Jones (Fielding), 136
Tomorrow's Eve (Villiers), 44, 45, 109, 243–246
Totemism (Lévi-Strauss), 306
Tour in the Manufacturing Districts of Lancashire (Taylor), 232
Towards a History of Religion and Philosophy in Germany (Heine), 220, 221–222
Trains, 239–242, 253, 276–277
Traité de l'espirit de l'homme (La Forge), 129
Trattato dell'arte della pittura (Lomazzo), 92–93
Treatise on Man (Descartes), 96–97, 117, 119–121, 129
Treatise on Sensations, A (Condillac), 134
Treatise on Systems, A (Condillac), 134
Treatise on the World (Descartes), 96–97, 117
Triumph of the Will (Riefenstahl), 276
Truitt, E. R., 65–66
Turing, Alan, 298
Turner, Victor, 35
Turriano, Juanelo, 81, 122
Twelfth-Century Renaissance, 58–59
Twenty Years After (Dumas), 171–172
Two Discourses Concerning the Soul of Brutes (Willis), 125

Übermensch. See Superman
"Uncanny, The" (Freud), 22–24
Uncanny, automaton and the theme of the, 22–24, 41–43, 47–55, 184, 187–188, 194, 195–198, 207–218, 219–220, 274–275
"Uncanny Valley, The" (Mori), 47–49, 317–318n75
Under Fire (Barbausse), 269–270
"Unterthängiste Vostellung unser, der sämtlicher Spieler und reddened Damen in Europa entgegen und wider die Einführung der Kemplischen Spiel-und Sprechmaschinen" (Jean Paul), 201
Ure, Andrew, 231–232

VanderMeer, Jeff, 46
Vartanian, Aram, 130
Vasari, Giorgio, 81, 93
Vasari, Ruggero, 268, 286–288, 296
Vaucanson, Jacques de, 7, 11, 103–111, 121, 139–143, 174, 175, 177, 181–183, 186, 187, 202, 214, 217, 228, 229
Veit, Dorothea, 203
Venel, Gabriel-François, 150, 151
"Venus of Ille, The" (Mérimée), 3
Verne, Jules, 27, 260
Vicar of Wakefield (Goldsmith), 153
Vigneul-Marville (Bonaventure d'Argonne), 123
Vila, Anne, 153
Villard de Honnecourt, 61
Villiers de l'Isle Adam, Auguste, 10, 27, 44, 45, 109, 235, 243–246, 257, 261, 267, 268
Vitalism, enlightenment, 149–157, 182–183, 188–190, 193
Vitalism, modernist, 39, 250
Vitalistic physiology, 149–153, 165, 182, 188–189, 193
Vitalistic political theory, 155–157, 182–184
Vitruvius Pollio, Marcus, 80
Vogt, Karl, 230
Voltaire, 110, 138, 146, 159, 160–161
"Von Kempelen and his Discovery" (Poe), 178
Voyage of Argo, The (Apollonius of Rhodes), 15–16
"Voyage pittoresque et industriel dans le Paraguay–Roux et la palingénésie austral" (Nodier), 27
Voyage to the Moon and the Sun (Bergerac), 137

Walker, D. P., 92
WALL-E, 302
Walpole, Horace, 194
Walpole, Robert, 154
Wandel, Torbjörn, 235
"War, the World's Only Hygiene." See "Multiplied Man and the Reign of the Machine"
Warwick, Kevin, 300
Watt, James, 108, 238
We (Zemyatin), 271–272
"We Grow out of Metal" (Gastev), 271
Wealth of Nations (Smith), 233
Webb, Jane, 218
Weeks, Thomas, 175, 229
Weichardt. Wilhelm, 247
Wellman, Kathleen, 130, 136
Wells, H. G., 237
Westworld, 44
"What is Enlightenment?" (Kant), 186
Whiston, William, 141
Whytt, Robert, 149, 152
Wiener, Norbert, 299
Wilkes, John, 154
Wilkins, John, 56, 78–79, 97–99, 115
William of Auvergne (of Paris), 58, 59, 77, 86, 93, 94
William of Malmesbury, 69
William of Rubruck, 64, 66
Williams, Elizabeth, 152
Willis, Thomas, 125, 128, 131, 151
Willis, Robert, 180
Windisch, Karl Gottlieb von, 176, 179–180
Winter, Allison, 167
Winter's Tale (Shakespeare), 68
Wise, Norton, 231
Wolfe, Jessica, 65
Wolff, Christian, 188
Wöllner, Johann Christof, 186
Wollstonecraft, Mary, 161
Womb-envy theory, 26–27
Wonders, 60–61, 80–81, 95–96, 113–115, 184
Wood, Gaby, 52, 123
Wordsworth, William, 196, 216
World War I, 10, 267, 268–270, 274, 282, 286, 293–294, 296
Wright, Wilbur, 254
Wunderkammer. See Wonders

Yates, Frances, 83, 115

Zamyatin, Yevgeny, 271–272
Zielke, Willy, 277
Zinner, Ernst, 90
Žižek, Slavoj, 181
Zola, Emile, 240–241

Harvard University Press is a member of Green Press Initiative (greenpressinitiative.org), a nonprofit organization working to help publishers and printers increase their use of recycled paper and decrease their use of fiber derived from endangered forests. This book was printed on recycled paper containing 30% post-consumer waste and processed chlorine free.